Computational Intelligence, Optimization and Inverse Problems with Applications in Engineering

Gustavo Mendes Platt • Xin-She Yang
Antônio José Silva Neto

Editors

Computational Intelligence, Optimization and Inverse Problems with Applications in Engineering

 Springer

Editors
Gustavo Mendes Platt
Universidade Federal do Rio Grande
Santo Antônio da Patrulha
Rio Grande do Sul, Brazil

Xin-She Yang
School of Science and Technology
Middlesex University
London, UK

Antônio José Silva Neto
Polytechnic Institute
Universidade do Estado do Rio de Janeiro
Nova Friburgo, Rio de Janeiro, Brazil

ISBN 978-3-030-07190-5   ISBN 978-3-319-96433-1   (eBook)
https://doi.org/10.1007/978-3-319-96433-1

This Springer imprint is published by the registered company Springer Nature Switzerland AG
The registered company address is: Gewerbestrasse 11, 6330 Cham, Switzerland

To Monique, Ana Luísa, and Luís Felipe
Gustavo Mendes Platt

To Gilsineida, Lucas, and Luísa
Antônio José Silva Neto

Foreword

This book covers recent developments and novel applications on computational intelligence to model and solve a variety of relevant and challenging engineering problems. These problems include interesting applications from thermodynamics, chemical engineering, civil engineering, nuclear engineering, fault diagnosis, environmental applications, heat transfer, and image processing. In particular, the use of metaheuristics is reviewed and analyzed using different engineering problems. Metaheuristics (or also known as stochastic global optimization methods) are widely recognized as powerful optimization tools that can be used to resolve a diversity of multivariable engineering problems with non-convex objective functions that can also be subjected to equality and inequality constraints. This type of optimization methods can handle discrete and continuous variables and discontinuities in the objective function or even can be incorporated with black-box models.

To date, there are challenges in the development and application of effective metaheuristics to resolve real-world optimization problems. In this book, several algorithms and their engineering applications are described including traditional metaheuristics (e.g., simulated annealing, genetic algorithm, particle swarm optimization and differential evolution) and novel methods such as multi-particle collision algorithm and lightning optimization algorithm. Objective functions analyzed in this book involve both benchmark and real-world application problems. Additionally, some chapters cover important aspects on the application of artificial neural network modeling and fuzzy logic, which are also computational intelligence approaches that can address complex real-world problems from the engineering context.

This book contains a collection of interesting contributions that are organized in 14 chapters. Specifically, Chap. 1 covers the application of some nature-inspired metaheuristics for solving two challenging phase equilibrium problems: calculation of reactive azeotropes and the calculation of dew point pressures in systems with double retrograde vaporization. Authors of this chapter compared the performance of differential evolution and harmony search in these thermodynamic calculations. Results showed that differential evolution was the best method for solving these problems.

In Chap. 2, a hybrid approach based on differential evolution, reliability analysis and effective mean concept was proposed to solve reliability-based robust optimization problems. This methodology was tested in five examples (two mathematical test cases and three engineering system design problems) proving that it was effective for reliability-based robust design of engineering systems.

Two approaches for generating an initial population in population-based methods have been tested and analyzed with differential evolution and challenging practical optimization problems, and results are reported in Chap. 3. The main conclusion reported in this chapter indicated that Mersenne Twister pseudorandom generator should be used for small populations, while the Sobol sequence was a better option for large population sizes.

Chapter 4 describes the implementation of a modified particle swarm optimization method to solve the Steiner tree problem, which is classified as a combinatorial optimization. Numerical results were statistically analyzed using problems with sets of 1000 and 10,000 points showing the robustness and efficiency of this improved method.

The vibration-based damage identification problem has been solved using a multi-particle collision algorithm with rotation-based sampling, and results are reported in Chap. 5. Performance of this metaheuristic has been tested with and without the application of Hooke-Jeeves pattern search method. Authors of this chapter concluded that this algorithm provided proper estimations with and without a low level of noise.

Chapter 6 provides a review of different metaheuristics for solving design optimization problems in civil engineering. Objective functions from civil engineering applications are challenging due to the characteristics of design variables and the presence of nonlinear constraints. Therefore, this chapter discusses different metaheuristics and their uses in the context of civil engineering.

Fault diagnosis in a nonlinear bioreactor benchmark problem with ant colony optimization with dispersion, differential evolution with particle collisions, and the covariance matrix adaptation evolution strategy is reported in Chap. 7. Description of the fault diagnosis problem was provided, and some examples were solved with these methods, and different noise levels, showing the capabilities of these metaheuristics.

Chapter 8 illustrates the application of cross-entropy algorithm, particle swarm optimization, artificial bee colonies, and population-based incremental learning to resolve a challenging real-world optimization problem from nuclear engineering, which is known as loading pattern in nuclear reactors or in-core fuel management optimization. Numerical results illustrated the advantages and disadvantages of these metaheuristics.

In Chap. 9, lightning optimization algorithm has been employed for parameter estimation in the context of anomalous diffusion phenomenon. Authors of this chapter concluded that lightning optimization algorithm is a novel and interesting method that is inspired by the observation of atmospheric discharges, which can also be improved in future works.

Chapter 10 reports the analysis of different complex topologies of neural networks used to predict two climatic variables: temperature and solar radiation. These topologies included complete, random, scale-free network and small-world networks. Results illustrated the capabilities and limitations of tested neural networks in this modeling problem.

In Chap. 11, genetic algorithm has been applied to resolve a heat transfer problem of an H-shaped cavity into a solid body. This metaheuristic was effective to solve the optimization problem concluding that the thermal resistance depended on the cavity volume fraction.

Chapter 12 describes the application and comparison of particle swarm optimization and genetic algorithm to improve the template matching for image processing. Overall, particle swarm optimization has a better performance to achieve real-time execution in template matching.

In Chap. 13, particle swarm optimization has been used in solving the state and parameter estimation using a particle filter algorithm. This hybrid method has been tested in the resolution of an inverse heat conduction problem in a thin plate. Results showed the reliability of this approach especially for those problems involving abrupt transition.

Finally, Chap. 14 covers the fault detection with kernel methods and the metaheuristics differential evolution and particle swarm optimization. The design of a fault detection system for the Tennessee Eastman process was used as a case of study. Results showed that differential evolution yielded a faster convergence to estimate the parameters in this fault detection problem.

Active researchers have contributed to the conformation of this book, and its content can be used in undergraduate and postgraduate courses of several engineering programs. The edition of this book has been leaded by Prof. G.M. Platt, Prof. X.-S. Yang, and Prof. A.J. Silva Neto whom are well-recognized authors due to their contributions to the application of computational intelligence in the solution of challenging engineering problems.

I am sure that the readers will find this book interesting and enlightening. Research on computational intelligence and its numerous applications will continue to be an active area in science and engineering, and this book will serve as a source of new research topics and contribute to further developments in this topic.

Instituto Tecnológico de Aguascalientes   Adrián Bonilla-Petriciolet
Aguascalientes, Mexico
May 2018

Preface

The initial ideas of this book emerged in 2017, during the preparation of the XX Brazilian National Meeting on Computational Modeling (XX ENMC), which was held in Nova Friburgo, RJ, Brazil. The increasing number of works in the Technical Sessions on *Optimization, Inverse Problems and Computational Intelligence* in the previous ENMCs—dealing mainly with applications of metaheuristic algorithms in engineering problems—inspired the editors to organize a book focusing on the application of these subjects in engineering and physical problems.

The editors invited a group of active researchers in this field to participate on this project. Authors from four countries and 18 different universities/research centers contributed with chapters to this book, covering the applications of optimization, inverse problems, and computational intelligence in chemical, mechanical, civil and nuclear engineering, meteorology, and combinatorial optimization.

The book is intended for use in undergraduate and graduate courses of chemical, mechanical, civil and nuclear engineering, computational modeling, applied mathematics, and related fields, where the use of state-of-the-art computational tools is encouraged.

Santo Antônio da Patrulha, Brazil   Gustavo Mendes Platt
London, UK   Xin-She Yang
Nova Friburgo, Brazil   Antônio José Silva Neto
May 2018

Acknowledgments

The editors acknowledge the support provided by the Springer representation in Brazil for the publication of this book.

We want also to extend our thanks to the Brazilian research supporting agencies: Fundação Coordenação de Aperfeiçoamento de Pessoal de Nível Superior (CAPES), Conselho Nacional de Desenvolvimento Científico e Tecnológico (CNPq), and Fundação Carlos Chagas Filho de Amparo à Pesquisa do Estado do Rio de Janeiro (FAPERJ).

We also thank the authors for their high-standard contributions and speedy response in the review process.

Finally, we must acknowledge the doctorate student Gustavo Barbosa Libotte (IPRJ-UERJ), for his efforts and dedication, which were fundamental in the LaTeX edition of the book and in the management of the repository.

Contents

1 An Overview of the Use of Metaheuristics in Two Phase Equilibrium Calculation Problems 1
Gustavo Mendes Platt, Lucas Venancio Pires de Carvalho Lima,
Gustavo Barbosa Libotte, and Vinícius Magno de Oliveira Coelho
1.1   Introduction 1
1.2   Analysis of Previous Works 2
1.3   Description of the Problems 3
    1.3.1   Calculation of a Reactive Azeotrope in a Ternary
        System (RA Problem) 3
    1.3.2   Calculation of Dew Point Pressures in a Binary
        System with Double Retrograde Vaporization
        (DRV Problem) 8
1.4   Description of the Metaheuristics and Statistical Analysis 13
    1.4.1   The Differential Evolution Algorithm (DE) 14
    1.4.2   The Harmony Search Algorithm (HS) 15
    1.4.3   Statistical Analysis 15
1.5   Results and Discussion 18
    1.5.1   Reactive Azeotrope Calculation (RA) 18
    1.5.2   Double Retrograde Vaporization Calculation 21
1.6   Conclusions 25
References 25

2 Reliability-Based Robust Optimization Applied to Engineering System Design 29
Fran Sérgio Lobato, Márcio Aurelio da Silva,
Aldemir Aparecido Cavalini Jr., and Valder Steffen Jr.
2.1   Introduction 29
2.2   Robust Design (RD) 31
2.3   Reliability-Based Optimization (RBO) 32
    2.3.1   First Order Reliability Method (FORM) 33
    2.3.2   Second-Order Reliability Method (SORM) 34

            2.3.3   Sequential Optimization and Reliability
                         Assessment (SORA).. 34
            2.3.4   Inverse Reliability Analysis (IRA)......................... 35
    2.4   Differential Evolution Algorithm (DE) 36
    2.5   Methodology 37
    2.6   Mathematical and Engineering Applications 39
            2.6.1   Nonlinear Limit State.. 40
            2.6.2   Highly Nonlinear Limit State 41
            2.6.3   Short Column Design 43
            2.6.4   Cantilever Beam Problem 45
            2.6.5   Three Bar Truss Problem................................... 48
    2.7   Conclusion... 50
    References.. 51

3   **On Initial Populations of Differential Evolution for Practical**
    Optimization Problems 53
    Wagner Figueiredo Sacco and Ana Carolina Rios-Coelho
    3.1   Introduction... 53
    3.2   The Initialization Schemes.. 54
            3.2.1   The Mersenne Twister Pseudorandom Generator 54
            3.2.2   Mersenne Twister and Opposition-Based Learning....... 54
            3.2.3   The Sobol Low-Discrepancy Sequence 55
    3.3   Numerical Comparisons 55
            3.3.1   The Practical Problems 55
            3.3.2   Implementation and Setup 57
            3.3.3   Computational Results 58
    3.4   Conclusions.. 59
    References.. 60

4   **Application of Enhanced Particle Swarm Optimization in**
    Euclidean Steiner Tree Problem Solving in R^N......................... 63
    Wilson Wolf Costa, Marcelo Lisboa Rocha, David Nadler Prata,
    and Patrick Letouzé Moreira
    4.1   Introduction... 63
    4.2   Definition of Euclidean Steiner Tree Problem (ESTP) 64
            4.2.1   Smith's Algorithm 65
    4.3   Particle Swarm Optimization 65
    4.4   Resolution of the Euclidean Tree Problem in R^N with the
            Particle Swarm Optimization 67
            4.4.1   Definition of a Particle 67
            4.4.2   Overview of the Proposed Resolution 67
            4.4.3   Use of the Minimum Spanning Tree and Geometry
                       to Obtain Topology 68
            4.4.4   A Modified Particle Swarm Optimization for ESTP
                       Resolution.. 70

4.5 Computational Results 73
    4.5.1 Computational Test Environment 73
    4.5.2 Test Methodology 74
4.6 Discussion on the Experimental Results 80
    4.6.1 Solution Quality 80
    4.6.2 Computational Effort 82
4.7 Conclusion and Future Work 83
References.. 84

5 Rotation-Based Multi-Particle Collision Algorithm with Hooke–Jeeves Approach Applied to the Structural Damage Identification 87
Reynier Hernández Torres, Haroldo Fraga de Campos Velho, and Leonardo Dagnino Chiwiacowsky
5.1 Overview.. 87
5.2 Hybrid Algorithm for SDI 88
    5.2.1 Rotation-Based Multi-Particle Collision Algorithm with Hooke-Jeeves... 90
    5.2.2 Multi-Particle Collision Algorithm 90
    5.2.3 Opposition-Based Learning and Some Derived Mechanisms... 93
    5.2.4 Hooke–Jeeves Pattern Search Method.................... 96
5.3 Vibration-Based Damage Identification Problem as an Optimization Problem... 98
5.4 Damage Identification in a Cantilevered Beam 99
    5.4.1 Cantilevered Beam 99
    5.4.2 Experimental Results: Damage Identification from a Full Dataset 100
    5.4.3 Experimental Results: Damage Identification from a Reduced Dataset 101
5.5 Final Remarks 105
References.. 107

6 Optimization in Civil Engineering and Metaheuristic Algorithms: A Review of State-of-the-Art Developments 111
Gebrail Bekdaş, Sinan Melih Nigdeli, Aylin Ece Kayabekir, and Xin-She Yang
6.1 Introduction.. 111
6.2 Optimization and Metaheuristic Algorithms....................... 112
    6.2.1 Genetic Algorithm.. 113
    6.2.2 Simulated Annealing 113
    6.2.3 Particle Swarm Optimization 114
    6.2.4 Harmony Search 114
    6.2.5 Firefly Algorithm 114

         6.2.6   Cuckoo Search 114
         6.2.7   Bat Algorithm 115
         6.2.8   Flower Pollination Algorithm 115
         6.2.9   Other Metaheuristic Algorithms Used in Civil
                 Engineering 115
    6.3 Applications and Optimization in Civil Engineering 116
         6.3.1   Truss Structures 117
         6.3.2   Reinforced Concrete Members 119
         6.3.3   Frame Structures 120
         6.3.4   Bridges 121
         6.3.5   Tuned Mass Damper 122
         6.3.6   Construction Management 122
         6.3.7   Hydraulics and Infrastructures 123
         6.3.8   Transportation Engineering 124
         6.3.9   Geotechnics 125
    6.4 Conclusions 125
    References 125

7   **A Bioreactor Fault Diagnosis Based on Metaheuristics** 139
    Lídice Camps Echevarría, Orestes Llanes-Santiago,
    and Antônio José Silva Neto
    7.1 Introduction 139
    7.2 Fault Diagnosis Formulated as an Optimization Problem 141
    7.3 Description of the Bioreactor Benchmark Problem 142
         7.3.1   FDI Formulation for the Benchmark Problem 143
    7.4 Description of the Metaheuristics for the Benchmark
         Problem Fault Diagnosis 144
         7.4.1   Ant Colony Optimization with Dispersion (ACO-d) 144
         7.4.2   Differential Evolution with Particle Collision
                 (DEwPC) 147
         7.4.3   Description of the Algorithm 148
         7.4.4   Covariance Matrix Adaptation Evolution Strategy
                 (μ_w, λ) CMA-ES 150
    7.5 Test Cases Considered and Parameters Used in the
         Metaheuristics 153
         7.5.1   Implementation of ACO-d 154
         7.5.2   Implementation of DEwPC 154
         7.5.3   Implementation of (μ_w, λ) CMA-ES 154
    7.6 Results 155
         7.6.1   Comparison with Other FDI Methods 160
    7.7 Conclusions 161
    References 162

**8   Optimization of Nuclear Reactors Loading Patterns with
     Computational Intelligence Methods** 165
Anderson Alvarenga de Moura Meneses, Lenilson Moreira Araujo,
Fernando Nogueira Nast, Patrick Vasconcelos da Silva,
and Roberto Schirru
   8.1   Introduction 165
   8.2   Related Work 167
         8.2.1   Particle Swarm Optimization 167
         8.2.2   Cross-Entropy Algorithm 168
         8.2.3   Population-Based Incremental Learning 168
         8.2.4   Artificial Bee Colonies 168
   8.3   Loading Pattern Optimization 168
   8.4   Some Optimization Metaheuristics Applied to the LP
         Optimization 171
         8.4.1   Particle Swarm Optimization (PSO) 171
         8.4.2   Cross-Entropy Algorithm (CE) 172
         8.4.3   Population-Based Incremental Learning (PBIL) 174
         8.4.4   Artificial Bee Colonies (ABC) 176
   8.5   Results and Discussion 178
         8.5.1   Short-Run Results (PSO and CE) 179
         8.5.2   Long-Run Results (PBIL and ABC) 180
   8.6   Conclusion 180
   Appendix 181
   References 181

**9   Inverse Problem of an Anomalous Diffusion Model Employing
     Lightning Optimization Algorithm** 185
Luciano Gonçalves da Silva, Diego Campos Knupp,
Luiz Bevilacqua, Augusto César Noronha Rodrigues Galeão,
and Antônio José Silva Neto
   9.1   Introduction 185
   9.2   Direct Problem Formulation and Solution 186
   9.3   Inverse Problem 188
         9.3.1   Maximum Likelihood 189
         9.3.2   Lightning Optimization Algorithm (LOA) 190
   9.4   Results and Discussion 195
   9.5   Conclusions 198
   References 199

**10 Study of the Impact of the Topology of Artificial Neural
     Networks for the Prediction of Meteorological Data** 201
Roberto Luiz Souza Monteiro, Hernane Borges de Barros Pereira,
and Davidson Martins Moreira
   10.1   Introduction 201
   10.2   Materials and Methods 202
          10.2.1   Artificial Neural Networks 202

10.3   Methods... 207
10.4   Results 207
10.5   Concluding Remarks.. 212
References.. 213

**11   Constructal Design Associated with Genetic Algorithm to
         Maximize the Performance of H-Shaped Isothermal Cavities**......... 215
Emanuel da Silva Dias Estrada, Elizaldo Domingues dos Santos,
Liércio André Isoldi, and Luiz Alberto Oliveira Rocha
11.1   Introduction.. 215
11.2   H-Shaped Construct: Constructal Design and Numerical
         Formulation 216
11.3   Optimal H-Shaped Cavities.. 220
11.4   Conclusions... 223
References.. 225

12   Co-design System for Tracking Targets Using Template Matching... 227
Yuri Marchetti Tavares, Nadia Nedjah,
and Luiza de Macedo Mourelle
12.1   Introduction.. 227
12.2   Template Matching.. 228
12.3   Software Development... 229
         12.3.1   Genetic Algorithms.. 230
         12.3.2   Particle Swarm Optimization 231
         12.3.3   Comparison Between GA and PSO........................ 233
12.4   Hardware Architecture 236
         12.4.1   Dedicated Co-processor 238
         12.4.2   Memory Controllers 241
12.5   Performance and Results.. 242
12.6   Conclusions... 244
References.. 245

**13   A Hybrid Estimation Scheme Based on the Sequential
         Importance Resampling Particle Filter and the Particle Swarm
         Optimization (PSO-SIR)**... 247
Wellington Betencurte da Silva, Julio Cesar Sampaio Dutra,
José Mir Justino da Costa, Luiz Alberto da Silva Abreu,
Diego Campos Knupp, and Antônio José Silva Neto
13.1   Introduction.. 247
13.2   Physical Problem and Mathematical Formulation 248
13.3   Inverse Problem Formulation and Solution 249
         13.3.1   Principle of the Sequential Monte Carlo Based
                     Estimation.. 250
         13.3.2   The Particle Filter Method 251

13.4   The Scheme Based on PSO and SIR Particle Filter   253
        13.4.1   The Particle Swarm Optimization Method (PSO)   253
        13.4.2   The PSO-SIR Filter Algorithm   254
13.5   Results and Discussions..   255
13.6   Conclusion...   260
References..   260

**14   Fault Detection Using Kernel Computational Intelligence
Algorithms**...   263
Adrián Rodríguez-Ramos, José Manuel Bernal-de-Lázaro,
Antônio José Silva Neto, and Orestes Llanes-Santiago
14.1   Introduction..   263
14.2   Preprocessing and Classification Tasks for Fault Diagnosis   264
        14.2.1   Preprocessing by Using Kernel ICA......................   264
        14.2.2   Kernel Fuzzy C-Means (KFCM)   265
        14.2.3   Optimization Algorithms and Kernel Function   266
14.3   Results and Discussion...   270
        14.3.1   Study Case: Tennessee Eastman Process   270
        14.3.2   Experimental Results   272
        14.3.3   Statistical Tests   277
14.4   Conclusions and Future Work   279
References..   279

Index...   283

Contributors

Luiz Alberto da Silva Abreu Department of Mechanical Engineering and Energy, Polytechnic Institute, IPRJ-UERJ, Nova Friburgo, Brazil

Lenilson Moreira Araujo Institute of Engineering and Geosciences, Federal University of Western Pará, Santarém, Brazil

Gebrail Bekdaş Department of Civil Engineering, Istanbul University, Avcılar, Istanbul, Turkey

José Manuel Bernal-de-Lázaro Department of Automation and Computing, Universidad Tecnológica de La Habana José Antonio Echeverría, Cujae, Habana, Cuba

Luiz Bevilacqua Federal University of Rio de Janeiro, COPPE-UFRJ, Rio de Janeiro, Brazil

Aldemir Aparecido Cavalini Jr. LMEst-Structural Mechanics Laboratory, School of Mechanical Engineering, Federal University of Uberlândia, Uberlândia, MG, Brazil

Leonardo Dagnino Chiwiacowsky Graduate Program in Industrial Engineering (PPGEP), University of Caxias do Sul (UCS), Bento Gonçalves, RS, Brazil

Vinícius Magno de Oliveira Coelho Department of Computational Modeling, Polytechnic Institute, Rio de Janeiro State University, Nova Friburgo, Brazil

Wilson Wolf Costa Postgraduate Program in Computational Modelling of Systems, Federal University of Tocantins, Palmas, Brazil

José Mir Justino da Costa Statistics Department, Federal University of Amazonas, Manaus, Brazil

Luciano Gonçalves da Silva Instituto Federal de Educação, Ciência e Tecnologia do Pará, IFPA, Paragominas, Brazil

Márcio Aurelio da Silva LMEst-Structural Mechanics Laboratory, School of Mechanical Engineering, Federal University of Uberlândia, Uberlândia, MG, Brazil

Patrick Vasconcelos da Silva Institute of Engineering and Geosciences, Federal University of Western Pará, Santarém, Brazil

Wellington Betencurte da Silva Chemical Engineering Program, Center of Agrarian Sciences and Engineering, Federal University of Espírito Santo, Alegre, Brazil

Antônio José Silva Neto Department of Mechanical Engineering and Energy, Polytechnic Institute, IPRJ-UERJ, Nova Friburgo, Brazil

Haroldo Fraga de Campos Velho Associated Laboratory for Computing and Applied Mathematics (LAC), National Institute for Space Research (INPE), São José dos Campos, SP, Brazil

Elizaldo Domingues dos Santos School of Engineering, Federal University of Rio Grande, Rio Grande, Brazil

Julio Cesar Sampaio Dutra Chemical Engineering Program, Center of Agrarian Sciences and Engineering, Federal University of Espírito Santo, Alegre, Brazil

Lídice Camps Echevarría Mathematics Department, Universidad Tecnológica de La Habana (Cujae), La Habana, Cuba

Emanuel da Silva Dias Estrada Centre for Computational Sciences, Federal University of Rio Grande, Rio Grande, Brazil

Augusto César Noronha Rodrigues Galeão Laboratório Nacional de Computação Científica, LNCC, Petrópolis, Brazil

Liércio André Isoldi School of Engineering, Federal University of Rio Grande, Rio Grande, Brazil

Aylin Ece Kayabekir Department of Civil Engineering, Istanbul University, Avcılar, Istanbul, Turkey

Diego Campos Knupp Department of Mechanical Engineering and Energy, Polytechnic Institute, IPRJ-UERJ, Nova Friburgo, Brazil

Gustavo Barbosa Libotte Department of Computational Modeling, Polytechnic Institute, Rio de Janeiro State University, Nova Friburgo, Brazil

Lucas Venancio Pires de Carvalho Lima Department of Mechanical Engineering and Energy, Polytechnic Institute, IPRJ-UERJ, Nova Friburgo, Brazil

Fran Sérgio Lobato NUCOP-Laboratory of Modeling, Simulation, Control and Optimization, School of Chemical Engineering, Federal University of Uberlândia, Uberlândia, MG, Brazil

Anderson Alvarenga de Moura Meneses Institute of Engineering and Geosciences, Federal University of Western Pará, Santarém, Brazil

Roberto Luiz Souza Monteiro Program of Computational Modeling, SENAI CIMATEC, Salvador, Brazil

Davidson Martins Moreira Program of Computational Modeling, SENAI CIMATEC, Salvador, Brazil

Patrick Letouzé Moreira Postgraduate Program in Computational Modelling of Systems, Federal University of Tocantins, Palmas, Brazil

Luiza de Macedo Mourelle Department of Systems Engineering and Computing, Rio de Janeiro State University, Rio de Janeiro, Brazil

Fernando Nogueira Nast Institute of Engineering and Geosciences, Federal University of Western Pará, Santarém, Brazil

Nadia Nedjah Department of Electronic Engineering and Telecommunications, Rio de Janeiro State University, Rio de Janeiro, Brazil

Sinan Melih Nigdeli Department of Civil Engineering, Istanbul University, Avcılar, Istanbul, Turkey

Hernane Borges de Barros Pereira Program of Computational Modeling, SENAI CIMATEC, Salvador, Brazil

Gustavo Mendes Platt Universidade Federal do Rio Grande, Santo Antônio da Patrulha, Rio Grande do Sul, Brazil

David Nadler Prata Postgraduate Program in Computational Modelling of Systems, Federal University of Tocantins, Palmas, Brazil

Ana Carolina Rios-Coelho Federal University of Western Pará, Institute of Engineering and Geosciences, Santarém, PA, Brazil

Luiz Alberto Oliveira Rocha UNISINOS, São Leopoldo, Brazil

Marcelo Lisboa Rocha Postgraduate Program in Computational Modelling of Systems, Federal University of Tocantins, Palmas, Brazil

Adrián Rodríguez-Ramos Department of Automation and Computing, Universidad Tecnológica de La Habana José Antonio Echeverría, Cujae, Habana, Cuba

Wagner Figueiredo Sacco Federal University of Western Pará, Institute of Engineering and Geosciences, Santarém, PA, Brazil

Orestes Llanes-Santiago Automatic and Computing Department, Universidad Tecnológica de La Habana (Cujae), La Habana, Cuba

Roberto Schirru Program of Nuclear Engineering, Federal University of Rio de Janeiro, Rio de Janeiro, Brazil

Valder Steffen Jr. LMEst-Structural Mechanics Laboratory, School of Mechanical Engineering, Federal University of Uberlândia, Uberlândia, MG, Brazil

Yuri Marchetti Tavares Department of Weapons, Navy Weapons Systems Directorate, Brazilian Navy, Rio de Janeiro, Brazil

Reynier Hernández Torres Associated Laboratory for Computing and Applied Mathematics (LAC), National Institute for Space Research (INPE), São José dos Campos, SP, Brazil

Xin-She Yang School of Science and Technology, Middlesex University, London, UK

Acronyms

ABC	Artificial Bee Colonies
ABC-AP	Artificial Bee Colony with adaptive penalty function
ACO	Ant colony optimization
ACO-d	Ant Colony Optimization with dispersion
ARMA	Auto Regressive Moving Average
BB-BC	Big bang big crunch algorithm
CE	Cross-Entropy Algorithm
CMA-ES	Covariance Matrix Adaptation Evolution Strategy
CP	Complete network
CS	Cuckoo search
DE	Differential evolution
DEwPC	Differential Evolution with Particle Collision
DHPSACO	Discrete heuristic particle swarm ant colony optimization
DRV	Double retrograde vaporization
EFPD	Effective Full Power Day
EOC	End-of-cycle
EOS	Equations of state
ES	Eagle strategy
ESO	Evolutionary structural optimization
ESTP	Euclidean Steiner Tree Problem
FA	Firefly algorithm
FAR	False Alarm Rate
FAs	Fuel Assemblies
FDI	Fault Detection and Isolation
FDR	Fault Detection Rate
FORM	First Order Reliability Method
FPA	Flower Pollination Algorithm
FPBIL	Parameter-Free PBIL
GA	Genetic algorithm
HJ	Hooke–Jeeves
HS	Harmony Search

HY	Hybrid network
IB	Isobutene
ICDE	Improved constrained differential evolution algorithm
IRA	Inverse reliability analysis
KFCM	Kernel Fuzzy C-means
KICA	Kernel Independent Component Analysis
LAN	Local area network
LOA	Lightning Optimization Algorithm
LP	Loading Pattern
LSA	Lightning Search Algorithm
MAE	Mean absolute error
MBA	Mine Blast Algorithm
MEC	Mean Effective Concept
MeOH	Methanol
MLP	Multilayer perceptron
MLR	Multiple linear regression
MO	Multi-objective
MPCA	Multi-Particle Collision Algorithm
MPP	Most probable point
MST	Minimum spanning tree
MTBE	Methyl-*tert*-butyl-ether
NFE	Fitness-function evaluations
NPP	Nuclear Power Plant
OBL	Opposition-based learning
PBIL	Population-Based Incremental Learning
PCA	Particle Collision Algorithm
PCC	Pearson's Correlation Coefficient
PR	Peng-Robinson
PSO	Particle Swarm Optimization
PSOPC	Particle Swarm Optimizer with passive congregation
PWR	Pressurized Water Reactor
RA	Reactive azeotrope
RBF	Radial base function networks
RBMPCA-HJ	Rotation-based multi-particle collision algorithm with Hooke–Jeeves
RBO	Reliability-based optimization
RBS	Rotation-based sampling
RC	Reinforced concrete
RD	Robust design, random network
RGD	Robust geotechnical design
RK	Random Keys
SA	Simulated Annealing
SDI	Structural Damage Identification
SF	Scale-free network
SHM	Structural health monitoring

SIR	Sampling Importance Resampling
SMRF	Seismic steel moment-resisting frame
SORA	Sequential Optimization and Reliability Assessment
SORM	Second-Order Reliability Method
SQP	Sequential Quadratic Programming
STP	Steiner Tree Problem
SVDK	Smart Vision Development Kit
SW	Small-world network
TE	Tennessee Eastman
TLBO	Teaching-learning based optimization
TMD	Tuned mass damper
VRPTW	Vehicle routing problem with time windows
WAM	Water allocation module
WWER	Water-Water Energetic Reactor

Chapter 1
An Overview of the Use of Metaheuristics in Two Phase Equilibrium Calculation Problems

Gustavo Mendes Platt, Lucas Venancio Pires de Carvalho Lima, Gustavo Barbosa Libotte, and Vinícius Magno de Oliveira Coelho

1.1 Introduction

Phase equilibrium calculations are commonly described as optimization problems or nonlinear algebraic systems (that can also be converted to scalar objective-functions). Furthermore, in many occasions, these problems show more than one solution (root), as pointed out by Platt [24]. In this context, metaheuristics arise as natural options to solve these challenging situations, even though some deterministic routines can also be used (for instance, see Guedes et al. [13]).

Nature-inspired algorithms—a special kind of metaheuristics—have been continuously proposed in the last years. This kind of optimization framework is usually tested in a set of benchmark functions (for instance, Himmelblau function, Rosenbrock's banana function, Ackley function). Thus, many authors compared the results (in terms of number of function evaluations, computation time, mean value of fitness, etc.) among metaheuristics. On the other hand, real engineering problems

G. M. Platt (✉)
Universidade Federal do Rio Grande, Santo Antônio da Patrulha, Rio Grande do Sul, Brazil
e-mail: gmplatt@furg.br

L. V. P. d. C. Lima
Department of Mechanical Engineering and Energy, Polytechnic Institute, IPRJ-UERJ, Nova Friburgo, Brazil
e-mail: lucas.lima@iprj.uerj.br

G. B. Libotte · V. M. d. O. Coelho
Department of Computational Modeling, Polytechnic Institute, Rio de Janeiro State University, Nova Friburgo, Brazil
e-mail: gustavolibotte@iprj.uerj.br; vcoelho@iprj.uerj.br

© Springer Nature Switzerland AG 2019
G. Mendes Platt et al. (eds.), *Computational Intelligence, Optimization and Inverse Problems with Applications in Engineering*,
https://doi.org/10.1007/978-3-319-96433-1_1

1

are not common scenarios for comparisons. As pointed out by Nesmachnow [20], "small toy problems" must be avoided as tools for comparing metaheuristics.

In this chapter, we present an overview of the application of some metaheuristics in two "hard" phase equilibrium problems: the calculation of reactive azeotropes, and the prediction of dew point pressures in a binary system with double retrograde vaporization (DRV). Moreover, we addressed a detailed comparison of two metaheuristic algorithms—Differential Evolution (DE) [35] and Harmony Search (HS) [12]—in these two problems.

The chapter is organized as follows: the next section (Sect. 1.2) contains a review regarding the application of metaheuristics in the studied problems. Sections 1.3 and 1.4 present the description of the problems (with thermodynamic details) and the metaheuristics employed, respectively, as well as a brief explanation regarding the statistical analysis. The results are depicted and discussed in Sect. 1.5. Finally, some conclusions are presented in Sect. 1.6.

1.2   Analysis of Previous Works

In this section we present a description of the use of metaheuristics in phase equilibrium problems, particularly in the calculation of reactive azeotropes, and in the prediction of double retrograde vaporization. The description of the physical phenomena will be detailed in the next section.

Bonilla-Petriciolet et al. [7] employed a new methodology—based on Simulated Annealing algorithm—devoted to the calculation of homogeneous azeotropes (reactive and non-reactive). This strategy demonstrated to be robust and computationally efficient.

Platt [23] presented a study of the application of the Flower Pollination Algorithm [43] in the prediction of the double retrograde vaporization phenomenon in the system ethane + limonene at $T = 307.4$ K. The Flower Pollination Algorithm exhibits an unusual movement for the elements of the population: the Lévy flights [21]. This characteristic is an important difference between the Flower Pollination Algorithm and other metaheuristics (the Lévy flight is a hyperdiffusive movement [19]). Platt [23] demonstrated that DRV is a very difficult problem to solve using metaheuristics. The neighborhood of the solutions was found, but the existence of a large amount of local optima and infinite global optima (trivial solutions, without physical utility, as explained in the next section) are, undoubtedly, features that make this problem a challenge for stochastic algorithms.

Platt et al. [27] employed the Firefly algorithm [42] and the Luus-Jaakola method [17] in the prediction of the two azeotropes in the reactive system isobutene + methanol + MTBE (methyl-*tert*-butyl-ether, an anti-knocking agent used in gasoline) at 8 atm. These authors demonstrated that the calculation of reactive azeotropes can be a difficult task when using metaheuristics.

Platt et al. [26] solved the prediction of the double retrograde vaporization in the system ethane + limonene using a hyperheuristic structure. The hyperheuristic algorithm can be viewed, *grosso modo*, as "a metaheuristic that chooses other metaheuristics." In this work, the Simulated Annealing was chosen as the master-metaheuristic. This algorithm then was capable to choose between the following techniques: Differential Evolution, Particle Swarm Optimization, Luus-Jaakola, Multimodal Particle Swarm Optimization and, obviously, Simulated Annealing.

Other metaheuristics, such as Harmony Search [6], Genetic Algorithms [2, 29], Simulated Annealing, [5] and Ant Colony Optimization [11] have also been successfully used in phase/chemical equilibrium problems (parameter estimation of thermodynamic models, azeotropy calculations, phase stability calculations, among others), but not specifically in the two instances addressed here.

1.3   Description of the Problems

In this section we present some details of the two problems approached in this chapter: (1) the calculation of a reactive azeotrope (henceforth referred to as RA problem) in a ternary system; and (2) the calculation of dew point pressures (and liquid phase compositions) in a binary system with double retrograde vaporization (represented by DRV problem).

These two phase/chemical equilibrium problems are "hard" tests for any methodology devoted to the solution of nonlinear algebraic systems or optimization techniques, by the following reasons: (1) high nonlinearity (as a consequence of the thermodynamic models); (2) existence of multiple solutions (global minima, from an optimization point of view); and (3) multiple local minima (in the case of DRV), that are not solutions of the problem. Moreover, the RA problem is in $\mathbb{R}^5$ (and many benchmarks usually employed are in lower dimensions).

1.3.1   Calculation of a Reactive Azeotrope in a Ternary System (RA Problem)

The concept of a reactive azeotrope was firstly introduced by Barbosa and Doherty [3]. These authors described a new set of conditions for azeotropes under a chemical equilibrium constraint, in opposition to the classical stationary points in bubble and dew point curves/surfaces. In a subsequent paper [4], the same authors presented a new set of compositional coordinates, which are equal in the reactive azeotropic condition.

The reactive azeotrope condition, considering only one chemical equilibrium reaction, is represented as [3]:

$$\frac{y_1 - x_1}{v_1 - v_T x_1} = \frac{y_i - x_i}{v_i - v_T x_i} \tag{1.1}$$

In this expression, y and x represent vapor and liquid molar fractions, respectively. The subscripts 1 and i are related to the indexes of components in the mixture. v_i is the stoichiometric coefficient for a component i and v_T is the sum of the stoichiometric coefficients for all reactants/products of the equilibrium reaction.

Another set of compositional coordinates in reactive systems with multiple equilibrium chemical reactions was proposed by Ung and Doherty [37]. In this case, liquid and vapor transformed coordinates are:

$$X_i = \left(\frac{x_i - \boldsymbol{v}_i^t \boldsymbol{N}^{-1} \boldsymbol{x}_{ref}}{1 - \boldsymbol{v}_{tot}^t \boldsymbol{N}^{-1} \boldsymbol{x}_{ref}} \right)$$

$$Y_i = \left(\frac{y_i - \boldsymbol{v}_i^t \boldsymbol{N}^{-1} \boldsymbol{y}_{ref}}{1 - \boldsymbol{v}_{tot}^t \boldsymbol{N}^{-1} \boldsymbol{y}_{ref}} \right) \tag{1.2}$$

In this last equation, $\boldsymbol{v}_i$ is a vector with the stoichiometric coefficients for the component i in all equilibrium chemical reactions in the system. $\boldsymbol{v}_{tot}$ represents the sum of the stoichiometric coefficients for all reacting components in all chemical reactions. Finally, $\boldsymbol{N}$ is a square matrix with the stoichiometric coefficients for the reference components (represented by the subscript ref) in the chemical reactions (the number of reference components is equal to the number of equilibrium reactions).

Remark There are two important matrices in the context of the calculation of vapor–liquid equilibrium using the Ung–Doherty coordinate transformation: the matrix of stoichiometric coefficients $\boldsymbol{v}$ and the matrix of the stoichiometric coefficients for the reference components $\boldsymbol{N}$. $\boldsymbol{v}$ can be represented as:

$$\boldsymbol{v} = \begin{bmatrix} v_{1,1} & v_{1,2} & \cdots & v_{1,r-1} & v_{1,r} \\ v_{2,1} & v_{2,2} & \cdots & v_{2,r-1} & v_{2,r} \\ \vdots & \vdots & \ddots & \vdots & \vdots \\ v_{c,1} & v_{c,2} & \cdots & v_{c,r-1} & v_{c,r} \end{bmatrix}$$

Moreover, $\boldsymbol{N}$ is a (square) submatrix of $\boldsymbol{v}$: the number of lines of $\boldsymbol{N}$ is equal to the number of equilibrium reactions r. As an example, we will detach the Ung–Doherty definitions for a system with five components and two equilibrium reactions. In this case:

(continued)

$$\nu = \begin{bmatrix} \nu_{1,1} & \nu_{1,2} \\ \nu_{2,1} & \nu_{2,2} \\ \nu_{3,1} & \nu_{3,2} \\ \nu_{4,1} & \nu_{4,2} \\ \nu_{5,1} & \nu_{5,2} \end{bmatrix}$$

Considering that the reference components are (for example) 2 and 4, the matrix N is then $N = \begin{bmatrix} \nu_{2,1} & \nu_{2,2} \\ \nu_{4,1} & \nu_{4,2} \end{bmatrix}$. Using this notation, $\nu_1{}^t = \begin{bmatrix} \nu_{1,1} & \nu_{1,2} \end{bmatrix}$.

The inverse matrix can be easily calculated (if it exists, as pointed out by Ung and Doherty [37]): $N^{-1} = \frac{1}{\nu_{2,1}\nu_{4,2} - \nu_{2,2}\nu_{4,1}} \begin{bmatrix} \nu_{4,2} & -\nu_{2,2} \\ -\nu_{4,1} & \nu_{2,1} \end{bmatrix}$.

The term $\nu_i^t N^{-1} x_{ref}$ that appears in Eq. (1.2) is then (for component 1):

$$\nu_1^t N^{-1} x_{ref} = \frac{1}{\nu_{2,1}\nu_{4,2} - \nu_{2,2}\nu_{4,1}} \begin{bmatrix} \nu_{1,1} & \nu_{1,2} \end{bmatrix} \begin{bmatrix} \nu_{4,2} & -\nu_{2,2} \\ -\nu_{4,1} & \nu_{2,1} \end{bmatrix} \begin{bmatrix} x_2 \\ x_4 \end{bmatrix}$$

A similar analysis was presented in a system representing the alkylation of xylenes with di-*tert*-butyl-benzene by Ung and Doherty [36].

The reactive system studied here is composed by isobutene (IB) (1), methanol (MeOH) (2), and methyl-*tert*-butyl-ether (3) (MTBE). In this system, an etherification reaction occurs: IB + MeOH $\rightleftharpoons$ MTBE. Since only one chemical equilibrium reaction takes place: $\nu = \begin{bmatrix} \nu_{1,1} \\ \nu_{2,1} \\ \nu_{3,1} \end{bmatrix} = \begin{bmatrix} -1 \\ -1 \\ 1 \end{bmatrix}$ and, therefore, $\nu_{tot} = \mathbf{1}^t \nu = -1 - 1 + 1 = -1$, where $\mathbf{1}^t = \begin{bmatrix} 1 & 1 & \cdots & 1 & 1 \end{bmatrix}$.

Considering the definition of Ung–Doherty coordinates, and using MTBE as the reference component, $N = N^{-1} = 1$. Thus, Ung–Doherty compositions for components 1 and 2 are given by:

$$X_1 = \frac{x_1 + x_3}{1 + x_3}, \quad X_2 = \frac{x_2 + x_3}{1 + x_3} \tag{1.3}$$

$$Y_1 = \frac{y_1 + y_3}{1 + y_3}, \quad Y_2 = \frac{y_2 + y_3}{1 + y_3} \tag{1.4}$$

The condition for azeotropy in a ternary equilibrium reactive system is [37]:

$$X_1 = Y_1 \tag{1.5}$$

In this system, the vapor–liquid equilibrium is represented by the modified Raoult's Law [38], according to:

$$P y_i = x_i \gamma_i P_i^{sat}, \quad i = 1, 2, 3 \tag{1.6}$$

In these equations, P is the system pressure, γ_i is the activity coefficient of component i (calculated by an appropriated model), and P_i^{sat} is the saturation pressure of a pure component.

Following the same thermodynamic modelling employed by Maier et al. [18], the results will be obtained using the Wilson model [41].

Since we are dealing with a system in a chemical equilibrium condition, the following constraint must be included:

$$K = \prod_{i=1}^{c} \hat{a}_i^{v_i} \tag{1.7}$$

where K is the chemical equilibrium constant and $\hat{a}_i$ is the activity of component i, calculated as $\hat{a}_i = \gamma_i x_i$.

A more usual form employed in simultaneous phase and chemical equilibrium calculation can be obtained as follows:

$$\log(K) = \log\left(\prod_{i=1}^{c} \hat{a}_i^{v_i}\right) \rightarrow \log(K) = \log\left(\hat{a}_1^{v_1}\right) + \log\left(\hat{a}_2^{v_2}\right) + \cdots + \log\left(\hat{a}_c^{v_c}\right) \tag{1.8}$$

Or, in a compact notation:

$$\log(K) - \boldsymbol{v}^t \begin{bmatrix} \log(\hat{a}_1) \\ \log(\hat{a}_2) \\ \vdots \\ \log(\hat{a}_c) \end{bmatrix} = 0 \tag{1.9}$$

The total number of equations in this system is: 3 [Eq. (1.6)] + 1 [Eq. (1.9)] + 2 (constraints over molar fraction vectors, i.e., $\sum_{i=1}^{c} x_i = 1$ and $\sum_{i=1}^{c} y_i = 1$). Thus, we obtain 6 equations.

In this system, the thermodynamic coordinates are: x_i ($i = 1, 2, 3$), y_i ($i = 1, 2, 3$), T and P. The system pressure is specified (8 atm); therefore, only 7 variables remain. Thus, the simultaneous chemical and phase equilibrium calculation can be done with only one specification for molar fractions (x_1 or y_1, for instance). In this particular problem, the unique degree of freedom is attended by the reactive azeotropy relation, Eq. (1.5).

In order to diminish the number of variables involved in the optimization procedure, the third expression in Eq. (1.6) can be expressed as $P(1 - y_1 - y_2) = (1 - x_1 - x_2)\gamma_3 P_3^{sat}$, due to the constraints $y_3 = 1 - y_1 - y_2$ and $x_3 = 1 - x_1 - x_2$. The final formulation of the nonlinear algebraic system is then represented as:

$$P y_i - x_i \gamma_i P_i^{sat} = 0, \quad i = 1, 2$$

$$P (1 - y_1 - y_2) - (1 - x_1 - x_2) \gamma_3 P_3^{sat} = 0$$

$$\log(K) - \boldsymbol{v}^t \begin{bmatrix} \log(\hat{a}_1) \\ \log(\hat{a}_2) \\ \vdots \\ \log(\hat{a}_c) \end{bmatrix} = 0$$

$$X_1 - Y_1 = 0 \tag{1.10}$$

In this situation, the variables are: x_1, x_2, y_1, y_2 and T, and the optimization procedure is a search in $\mathbb{R}^5$.

The nonlinear algebraic system must be converted to a merit function (since the algorithms to be evaluated are optimization procedures). A simple and obvious proposal is the sum of squares of the residues of the nonlinear equations: $f = \sum_{k=1}^{5} F_k^2$, where F_k represents each residue of the equations expressed in Eq. (1.10).

Figure 1.1 presents a pictorial representation of the azeotropic condition in the reactive system IB + MeOH + MTBE at 8 atm, with $K = 49$. The two azeotropes are represented, in the Ung–Doherty coordinate system, by the tangency between the bubble and dew point curves. Obviously, the problem is not in the plane (as it might seen at first sight), but the projection in the Ung–Doherty plane produces a useful and pictorial representation of the azeotropes. Moreover, the curves of bubble and dew point temperatures are not obtained in the search for azeotropes, and are generated by a specific computational routine. The coordinates of the reactive azeotropes are presented in Table 1.1.

An amplification of Fig. 1.1, focusing on the region of the reactive azeotropy, can be viewed in Fig. 1.2. The two azeotropes (a minimum and a maximum in the tangency points) are very clear in this figure.

Table 1.1 Compositions and temperatures in the RA problem

Root	x_1	x_2	y_1	y_2	$T(K)$
1	0.0138208	0.4037263	0.0748231	0.4406101	391.1756
2	0.0446038	0.1197971	0.1727712	0.2378773	392.2444

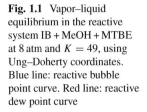 **Fig. 1.1** Vapor–liquid equilibrium in the reactive system IB + MeOH + MTBE at 8 atm and $K = 49$, using Ung–Doherty coordinates. Blue line: reactive bubble point curve. Red line: reactive dew point curve

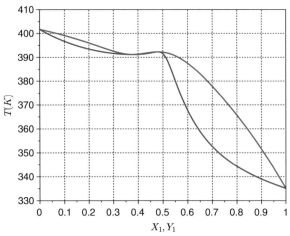

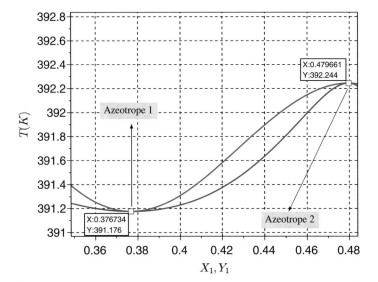

Fig. 1.2 Amplification of the vapor–liquid equilibrium curves in the vicinities of the azeotropes. Blue line: reactive bubble point curve. Red line: reactive dew point curve

1.3.2 Calculation of Dew Point Pressures in a Binary System with Double Retrograde Vaporization (DRV Problem)

The phenomenon known as "double retrograde vaporization" (DRV) was firstly identified in the binary systems formed by methane + n-butane [8] and methane + n-pentane [9] at temperatures close to the critical temperature of methane. The DRV is characterized by the existence of three or four dew point pressures in a small composition range (close to the more volatile pure component of the system).

Moreover, this phenomenon only occurs at temperatures very close to the critical temperature of the system. Thus, dew point calculations in the vicinities of DRV can be considered as difficult problems from a numerical point of view.

Some authors employed cubic equations of state in order to predict dew point pressures and liquid phase compositions in systems with DRV [28].

Following the approach of Raeissi and Peters [28], we used, in this work, the Peng–Robinson equation of state (PR EOS) [22]:

$$P = \frac{RT}{\bar{V} - b} - \frac{a_m}{\bar{V}(\bar{V} - b_m) + b_m(\bar{V} - b_m)} \tag{1.11}$$

In this equation, P is the system pressure, T is the absolute temperature, and $\bar{V}$ is the molar volume. The quantities a_m and b_m are parameters of the mixture. Moreover, R is the Universal gas constant.

Given a (T, P) pair, the Peng–Robinson equation of state can also be represented by a cubic polynomial in terms of the compressibility factor (Z), as follows [38]:

$$Z^3 - (1 - B)Z^2 + (A - 2B - 3B^2) - (AB - B^2 - B^3) = 0 \tag{1.12}$$

The mixture parameters A and B are defined as

$$A = \frac{a_m P}{R^2 T^2}; \quad B = \frac{b_m P}{RT} \tag{1.13}$$

Moreover, the following quantities are necessary, for pure components [22]:

$$a_i = 0.45724 \frac{R^2 T_{ci}^2}{P_{ci}}; \quad b_i = 0.07780 \frac{RT_{ci}}{P_{ci}} \tag{1.14}$$

Parameters T_{ci} and P_{ci} represent, respectively, the critical temperature and the critical pressure for such a component. Since we are dealing with mixtures (and not pure substances), some combination and mixing rules are necessary. The combination rule employed here uses [38]

$$a_{ij} = \sqrt{a_i \alpha_i a_j \alpha_j}(1 - k_{ij}) \tag{1.15}$$

where the parameter α_i is calculated by

$$\alpha_i^{0.5} = 1 + \left(0.37464 + 1.54226\omega_i - 0.26992\omega_i^2\right)\left(1 - T_{Ri}^{0.5}\right) \tag{1.16}$$

and k_{ij} is called "binary interaction parameter" (obtained from vapor–liquid equilibrium data; here, we used $k_{ij} = 0$, $\forall i, j$). The combination rule for parameter b is not necessary, i.e., $b_i = b_{ij}$, $\forall j$. Moreover, ω_i is called "acentric factor" for a pure substance and the reduced temperature is calculated by $T_{Ri} = \frac{T}{T_{ci}}$.

By definition, the compressibility factor is $Z = \frac{P\bar{V}}{RT}$, where $\bar{V}$ is the molar volume of the mixture. Thus, for a pair (T, P), the cubic polynomial represented by Eq. (1.12) can be solved and three roots obtained. If one obtains three real roots, the smallest root is the liquid phase compressibility factor, whereas the largest root corresponds to vapor phase. These quantities are necessary in order to evaluate the fugacity coefficients for both phases.

The fugacity coefficient for component i with the PR EOS and using any set of mixing rules can be calculated by Haghtalab and Espanani [14]:

$$
\log\left(\hat{\phi}_i\right) = \frac{1}{b_m}\left[\frac{\partial(nb_m)}{\partial n_i}\right](Z-1) - \log(Z-B)
$$

$$
- \frac{a_m}{2\sqrt{2}b_m RT}\left[\frac{1}{a_m}\left(\frac{1}{n}\frac{\partial(n^2 a_m)}{\partial n_i}\right) - \frac{1}{b_m}\left(\frac{\partial(nb_m)}{\partial n_i}\right)\right]
$$

$$
\log\left(\frac{Z+2.414B}{Z-0.414B}\right) \tag{1.17}
$$

The (classical) mixing rules employed here are calculated by Walas [38]:

$$
a_m = \sum_i \sum_j a_{ij} z_i z_j; \quad b_m = \sum_i b_i z_i \tag{1.18}
$$

Then, the derivatives $\frac{\partial(nb_m)}{\partial n_i}$ and $\frac{\partial(n^2 a_m)}{\partial n_i}$ that appear in Eq. (1.17) can be obtained as follows:

$$
nb_m = n\sum_i b_i z_i = \sum_i b_i n_i \tag{1.19}
$$

and, clearly:

$$
\frac{\partial(nb_m)}{\partial n_i} = b_i \tag{1.20}
$$

In a similar way:

$$
n^2 a_m = n^2 \sum_i \sum_j a_{ij} z_i z_j = \sum_i \sum_j a_{ij} n_i n_j \tag{1.21}
$$

and, thus:

$$
\frac{\partial(n^2 a_m)}{\partial n_i} = 2\sum_j a_{ij} n_j \rightarrow \frac{1}{n}\frac{\partial(n^2 a_m)}{\partial n_i} = 2\sum_j a_{ij} z_j \tag{1.22}
$$

Finally, the fugacity coefficient for the component i in the mixture with this set of mixing rules is obtained by:

$$\log\left(\hat{\phi}_i\right) = \frac{b_i}{b_m}(Z-1) - \log(Z-B)$$

$$-\frac{a_m}{2\sqrt{2}b_m RT}\left[\frac{2\sum_j a_{ij}z_j}{a_m} - \frac{b_i}{b_m}\right]\log\left(\frac{Z+2.414B}{Z-0.414B}\right) \quad (1.23)$$

Or, in a more convenient way:

$$\log\left(\hat{\phi}_i\right) = \frac{b_i}{b_m}(Z-1) - \log(Z-B)$$

$$-\frac{A}{2\sqrt{2}B}\left[\frac{2\sum_j a_{ij}z_j}{a_m} - \frac{b_i}{b_m}\right]\log\left(\frac{Z+2.414B}{Z-0.414B}\right) \quad (1.24)$$

The vapor–liquid coexistence problem is represented by the isofugacity equations [38]:

$$\hat{f}_i^L = \hat{f}_i^V \quad i = 1, \ldots, c \quad (1.25)$$

In this last expression, $\hat{f}_i$ represents the fugacity of a component i in the mixture. The superscripts L and V indicate the liquid and vapor phases, respectively. The fugacity of a component in the mixture for a phase α is expressed by: $\hat{f}_i^\alpha = \hat{\phi}_i^\alpha P z_i$.

Thus, we can formulate the dew point problem in a binary mixture as:

$$\hat{\phi}_1^L x_1 = \hat{\phi}_1^V y_1$$
$$\hat{\phi}_2^L x_2 = \hat{\phi}_2^V y_2 \quad (1.26)$$

In this situation, the problem is described by four equations: two equations describing the vapor–liquid equilibrium, Eq. (1.26), and the two constraints regarding the molar fractions, $x_1 + x_2 = 1$ and $y_1 + y_2 = 1$. Here, the variables are: T, P, x_1, x_2, y_1, y_2. Considering that y_1 (or y_2) and system temperature are specified, the variables become x_1, x_2 and P. Since $x_2 = 1 - x_1$, the dew point problem can be viewed as an inversion of a function from the plane to the plane (or, in other words, the solution of a two-dimensional nonlinear algebraic system):

$$F_1 = \hat{\phi}_1^L x_1 - \hat{\phi}_1^V y_1 = 0$$
$$F_2 = \hat{\phi}_2^L(1-x_1) - \hat{\phi}_2^V y_2 = 0 \quad (1.27)$$

The vector of variables is then represented as $\theta = \begin{bmatrix} x_1 \\ P \end{bmatrix}$.

Table 1.2 Dew point pressures and liquid compositions in the DRV problem

Root	Pressure (kPa)	x_1
1	619.24	0.1567
2	4859.5	0.9829
3	4931.2	0.9918
4	5007.3	0.9979

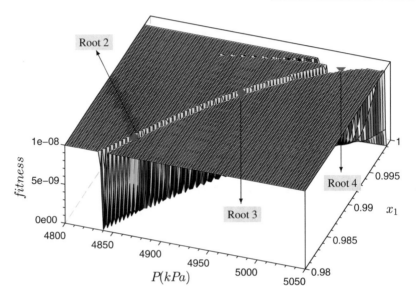

Fig. 1.3 Surface of fitness function in the DRV problem. The roots of the problem are pictorially represented by red triangles (with fitness equal to zero)

The final step is to convert the nonlinear system represented by Eq. (1.27) in a merit function (here, represented as "fitness," a term commonly used in bioinspired algorithms). A very usual approach is to sum squares of the residues of each nonlinear equation, in the form: $f = \sum_{i=1}^{2} F_i^2$. Table 1.2 presents the dew points for this problem.

Figure 1.3 exhibits the fitness function for the problem at hand. A similar figure was presented by Platt et al. [26], but using other ranges for the variables; here, we present a closer view of the multiple minima in this function. A basic analysis of this figure immediately indicates a large quantity of minima (many of them are local minima). Therefore, it does not seem a good idea to convert the nonlinear system in an optimization problem in this case. Indeed, it is not and, thus, this problem is a very severe test for new metaheuristics. On the other hand, even when formulated as a nonlinear algebraic system, to solve this system is a very hard task, since the basins of attraction of some roots are small, as pointed out by Platt et al. [26].

Figure 1.4 presents an upper view of the fitness function in the DRV problem. Root 1 is the low-pressure root and, then, this root does not appear in the figure.

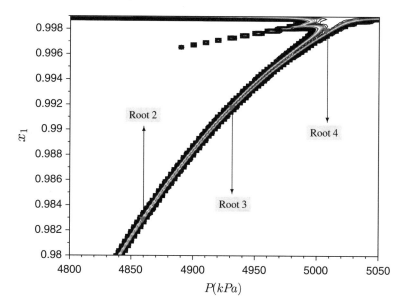

Fig. 1.4 Upper view of the fitness function in DRV problem, with emphasis to the high-pressure roots

Figure 1.5 presents in more detail the region in the neighborhood of Root 4. Although this solution presented the highest pressure in this problem, it is not the "hardest" root to determine, for many root-solving techniques (such as Newton-like methods), as explained by Libotte and Platt [16]. In fact, the high quantity of local minima diminishes when the zone of trivial solutions is approached. This situation can also be seen through Fig. 1.3.

1.4   Description of the Metaheuristics and Statistical Analysis

In this section, the two metaheuristics that will be used in the case study—Differential Evolution and Harmony Search—are presented and discussed. A brief description of other metaheuristics used (in previous works) in the thermodynamic problems addressed here is also depicted. Finally, the statistical methodology employed for results comparison is introduced. One should keep in mind that the comparison done in this chapter is generic, in the sense that it may be applied to any stochastic optimization algorithm. Furthermore, all techniques are employed without any further improvement in order to obtain multiple optima (such as niching techniques, for instance).

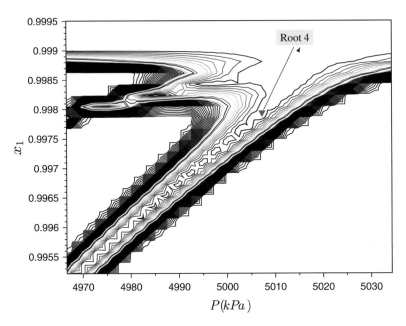

Fig. 1.5 Upper view of the fitness function in DRV problem: highlight of the vicinities of the highest pressure root

1.4.1 The Differential Evolution Algorithm (DE)

The Differential Evolution (DE) was proposed by Storn and Price [35], and is an extremely popular stochastic algorithm. Some authors claim that DE is among the most efficient stochastic algorithms in the present days [10]. A very useful review addressing many variants of DE (including applications to multiobjective optimization, large-scale problems, constrained problems) is presented by Das and Suganthan [10]. Our goal in this chapter is to present the applications of metaheuristics in some particular thermodynamic problems, and not to present comments/criticisms regarding the characteristics of the algorithms.

As described in the original paper by Storn and Price [35], the three steps of DE are: (1) mutation; (2) crossover; and (3) selection. The control parameters of DE are: F (employed to control the mutation step size); CR (the crossover constant); and the size of the population.

In this chapter we use the DE/rand/bin/1 implementation [35]. The pseudocode is presented in Storn and Price [35], for instance.

1.4.2   The Harmony Search Algorithm (HS)

Harmony Search (HS) is a metaheuristic [12] based on the process of improvisation of a melody by a jazz musician. According to Geem et al. [12], the global optimum of mathematical problems can be compared—using this metaphor—to a "fantastic harmony" and, thus, the optimization is a search for a beautiful harmony.

The main idea of Harmony Search is based on the decisions of a musician, which improvises a piece of a music using three strategies:

- play a piece of a known harmony, using his/her memory (called as harmony memory);
- if the musician uses his/her memory, he/she can promote a pitch adjustment;
- the musician can use an absolutely random piece of melody.

There are some controversies regarding the Harmony Search as a new optimization algorithm. For instance, Weiland [39] advocates that Harmony Search is a particular case of $(\mu + 1)$ evolution strategy [31]. In the same context, Sorensen [34] presented some criticisms about "new metaphors" in metaheuristics. More recently, Saka et al. [33] refuted the arguments of Weiland [39], pointing out some differences of Harmony Search and $(\mu + 1)$ strategy. This discussion—extremely relevant in the context of the proposal of new metaheuristics—is, however, out of the scope of this chapter. The proposal here is to apply a rigorous statistical analysis in the comparison of metaheuristics in problems arising from real-world situations. For this reason, we stay way from the previously mentioned discussion.

The parameters for Harmony Search are: (1) the harmony memory (represented by hm); (2) the size of harmonic memory (hms); the probability to accept an information from the harmonic memory (also called harmony considering rate) ($hmcr$); the probability to adjust the pitch (par); and, finally, the maximum change in pitch adjustment (b_{range}).

The harmonic memory hm represents the "population," and has dimension ($d \times hms$), where d is the dimension of the problem.

1.4.3   Statistical Analysis

1.4.3.1   Comparison of Different Sets of Data

When comparing two different sets of data, as the ones derived from the application of two different metaheuristics, one approach is to compare a set of n independent runs for each algorithm. Evaluating only the value of the mean (or the median) can lead to wrong conclusions, because even if two samples present different means (or medians), it is still possible to find no significantly differences between them; i.e. they can be two samples from the same population.

In that sense, it is important to evaluate if statistically the two evaluated samples are different from each other, before any conclusion. Hypothesis method can be used to evaluate this issue. The hypotheses are stated:

- H_0: No significant differences between the two studied samples were found; they come from the same population;
- H_1: Significant differences were found from the two populations.

H_0 is called the null hypothesis, and H_1 is called the alternative hypothesis. Statistical tests evaluate the probability of rejecting the null hypothesis. If the null hypothesis is rejected, the alternative hypothesis is considered reasonable. On the contrary, if the test fails to reject the H_0, the null hypothesis is considered reasonable. Since the two hypotheses have been defined, it is important to choose a statistical test to evaluate these hypotheses. Many different tests are proposed for this kind of comparison; in order to not choose an incorrect, some features have to be evaluated:

1. Is the test for paired or unpaired data?
2. Do we use parametric or non-parametric test?

Another issue that is important to understand when performing statistical tests is the possible types of errors. A type I error consists in rejecting the null hypothesis when H_0 is correct, and type II error consists in accepting the null hypothesis when it is false. The probability of committing a type I error is defined as α, and to commit a type II error is defined as β. The power of a test is defined as $1 - \beta$, or the chance of rejecting the null hypothesis when it is false.

Other important factor to understand is the p-value, which is the probability of committing a type I error, meaning that for a test to approve the null hypothesis, the p-value should be lower than a significance level arbitrarily chosen (equals to α when the hypothesis test is two-sided). For scientific purposes, it can be recommended an α value of 0.05. If the p-value found is lower than the arbitrarily chosen α, the test fails to reject the null hypothesis, and for our specific case, the two samples are said as coming from the same population.

1.4.3.2 Paired or Unpaired Data Tests

Pair-wise tests have, in their mathematical formulation, the hypothesis that samples data order are necessarily related to each other. Since, in our comparison of meta-heuristic methods, the objective is to compare independent results, it is necessary to choose statistical tests for unpaired samples.

1.4.3.3 Parametric or Non-parametric Tests

Parametric tests use, in their mathematical formulation, the hypothesis that samples come from a normal distribution (Gaussian), and it depends on parameters like mean and/or standard deviation. If the type of distribution is not clear, it is possible to use a normality test to evaluate this feature. Between those, it is possible to use Shapiro–Wilk test [32], Kolmogorov–Smirnov test or to use graphical methods as the quantile–quantile plot. Normality tests evaluate if the hypothesis of normality is reasonable to the obtained data. Razali and Wah [30] compared the power of different normality tests: Shapiro–Wilk, Kolmogorov–Smirnov, Lilliefors and Anderson-Darling tests. For those, they found that Shapiro–Wilk presents the greater power for all evaluated distributions and sample sizes. After evaluating or choosing that the normality hypothesis does not apply to our sample, it is better to use robust statistics in order to diminish the effect of the outliers. In this way, the median value is better because it is less sensitive to outliers than the mean value. Non-parametric tests are also less sensitive to outliers because they are based in a rank of the values in samples, and not in the value itself.

1.4.3.4 Using Non-parametric Test for Parametric Data

One question that can be raised is if it is possible to use non-parametric tests in parametric samples. There is no problem in doing it; the only disadvantage of using a non-parametric test is that the power of those tests is lower than the parametric ones.

1.4.3.5 Non-parametric Unpaired Data Tests

Wilcoxon rank sum test [15], also known as Mann–Whitney U test, is a non-parametric test to compare unpaired data. The basic hypothesis is that the two groups can be assumed as samples of one unique population [15]. This test evaluates if there is a shift between the two samples [40], by ranking all observations into a unique vector, and the test is performed in those ranks.

1.4.3.6 Representing Results

After evaluating if the samples are significantly different, it can be important to verify how different they are. In parametric statistics, this is classically performed by showing the mean value alongside the standard deviation. This is usually performed because, for normal distributions, the mean $\pm 1.96\times$ standard deviation means the interval of 95% of data for a two-sided interval of confidence. When data is not a sample of a Gaussian population, it is still possible to evaluate a confidence interval for non-parametric samples. This method is also based in ranking the values of the distribution. In order to better evaluate this distance, the median of each sample and the corresponding 95% confidence interval were computed for non-parametric distributions, using defined values [1]: $r = \frac{n}{2} - 1.96 \times \frac{\sqrt{n}}{2}$ $s = 1 + \frac{n}{2} + 1.96 \times \frac{\sqrt{n}}{2}$, where n is the number of samples in the study. In the present work, $n = 50$, resulting in $r = 18.07$ and $s = 32.93$. Thus, the confidence interval for the population median will be obtained from the 18th and the 33rd observations in increasing order for each studied sample. In this way, having the median and the confidence interval it is possible to make box-plots of the results, showing the outliers (values outside the confidence interval) as whiskers in the graph. This method shows clearly the differences between compared samples.

1.5   Results and Discussion

The first step of the procedure is to adjust the parameters of the metaheuristics, in order to produce useful results. In this chapter, we use, for the Differential Evolution: $F = 0.8$, $CR = 0.5$, and $NP = 20$ (NP is the size of the population). The maximum number of generations employed was 300. For the Harmony Search, the control parameters used are: $hms = 200$, $hmcr = 0.95$, $par = 0.3$, and $b_{range}^t = [0.01\ 0.01\ 0.01\ 0.01\ 1]$. The maximum number of iterations was 5000. The adjustment of parameters for both algorithms was conducted in order to obtain low standard deviations in the solutions. For instance, the difference between the maximum number of iterations was necessary, since HS consumed a higher number of iterations to obtain good solutions. This is not a drawback of HS—but only a characteristic for this particular application (phase equilibrium problems), when compared to DE. Furthermore, the "correct" adjustment of the control parameters is, per se, an optimization task (possibly using a hyperheuristic structure). That was not the objective in this chapter. A detailed description of the tuning process of metaheuristic parameters can be found in Ref. [25].

1.5.1   Reactive Azeotrope Calculation (RA)

1.5.1.1   Fitness Evaluation

In order to compare the performance of the two algorithms—Differential Evolution (DE) and Harmony Search (HS)—the fitness function (for the best element of the population) was analyzed for 50 runs each. A good approach is to start by verifying if data follow a normal distribution to choose a parametric or non-parametric statistical test. A qualitative overview of the data can be performed by evaluating its histogram, presented in Fig. 1.6.

As clearly observed in the histogram, data seem to be very far from a normal distribution. In order to evaluate this using a hypothesis test, a Shapiro–Wilk test was performed. If any of the two samples does not present a normal distribution, non-parametric test for comparison can be performed, such as the Wilcoxon rank-sum test (as an option for the parametric one, like the T-test).

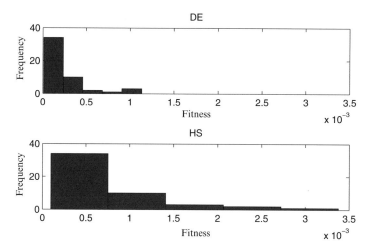

Fig. 1.6 Histogram of the fitness function (50 runs)—RA problem

Table 1.3 Fitness analysis—RA problem

Shapiro–Wilk parametric hypothesis test		
p-Value		
DE	1.20×10^{-7}	Non-normal
HS	2.28×10^{-6}	Non-normal
Non-parametric unpaired comparing samples		
Wilcoxon rank sum test		
p-Value		
DE_vs_HS	1.26×10^{-8}	Significantly different

Median and confidence interval for the median			
Method	Median	CI min(95%)	CI max(95%)
DE	1.57×10^{-4}	1.10×10^{-4}	2.30×10^{-4}
HS	4.42×10^{-4}	3.17×10^{-4}	7.14×10^{-4}

Results of the Shapiro–Wilk and Wilcoxon rank sum tests are shown in Table 1.3. As seen by the test and the confidence intervals, the DE presents a lower fitness value and a more confined variability, meaning that DE presents a better performance than the HS algorithm. This can be illustrated by a box plot, as presented in Fig. 1.7.

The median value and the confidence interval can also be observed by a more simple graphic representation (Fig. 1.8).

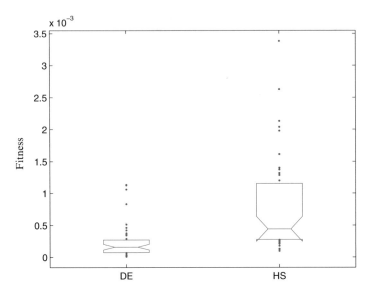

Fig. 1.7 Box plots of the fitness function for DE and HS algorithms in the RA problem

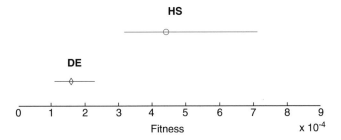

Fig. 1.8 Median value and confidence interval of the fitness function for DE and HS algorithms in the RA problem

1.5.1.2 Solution Distance

As in the previous analysis, the solution distance was evaluated as the lower Euclidean distance between the azeotrope solutions and the algorithm results, for the same 50 runs as presented in the fitness analysis (considering the best element of the population for each run, for both algorithms). Since the possible solutions present different dimensions (sizes), the computed distance was computed as a relative distance (estimated vectors difference were divided by the solutions vectors).

Table 1.4 shows the results for Shapiro–Wilk normality evaluation, and Wilcoxon rank sum evaluation, as well as the median and confidence intervals. It can be seen that the distance for the HS algorithm is higher than the one for DE algorithm.

Table 1.4 Solution distance—RA problem

Shapiro–Wilk parametric hypothesis test		
p-Value		
DE	4.29×10^{-5}	Non-normal
HS	2.00×10^{-3}	Non-normal
Non-parametric unpaired comparing samples		
Wilcoxon rank sum test		
p-Value		
DE_vs_HS	3.08×10^{-4}	Significantly different

Median and confidence interval for the median			
Method	Median	CI min(95%)	CI max(95%)
DE	0.205	0.139	0.297
HS	0.393	0.264	0.552

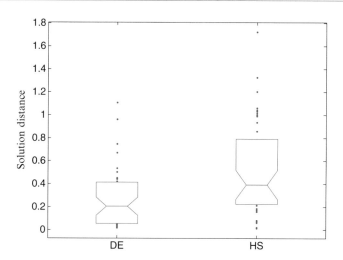

Fig. 1.9 Box plots of the solution distance for DE and HS algorithms in the RA problem

The box plots (Fig. 1.9) show that the HS presents higher outliers leading to the conclusion that this algorithm is less reliable than the DE algorithm, for predicting azeotropes. The median value and confidence interval are shown in Fig. 1.10.

1.5.2   Double Retrograde Vaporization Calculation

A similar analysis was performed for the DRV problem, considering the fitness evaluation and the distances to the solutions. In this case, the values for the fitness function are lower (when compared to RA results), as a consequence of the trivial

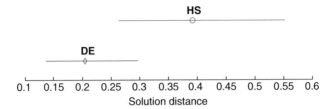

Fig. 1.10 Median value and confidence interval of the solution distances for DE and HS algorithms in the RA problem

Table 1.5 Fitness analysis—DRV problem

Shapiro–Wilk parametric hypothesis test		
p-Value		
DE	3.25×10^{-10}	Non-normal
HS	3.35×10^{-13}	Non-normal
Non-parametric unpaired comparing samples		
Wilcoxon rank sum test		
p-Value		
DE_vs_HS	4.25×10^{-18}	Significantly different

Median and confidence interval for the median			
Method	Median	CI min(95%)	CI max(95%)
DE	7.17×10^{-43}	0	2.87×10^{-42}
HS	1.18×10^{-13}	2.51×10^{-14}	3.544×10^{-13}

solutions (with null values for fitness). Obviously, from the mathematical point of view, these trivial solutions cannot be avoided by the metaheuristics.

1.5.2.1 Fitness Evaluation

For the fitness evaluation, results are presented in Table 1.5. Graphically, box plots (Fig. 1.11) and interval solutions (Fig. 1.12) can represent the performance of both algorithms. Once again, the results using DE proved to be better than the ones obtained by HS.

1.5.2.2 Solution Distance

Table 1.6 contains the results for the solution distance. The box plots are presented in Fig. 1.13, and a graphical representation of medians and confidence intervals is displayed in Fig. 1.14. Again, we observe a better performance of DE in comparison to HS (Table 1.7).

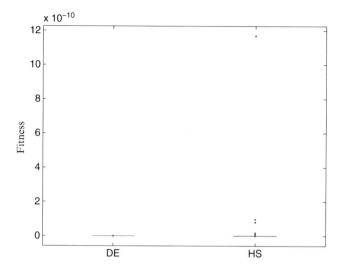

Fig. 1.11 Box plots of the fitness function for DE and HS algorithms in the DRV problem

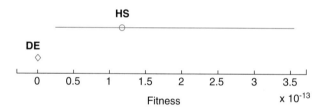

Fig. 1.12 Median value and confidence interval of the fitness function for DE and HS algorithms in the DRV problem

Table 1.6 Solution distance—DRV problem

Shapiro–Wilk parametric hypothesis test		
p-Value		
DE	4.29×10^{-5}	Non-normal
HS	2.00×10^{-3}	Non-normal
Non-parametric unpaired comparing samples		
Wilcoxon rank sum test		
p-Value		
DE_vs_HS	3.08×10^{-4}	Significantly different

Median and confidence interval for the median			
Method	Median	CI min(95%)	CI max(95%)
DE	2.05×10^{-1}	1.39×10^{-1}	2.97×10^{-1}
HS	3.93×10^{-1}	2.64×10^{-1}	5.52×10^{-1}

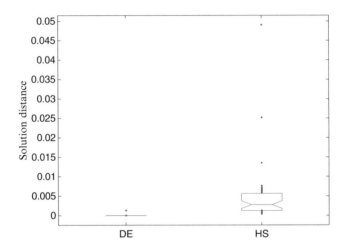

Fig. 1.13 Box plots of the solution distance for DE and HS algorithms in the DRV problem

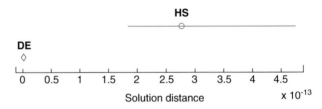

Fig. 1.14 Median value and confidence interval of the solution distance for DE and HS algorithms in the DRV problem

Table 1.7 Solution distance

Shapiro–Wilk parametric hypothesis test		
p-Value		
DE	4.39×10^{-10}	Non-normal
HS	1.26×10^{-10}	Non-normal
Non-parametric unpaired comparing samples		
Wilcoxon rank sum test		
p-Value		
DE_vs_HS	1.96×10^{-14}	Significantly different

Median and confidence interval for the median			
Method	Median	CI min(95%)	CI max(95%)
DE	9.24×10^{-6}	9.24×10^{-6}	2.67×10^{-5}
HS	2.77×10^{-3}	1.85×10^{-3}	4.75×10^{-3}

1.6 Conclusions

In this work we presented the application of two metaheuristics (Differential Evolution and Harmony Search) in two phase equilibrium problems: the calculation of a reactive azeotrope and the calculation of dew point pressures in a system with double retrograde vaporization.

An important conclusion that could be verified in this work is that the task of comparing metaheuristics is extremely complex. The most traditional comparisons such as means, better results, worst results, computing times, and number of evaluations of the objective function are usually on opposite sides of the scale (for example, shorter computing time may imply worst means). We chose not to evaluate computation times or number of evaluations of the objective function, focusing only on the quality of the solutions. In spite of this—as always happens with metaheuristics—the choice of the control parameters of each of the algorithms represents a weakness of this methodology of analysis.

Acknowledgements The authors acknowledge the financial support provided by FAPERJ–Fundação Carlos Chagas Filho de Amparo à Pesquissa do Estado do Rio de Janeiro, CNPq–Conselho Nacional de Desenvolvimento Científico e Tecnológico, and CAPES–Coordenação de Aperfeiçoamento de Pessoal de Nível Superior, research supporting agencies from Brazil.

References

1. Altman, D.G., Machin, D.T., Bryant, T.N., Martin, J.G.: Statistics with Confidence, 2nd edn. British Medical Journal Books, London (2000)
2. Alvarez, V.H., Larico, R., Ianos, Y., Aznar, M.: Parameter estimation for VLE calculation by global minimization: the genetic algorithm. Braz. J. Chem. Eng. **25**(2), 409–418 (2008)
3. Barbosa, D., Doherty, M.F.: Theory of phase diagrams and azeotropic conditions for two-phase reactive systems. Proc. R. Soc. Lond. A **413**, 443–458 (1987)
4. Barbosa, D., Doherty, M.F.: The simple distillation of homogeneous reactive mixtures. Chem. Eng. Sci. **43**, 541–550 (1988)
5. Bonilla-Petriciolet, A., Bravo-Sánchez, U.I., Castillo-Borja, F., Zapiain-Salinas, J.G., Soto-Bernal, J.J.: The performance of simulated annealing in parameter estimation for Vapor-Liquid equilibrium modeling. Braz. J. Chem. Eng. **24**(1), 151–162 (2007)
6. Bonilla-Petriciolet, A., Segovia-Hernández, J.G., Soto-Bernal, J.J.: Harmony search for parameter estimation in Vapor-Liquid equilibrium modeling. In: Proceeding of the 5th PSE Asia, Singapore, pp. 719–726 (2010)
7. Bonilla-Petriciolet, A., Iglesias-Silva, G.A., Hall, K.R.: Calculation of homogeneous azeotropes in reactive and non-reactive mixtures using a stochastic optimization approach. Fluid Phase Equilib. **281**(1), 22–31 (2011)
8. Chen, R.J.J., Chappelear, P.S., Kobayashi, R.: Dew point loci for Methane-n-butane binary system. J. Chem. Eng. Data **19**, 53–58 (1974)
9. Chen, R.J.J., Chappelear, P.S., Kobayashi, R.: Dew point loci for Methane-n-pentane binary system. J. Chem. Eng. Data **19**, 58–61 (1974)
10. Das, S., Suganthan, P.N.: Differential evolution: a survey of the state-of-the-art. IEEE Trans. Evol. Comput. **15**(1), 4–31 (2011)

11. Fernández-Vargas, J.A., Bonilla-Petriciolet, A., Segovia-Hernández, J.G.: An improved ant colony optimization method and its application for the thermodynamic modeling of phase equilibrium. Fluid Phase Equilib. **353**, 121–131 (2013)
12. Geem, Z.W., Kim, J.H., Loganathan, G.V.: A new heuristic optimization algorithm: harmony search. Simulation **76**(2), 60–68 (2001)
13. Guedes, A.L., Moura Neto, F.D., Platt, G.M.: Prediction of azeotropic behaviour by the inversion of functions from the plane to the plane. Can. J. Chem. Eng. **93**, 914–928 (2015)
14. Haghtalab, A., Espanani, R.: A new model and extension of Wong–Sandler mixing rule for prediction of (Vapour + Liquid) equilibrium of polymer solutions using EOS/G^E. J. Chem. Thermodyn. **36**, 901–910 (2004)
15. Hollander, M., Wolfe, D.A.: Nonparametric Statistical Methods. Wiley, Hoboken (1999)
16. Libotte, G.B., Platt, G.M.: The relationship among convergence in basins of attraction and the critical curve in the double retrograde vaporization calculation. In: XVIII Brazilian Meeting on Computational Modelling, Salvador, 13–16 October 2016 (in Portuguese)
17. Luus, R., Jaakola, T.H.I.: Optimization by direct search and systematic reduction of the size of search region. AIChE J. **19**(4), 760–766 (1973)
18. Maier, R.W., Brennecke, J.F., Stadtherr, M.A.: Reliable computation of reactive azeotropes. Comput. Chem. Eng. **24**(8), 1851–1858 (2000)
19. Mantegna, R.N.: Fast, Accurate algorithm for numerical simulation of Lévy stable stochastic processes. Phys. Rev. E **49**, 4677–4683 (1994)
20. Nesmachnow, S.: Using metaheuristics as soft computing techniques for efficient optimization. In: Khosrow-Pour, M. (ed.) Encyclopedia of Information Science and Technology, pp. 7390–7399 (2014)
21. Pavlyukevich, I.: Lévy flights, non-local search and simulated annealing. J. Comput. Phys. **226**, 1830–1844 (2007)
22. Peng, D.Y., Robinson, D.B.: A new two-constant equation of state. Ind. Eng. Chem. Fundam. **15**, 59–64 (1976)
23. Platt, G.M.: Computational experiments with flower pollination algorithm in the calculation of double retrograde dew points. Int. Rev. Chem. Eng. **6**, 95–99 (2014)
24. Platt, G.M.: Numerical experiments with new metaheuristic algorithms in phase equilibrium problems. Int. J. Math. Model. Numer. Optim. **7**(2), 189–211 (2016)
25. Platt, G.M., Lima, L.V.P.C.: Azeotropy in a refrigerant system: an useful scenario to test and compare metaheuristics. Int. J. Metaheuristics **7**(1), 43–66 (2018)
26. Platt, G.M., Bastos, I.N., Domingos, R.P.: Use of hyperheuristics to solve nonlinear algebraic systems: application to double azeotrope calculation. Int. Rev. Chem. Eng. **2**(3), 371–377 (2010)
27. Platt, G.M., Domingos, R.P., de Andrade, M.O.: Application of the firefly and Luus-Jaakola algorithms in the calculation of a double reactive azeotrope. Comput. Sci. Discov. **7**, 015002 (2014)
28. Raeissi, S., Peters, C.J.: Simulation of double retrograde vaporization using the Peng-Robinson equation of state. J. Chem. Thermodyn. **35**, 573–581 (2003)
29. Rashtchian, D., Ovaysi, S., Taghikhani, V., Ghotbi, C.: Application of the genetic algorithm to calculate the interaction parameters for multiphase and multicomponent systems. Iran. J. Chem. Chem. Eng. **26**(3), 89–102 (2007)
30. Razali, N.M., Wah, Y.B.: Power comparisons of Shapiro-Wilk, Kolmogorov-Smirnov, Lilliefors and Anderson-Darling tests. J. Stat. Model. Anal. **2**(1), 21–33 (2011)
31. Rechenberg, I: Evolutionsstrategie: Optimierung Technischer Systeme nach Prinzipien der Biologischen Evolution. Frommann-Holzboog, Stuttgart (1974) (in German)
32. Royston, P.: Approximating the Shapiro-Wilk W-test for non-normality. Stat. Comput. **2**, 117–119 (1992)
33. Saka, M.P., Hasançebi, O., Geem, Z.W.: Metaheuristics in structural optimization and discussions on harmony search algorithm. Swarm Evol. Comput. **28**, 88–97 (2016)
34. Sorensen, K.: Metaheuristics – the metaphor exposed. Int. Trans. Oper. Res. **22**(1), 3–18 (2015)

35. Storn, R., Price, K.: Differential evolution – a simple and efficient heuristic for global optimization over continuous spaces. J. Global Optim. **11**, 341–359 (1997)
36. Ung, S., Doherty, M.F.: Calculation of residue curve maps for mixtures with multiple equilibrium chemical reactions. Ind. Eng. Chem. Res. **34**, 3195–3202 (1995)
37. Ung, S., Doherty, M.F.: Necessary and sufficient conditions for reactive azeotropes in multireaction mixtures. AIChE J. **41**(11), 2383–2392 (1995)
38. Walas, S.M.: Phase Equilibria in Chemical Engineering. Butterworth, Stoheham (1985)
39. Weiland, D.: A critical analysis of the harmony search algorithm – how not to solve Sudoku. Oper. Res. Perspect. **2**, 97–105 (2015)
40. Wild, C.J., Seber, G.: Chance Encounters: A First Course in Data Analysis and Inference. Wiley, New York (2000)
41. Wilson, G.M.: Vapor-liquid Equilibrium. XI. A new expression for the excess free energy of mixing. J. Am. Chem. Soc. **86**(2), 127–130 (1964)
42. Yang, X.-S.: Nature-Inspired Metaheuristic Algorithms. Luniver Press, Frome (2008)
43. Yang, X.-S.: Flower pollination algorithm for global optimization. Lect. Notes Comput. Sci. **7445**, 240–249 (2012)

Chapter 2
Reliability-Based Robust Optimization Applied to Engineering System Design

Fran Sérgio Lobato, Márcio Aurelio da Silva, Aldemir Aparecido Cavalini Jr., and Valder Steffen Jr.

2.1 Introduction

Along the last decades, various strategies to evaluate reliability and robustness in engineering system design have been proposed. The application of these approaches has become necessary since the mathematical models and the vector of design variables are commonly influenced by uncertainties. According to Ritto [26], the modeling process of engineering systems deals with uncertainties related to both the parameters and the simplifications adopted in the formulation of the model. In addition, deterministic solutions are most of the time infeasible in real-world applications. In this context, there are uncertainties that are associated both with manufacturing process and with high accuracy requirements that can be difficult to achieve or even economically unaffordable [16].

Probabilistic methods were proposed as an alternative to deal with uncertainties. These methods can be classified according to two main categories, namely the reliability-based design (RBO) [3, 19] and the robust design (RD) [22, 33]. Although these approaches are based on probabilistic methods, they are both different in their conception and applicability. RBO emphasizes high reliability in the design by ensuring the probabilistic achievement of the considered constraint at a desired level [8, 9]. According to Ravichandran [25], RBO can be defined as the ability of a

F. S. Lobato (✉)
NUCOP-Laboratory of Modeling, Simulation, Control and Optimization, School of Chemical Engineering, Federal University of Uberlândia, Uberlândia, MG, Brazil
e-mail: fslobato@ufu.br

M. A. da Silva · A. A. Cavalini Jr. · V. Steffen Jr.
LMEst-Structural Mechanics Laboratory, School of Mechanical Engineering, Federal University of Uberlândia, Uberlândia, MG, Brazil
e-mail: aacjunior@ufu.br; vsteffen@ufu.br

© Springer Nature Switzerland AG 2019
G. Mendes Platt et al. (eds.), *Computational Intelligence, Optimization and Inverse Problems with Applications in Engineering*,
https://doi.org/10.1007/978-3-319-96433-1_2

system or component to perform its required functions under given conditions. The RBO approach in the context of engineering systems is based on the probability of the desired system performance to fail. Differently, RD produces a solution that presents small sensitivity to changes in the vector of design variables [15, 25, 33]. In both approaches, Monte Carlo simulation is commonly used to perform an analysis around the deterministic solution [8]. Consequently, the main difficulty on the application of these two approaches is associated with their intrinsic high computational cost [3, 9, 19].

Various techniques have been proposed to determine the robust design and the probability of failure [3, 6, 9, 19]. Regarding the RD approach, new constraints and/or objectives are required (i.e., relationship between the mean and standard deviation of the vector of objective functions). Additionally, the probability distribution of the design variables and/or objective functions must be known. In this sense, probability tools are used to model uncertainties [21, 26, 28, 30]. Uncertain parameters are represented by random variables and their probability density functions are determined. Deb and Gupta [5] proposed an alternative formulation, in which the Mean Effective Concept (MEC)—originally developed for mono-objective design problems—was extended to the multi-objective context. No additional restrictions are considered in the optimization problem, which is rewritten as the mean vector of the original objective functions. Moreira et al. [20] and Souza et al. [31] applied successfully this methodology to solve design problems on chemical and mechanical engineering systems. In the reliability context, the stochastic variables are converted from the physical space to the standard normal space. The optimization problem is reformulated to compute the largest probability of failure, which is simultaneously equated to the desired value [6]. Thus, the RBO procedure can be classified into four main categories based on the approach used to determine the probability of failure [1]: (1) simulation methods; (2) double-loop methods; (3) decoupled methods; and (4) single-loop methods.

RD and RBO approaches have been applied simultaneously to various engineering design problems. Lagaros et al. [14] studied the influence of probabilistic constraints in a robust optimization problem devoted to steel structures. Uncertainties on the loads, material properties, and cross section geometry were taken into account by using the Monte Carlo simulation method combined with the Latin Hypercube sampling. Ravichandran [25] developed an integrated design methodology for reliability and robustness, in which the RBO and RD were combined into a single approach to minimize both the probability of failure and the variance of the design variables. Jeong and Park [12] proposed a technique based on a robustness index in association with the Enhanced Single Loop Single Vector and the Conjugate Gradient methods. Numerical examples and structural applications were used to compare the proposed approach with well-established techniques. Shahraki and Noorossana [29] combined the concepts of robust design, reliability based design, and multi-objective optimization in a single algorithm. The optimal parameters of the implemented algorithm were determined in order to increase the efficiency and decrease the computational cost of the reliability analysis. Wang et al. [36] proposed a hybrid approach to find reliable and robust solutions under uncertainty in the design variables considering a hook structure. The strategy combines a single-loop approach and a mono-objective robust design method (weighted sum) considering a global approximation (simple models are used to represent complex models) and global sensitivity analysis. Gholaminezhad et al. [11] proposed a multi-objective reliability-based robust design approach for dimensional synthesis of robot gripper mechanisms using the DE approach and the Monte Carlo simulation. The mean and standard deviation of the difference between the maximum and minimum gripping forces and the transmission ratio of actuated and experienced gripping forces were considered as objective functions. Keshtegar and Chakraborty [13] proposed an efficient-robust structural reliability method by adaptive finite-step length based on Armijo line search approach. Wang et al. [37] studied the reliability-based robust dynamic positioning for a turret-moored floating production storage and offloading vessel with unknown time-varying disturbances and input saturation.

It is worth mentioning that the number of contributions involving the simultaneous application of reliability and robustness is limited. In this contribution, the DE algorithm is associated with the IRA and MEC approaches to deal with reliability and robustness in engineering system design. The proposed methodology is evaluated by considering two mathematical test cases (nonlinear limit state and highly nonlinear limit state) and three engineering system design applications (short column design, cantilever beam problem, and three bar truss problem). The organization of this chapter is the following. Sections 2.2 and 2.3 introduce briefly the RD and RBO concepts, respectively. Section 2.4 presents a review about the DE algorithm. The proposed methodology is presented in Sect. 2.5. For illustration purposes, Sect. 2.6 brings applications both involving mathematical functions and design of classical engineering systems. Finally, the conclusions are outlined in the last section of the present contribution.

2.2 Robust Design (RD)

The solution of optimization problems is commonly determined considering reliable mathematical models, variables, and parameters, i.e., no errors of modeling and estimation are taken into account [5]. However, most engineering systems are sensitive to small changes in the vector of design variables. The concept of robust optimization should be used to minimize this effect. According to Taguchi [33], this approach is used to find decisions that are optimal for the worst-case realization of the uncertainties within a given set of values.

Traditionally, new constraints and/or objective functions are required to introduce robustness both in mono- and multi-objective applications. Deb and Gupta [5] proposed an extension of the MEC approach to overcome this difficulty. No additional constraint is considered in the original problem, which is rewritten to consider the mean value of the original objectives. The robustness measure is based on the following definition [5]:

Definition 2.1 Being $f : \Omega \subset \mathbb{R}^n \rightarrow \mathbb{R}$ an integrable function, the MEC of f associated with the neighborhood δ of the variable x is a function $f^{eff}(x, \delta)$ given by:

$$f^{eff}(x, \delta) = \frac{1}{|B_\delta(x)|} \int_{y \in B_\delta(x)} f(y) dy \qquad (2.1)$$

subject to

$$g_1^{eff}(x, \delta) \leq 0, \quad j = 1, \ldots, k; \ x \in \Omega \subset \mathbb{R}^n \qquad (2.2)$$

$$x_i^l \leq x \leq x_i^u, \quad i = 1, \ldots, n \qquad (2.3)$$

where f is the objective function, g is the vector of inequality constraints, n is the number of design variables, $|B_\delta(x)|$ is the hypervolume of the neighborhood, and x^l and x^u represent the lower and upper limits of the vector of design variables, respectively. N candidate solutions are generated randomly by using the Latin Hypercube method to estimate the integral defined by Eq. (2.1). It is worth mentioning that this procedure increases the computational cost due to additional evaluations of the objective functions required to solve the optimization problem [5, 20, 31].

2.3 Reliability-Based Optimization (RBO)

The solution of deterministic optimization problems is obtained on a given constraint boundary (or at the intersection of more than one constraint boundary). In this case, any perturbation in the vector of design variables results in an infeasible solution with a probability distribution around the optimal solution. As mentioned by Deb et al. [6], in order to find a reliable solution (i.e., small associated probability to produce an infeasible solution), the true optimal solution must be penalized and an interior solution can be chosen within the feasible region. From a given reliability measure R, it is desired to find a feasible solution that ensures that the probability of having an infeasible solution is given by $(1 - R)$. Mathematically, the RBO problem can be formulated as [6]:

$$\min f(x_d, x_r) \qquad (2.4)$$

$$P(G_j(x_d, x_r) \leq 0) \geq R_j, j = 1, \ldots, n_g \qquad (2.5)$$

$$x_{di}^l \leq x_d \leq x_{di}^u, \quad i = 1, \ldots, n_d \ \text{and} \ x_{rk}, \quad k = 1, \ldots, n_r \qquad (2.6)$$

where f and G_j are the objective and constraint functions, respectively, x_d is the vector of design variables (deterministic values associated with the lower x_{di}^l and upper x_{di}^u limits), x_r is the vector of random variables, n_g is the number of probabilistic constraints, n_d is the number of design variables, n_r is the number of random variables, and R_j is the desired reliability. P is the probability of G_j to be less than or equal to zero ($Gj \geq 0$ indicates failure). The probability of failure can be defined by a cumulative distribution function, as follows [6]:

$$P(G_j(x_d, x_r) \leq 0) = \int \cdots \int_{G_j \leq 0} f_X(x_r) dx_r \qquad (2.7)$$

where f_X is a joint probability density function.

The integral defined by Eq. (2.7) should be evaluated along the design space specified by the vector of inequality constraints to determine the probability of failure P. The analytical (or numerical) evaluation of Eq. (2.7) is expensive due to the specified domain. In order to overcome this difficulty, the original problem with random variables x_r can be transformed into a similar problem with new random variables (u) by using the so-called Rosenblatt transformation [27]. In this approach, considering the new search space (u-space), the most probable point (MPP) for failure is found by locating the minimum distance between the origin and the limit-state (or the constraint function), which is defined by the reliability coefficient β. The probabilistic constraint in Eq. (2.7) can be further expressed through an inverse transformation, as given by Eq. (2.8).

$$P(G_j(x_d, x_r) \leq 0) = \Phi(\beta) \qquad (2.8)$$

where Φ is the standard normal distribution function. This transformation is used by different approaches to measure the associated reliability, and avoid the evaluation of the integral given by Eq. (2.7). In the reliability coefficient approach, this parameter is used to describe the probabilistic constraint in Eq. (2.7).

As mentioned earlier, the RBO procedures are classified according to four categories. Basically, the idea is to overcome the computational difficulties by simplifying the integral of $f_X(x_r)$ (see Eq. (2.7)) and approximating the function $G_j(x_d, x_r)$ [6]. Next, three of the most commonly used reliability analysis methods will be briefly presented.

2.3.1   First Order Reliability Method (FORM)

In this approach, the constraint $G_j(x_d, x_r)$ is approximated by a first order Taylor expansion (linearization procedure), being commonly used to evaluate the probability of failure presented by Eq. (2.7). The reliability coefficient β is interpreted as the minimum distance between the origin of the limit state surface in the standardized normal space (u-space) and the MPP for failure (design point searched by using computational methods) [39]. Therefore, β is determined by solving the following optimization problem [6, 39]:

$$\min_{u} G_j(x_d, u) j = 1, \ldots, n_g \qquad (2.9)$$

subject to

$$(u^T u)^{0.5} = \beta \qquad (2.10)$$

As mentioned by Tichy [35], accuracy problems occur when the performance function is strongly nonlinear, since the performance function is approximated by a linear function in the u-space at the design point.

2.3.2 Second-Order Reliability Method (SORM)

SORM was proposed as an attempt to improve the accuracy of FORM, which is obtained by approximating the limit state surface in the u-space at the design point by a second-order surface [10]. The SORM algorithm is time consuming due to the computation of second-order derivatives (Hessian matrix). In order to overcome this difficulty, various algorithms have been proposed. Der-Kiureghian and De Stefano [7] proposed an efficient point-fitting algorithm. In this approach, the major principal axis of the limit state surface and the corresponding curvature are obtained simultaneously with the design point, without computing the Hessian matrix. An alternative point-fitting SORM algorithm was presented by Zhao and Ono [40], in which a performance function is directly point-fitted by using a general form of the second-order polynomial for the standard normal random variables.

2.3.3 Sequential Optimization and Reliability Assessment (SORA)

The SORA method is based on a single-loop strategy, in which optimization cycles and reliability assessment are carried out sequentially [9]. The optimization and reliability assessment are decoupled from each other in each cycle. No reliability assessment is required within the optimization and the reliability assessment is only conducted after the optimization. According to Du and Chen [9], the idea behind this approach is to shift the boundaries of violated deterministic constraints to a feasible direction. This procedure is based on the reliability information obtained in the previous cycle. The design is improved from cycle to cycle and, consequently, the computation efficiency is improved significantly.

2.3.4 Inverse Reliability Analysis (IRA)

The IRA approach was proposed by Du [8], and consists in formulating an inverse problem for reliability analysis to find the value of G^p, given the following probability:

$$P(G(x_d, x_r) \leq G^p) = R \qquad (2.11)$$

where R is known (defined by the user).

Equation (2.11) indicates that the probability of the performance function $G(x_d, x_r)$ is equal to a given reliability measure R if $G(x_d, x_r) \leq G^p$, in which G^p is estimated by using FORM. In order to apply the FORM approach, the function $G(x_d, x_r)$ is used (see Eq. (2.12)), and the MPP for $P(G'_j(x_d, x_r) \leq 0) = P(G_j(x_d, x_r) \leq G^p)$ is considered as being u^*.

$$G'(x_d, x_r) = G(x_d, x_r) - G^p \qquad (2.12)$$

From FORM, if the probability R is known, the reliability coefficient is given by:

$$\beta = \left| \Phi^{-1}(R) \right| \qquad (2.13)$$

where the absolute value is used since the reliability coefficient is the minimum distance between the origin and the limit state (or the constraint function) and is always nonnegative.

Figure 2.1 shows that the MPP u^* is a tangent point of the circle with radius β and the performance function $G(x_d, x_r) - G^p = 0$. Additionally, u^* is a point associated with the minimum value of $G(x_d, x_r)$ on the circle.

Therefore, the MPP search for an inverse reliability analysis problem becomes: find the minimum value of $G(x_d, x_r)$ on the β-circle (or β-sphere for a 3-D problem, or β-hyper sphere for a higher dimensional problem). The mathematical model

Fig. 2.1 The inverse MPP search [8]

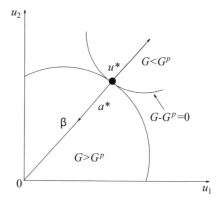

for the MPP search is stated as follows: find the MPP u^* where the performance function $G(u)$ is minimized while u^* remains on the surface of the β-circle considering the search direction a^*.

The described procedure can be summarized according to the following steps [8]:

1. Inform the starting point ($k = 0$, $u_0 = 0$, β, and distribution type);
2. Evaluate $G_j(u^k)$ and $\nabla G_j(u^k)$. Compute a as follows:

$$a^k = \frac{\nabla G_j(u^k)}{\left\| \nabla G_j(u^k) \right\|} \tag{2.14}$$

3. Update $u^{k+1} = -\beta a^k$ and $k = k + 1$;
4. Repeat steps 2 and 3 until convergence (tolerance) is achieved.

2.4 Differential Evolution Algorithm (DE)

The DE algorithm is an optimization technique proposed by Storn and Price [32] that belongs to the family of evolutionary computation, which differs from other evolutionary algorithms in the mutation and recombination schemes. Basically, DE executes its mutation operation by adding a weighted difference vector between two individuals to a third one. Then, the mutated individuals will perform discrete crossover and greedy selection with the corresponding individuals from the last generation to produce the offspring. The key control parameters of DE are the population size (NP), the crossover constant (CR), and the so-called perturbation rate (F). The canonical pseudo-code of DE algorithm is presented below.

Differential Evolution Algorithm
1: Initialize and evaluate population P
2:   while (not done) {
3:    for ($i = 0$; $i < NP$; i++) {
4:     Create candidate $C[i]$
5:     Evaluate candidate
6:    if ($C[i]$ is better than $P[i]$)
7:     $P_0[i] = C[i]$
8:     else
9:     $P_0[i] = P[i]$}
10:    $P = P_0$ }
11: *Create candidate* $C[i]$
12: Randomly select parents $P[i_1]$, $P[i_2]$, and $P[i_3]$
13: where i, i_1, i_2, and i_3 are different.
14: Create initial candidate
15: $C'[i] = P[i_1] + F \times (P[i_2] - P[i_3])$.

16:    Create final candidate $C[i]$ by crossing over the genes of $P[i]$ and $C'[i]$ as follows:

17:      for $(j = 0; j < NP; j{+}{+})$ {
18:      if $(r < CR)$
19:        $C[i][j] = C'[i][j]$
20:      else
21:        $C[i][j] = P[i][j]$}

P is the population of the current generation, P' is the population to be constructed for the next generation, $C[i]$ is the candidate solution with population index i, $C[i][j]$ is the j-th entry in the solution vector of $C[i]$, and r is a random number between 0 and 1.

Storn and Price [32] have given some simple rules for choosing the key parameters of DE for general applications. Normally, NP should be about 5–10 times the dimension of the problem (i.e., the number of design variables). As for F, it lies in the range between 0.4 and 1.0. Initially, $F = 0.5$ can be tried, then F and/or NP can be increased if the population converges prematurely. Storn and Price [32] proposed various mutation schemes for the generation of new candidate solutions by combining the vectors that are randomly chosen from the current population, as shown below.

$$\text{rand}/1 \rightarrow x = x_{r_1} + F\left(x_{r_2} - x_{r_3}\right) \tag{2.15}$$

$$\text{rand}/2 \rightarrow x = x_{r_1} + F\left(x_{r_2} - x_{r_3} + x_{r_4} - x_{r_5}\right) \tag{2.16}$$

$$\text{best}/1 \rightarrow x = x_{\text{best}} + F\left(x_{r_2} - x_{r_3}\right) \tag{2.17}$$

$$\text{best}/2 \rightarrow x = x_{\text{best}} + F\left(x_{r_2} - x_{r_3} + x_{r_4} - x_{r_5}\right) \tag{2.18}$$

$$\text{rand/best}/1 \rightarrow x = x_{r_1} + F\left(x_{\text{best}} - x_{r_1} + x_{r_1} - x_{r_2}\right) \tag{2.19}$$

$$\text{rand/best}/2 \rightarrow x = x_{r_1} + F\left(x_{\text{best}} - x_{r_1}\right) + F\left(x_{r_1} - x_{r_2} + x_{r_3} - x_{r_4}\right) \tag{2.20}$$

where x_{r1}, x_{r2}, x_{r3}, x_{r4}, and x_{r5} are candidate solutions randomly chosen and x_{best} is the candidate solution associated with the best fitness value; all of them present in the population of the current generation (the vector x is denoted by C').

2.5 Methodology

The proposed MEC + IRA + DE methodology starts by evaluating the candidates generated by the DE algorithm according to MEC, in which the N neighbors related to each candidate are determined by using the Latin Hypercube sampling method. Then, the IRA approach is applied to find the point with the highest associated

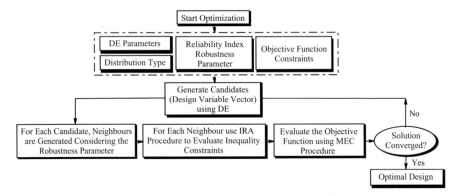

Fig. 2.2 Flowchart of the MEC + IRA + DE strategy proposed to solve reliability-based robust optimization problems

probability value. Finally, the integral of the objective function and constraints is evaluated considering the robustness parameter. This procedure continues until the stopping criterion is satisfied (maximum number of objective function evaluations). In this case, the MEC procedure is used to introduce robustness on the solution and the IRA analysis is performed to introduce reliability. The obtained solution considers both strategies simultaneously. Figure 2.2 presents the flowchart regarding the MEC + IRA + ED strategy proposed in this work to solve reliability-based robust optimization problems.

The MEC + IRA + DE strategy can be implemented as follows:

1. Initially, the reliability-based robust optimization problem (objective function, constraints, number of variables, reliability index, robustness parameter, number of samples, and distribution type) and the DE parameters (population size, number of generations, perturbation rate, crossover probability, and the strategy for the generation of potential candidates) are defined by the user;
2. The population of candidates is generated by using the DE algorithm. In this case, only the vector of design variables (x_d—deterministic values) is generated;
3. For each candidate, N neighbors are generated considering the robustness parameter δ;
4. For each neighbor, the IRA approach is applied to determine the value of the inequality constraint from the vector of random variables (x_r), i.e., the uncertain variables;
5. The values of x_d (defined by DE) and x_r (defined by IRA) are used to evaluate the vector of objective functions and the vector of inequality constraints (objective function defined by MEC);
6. The treatment of the constraints is performed through the Static Penalization method proposed by Castro [4]. This approach consists in attributing limited values to each objective function to play the role of penalization parameters, which guarantees that any non-dominated solution dominates any solution that

violates at least one of the given constraints. In the same way, any solution that violates a constraint will dominate any solution that presents two constraint violations, and so on. Therefore, layers of solutions are obtained and, consequently, the number of constraint violations corresponds to the rank of the solution. For a given problem, the vector containing the considered objective function is given by:

$$f(x) \equiv f(x) + r_p n_{\text{viol}} \qquad (2.21)$$

where $f(x)$ is the vector of objective functions, r_p is the vector of penalty parameters that depends on the considered problem, and n_{viol} is the number of violated constraints;

7. The iteration process is repeated until convergence is achieved (maximum number of generations).

2.6   Mathematical and Engineering Applications

In this section, the proposed methodology is evaluated through different mathematical and engineering test cases, namely: (1) nonlinear limit state, (2) highly nonlinear limit state, (3) short column design, (4) cantilever beam problem, and (5) three bar truss problem. In the context of the mentioned applications, the following points should be taken into account:

- The parameters used by the DE algorithm are the following [32]: 25 individuals (NP), 250 generations (GEN), perturbation rate (F) and crossover probability (CR) equal to 0.8, and DE/rand/1/bin strategy for the generation of potential candidates;
- In all test cases, the following values for the robustness parameter (δ) were considered: [0 2.5 5 7.5 10 15 20] (%);
- The number of samples (N) considered in MEC was 50;
- The random variables u are considered as being equal to zero in the first iteration of IRA;
- The derivatives of constraints (G_j) required by IRA were obtained analytically due to its simplicity;
- The stopping criterion used to finish the IRA strategy was the Euclidean norm of the vector u smaller than 1.0×10^{-5} along two consecutive iterations;
- The number of objective function evaluations is equal to $NP + NP \times GEN \times N \times N_{\text{IRA}}$, where N_{IRA} is the number of iterations required by IRA for each neighbor in each generation;
- The stopping criterion used to finish the DE algorithm is the maximum number of generations;
- All case studies were run 10 times to obtain the upcoming average values.

2.6.1 Nonlinear Limit State

This problem was originally proposed and solved by Aoues and Chateauneuf [2] in the reliability context. The associated mathematical formulation is given by:

$$\min_{x_d} f(x_{d1}, x_{d2}) = x_{d1}^2 + x_{d2}^2 \tag{2.22}$$

subject to

$$\begin{cases} P(0.2x_{d1}x_{d2}x_{r2}^2 - x_{r1} \leq 0) \geq R \\ 0 \leq x_{di} \leq 15, i = 1, 2 \end{cases} \tag{2.23}$$

This test case contains two design variables (x_{d1} and x_{d2}) and two random variables (x_{r1} and x_{r2}), which are normally distributed with mean values 5 and 3, respectively. The variation coefficients are equal to 0.3 for both variables. This RBO problem was solved considering different optimization strategies and target reliability R equal to 98.9830% (corresponding to $\beta = 2.32$), as presented in Table 2.1 (n_{eval} is the number of objective function evaluations required for each classical strategy). In this case, the initial condition was written as $(x_{d1}, x_{d2}) = (12, 12)$.

Table 2.2 shows the results obtained by using the MEC + IRA + DE approach considering the target reliability R equal to 98.9830% (corresponding to $\beta = 2.32$) and considering different values of δ.

Note that the results obtained by using the proposed strategy are similar to the ones determined by the classical approaches when robustness is disregarded ($\delta = 0$

Table 2.1 RBO results associated with the nonlinear limit state problem [2]

Strategy	f – Eq. (2.22)	n_{eval}
Reliability index approach	63.88	466
Performance measure approach	63.88	220
Karush–Kuhn–Tucker	63.88	211
Single loop approach	63.88	42
Sequential optimization and reliability assessment	63.88	200

Table 2.2 RBO results associated with the nonlinear limit state problem [2]

δ (%)	f – Eq. (2.22)	x_{d1}	x_{d2}	n_{eval}
0	63.09484[a]/1.47 × 10⁻⁵ [b]	5.61670[a]/2.2 × 10⁻⁵ [b]	5.61670[a]/1.9 × 10⁻⁵ [b]	1,875,025[a]/6[c]
2.5	64.43813/1.69 × 10⁻⁵	5.76905/1.93 × 10⁻⁵	5.58056/1.03 × 10⁻⁵	2,500,025/8
5	66.12401/1.98 × 10⁻⁵	5.75229/1.22 × 10⁻⁵	5.74283/1.02 × 10⁻⁵	3,125,025/10
7.5	67.25552/1.23 × 10⁻⁵	5.58285/1.11 × 10⁻⁵	5.99680/1.03 × 10⁻⁵	2,500,025/8
10	69.22563/2.34 × 10⁻⁵	5.84983/2.69 × 10⁻⁵	5.89708/2.99 × 10⁻⁵	2,812,525/9
15	71.78167/5.41 × 10⁻⁵	5.84511/3.44 × 10⁻⁵	6.08960/3.44 × 10⁻⁵	3,437,525/11
20	76.54139/1.69 × 10⁻⁵	5.80297/3.97 × 10⁻⁵	6.47010/3.94 × 10⁻⁵	2,187,525/7

[a] Mean value
[b] Standard deviation
[c] Average value required by IRA for each candidate in each run (N_{IRA})

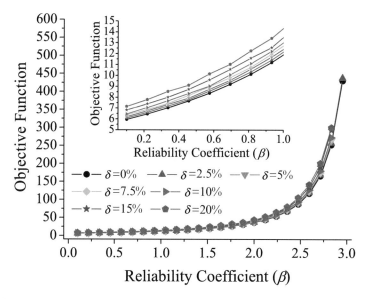

Fig. 2.3 Trade-off frontier between the reliability coefficient, the robustness parameter and the optimal solution for the nonlinear limit state problem

in Table 2.1). Considering $\delta > 0$, it can be observed that the value of the objective function increases according to δ. The value of the objective function worsened according to the robustness. The total number of evaluations n_{eval} required by the proposed methodology is higher than the one used by the classical approaches (that considers only RBO analysis). This is an expected behavior due to the characteristics of the DE algorithm. Remember that a population of candidates is determined in each generation of DE, increasing n_{eval}. It is worth mentioning that in all runs of the DE algorithm the problem always converged to the global minima solution.

Figure 2.3 presents the influence of β (0.1 (53.9828%)$\leq$ $\beta(R)$ $\leq$ 2.95 (99.8411%)) and $\delta \in$ [0 2.5 5 7.5 10 15 20] (%) by using the MEC + IRA + DE algorithm to solve the presented mathematical problem. Note that increasing reliability and robustness implies to increase the value of the objective function. In addition, the value of the objective function increases significantly for β higher than 2.

2.6.2 Highly Nonlinear Limit State

This mathematical problem was originally proposed and solved by Aoues and Chateauneuf [2] in the reliability context. Mathematically, this test case can be expressed as follows:

$$\min_{x_d} f(x_{d1}, x_{d2}) = x_{d1}^2 + x_{d2}^2 \tag{2.24}$$

Table 2.3 RBO results associated with the highly nonlinear limit state problem [2]

Strategy	f – Eq. (2.24)	n_{eval}
Reliability index approach	–	NC[a]
Performance measure approach	3.67	210
Karush–Kuhn–Tucker	–	NC
Single loop approach	3.67	60
Sequential optimization and reliability assessment	3.67	136

[a]No convergence

Table 2.4 Results associated with the highly nonlinear limit state problem considering different values of δ

δ (%)	f – Eq. (2.24)	x_{d1}	x_{d2}	n_{eval}
0	3.65327[a]/2.34 × 10^{-5} [b]	1.35153[a]/1.23 × 10^{-5} [b]	1.35153[a]/1.22 × 10^{-5} [b]	2,500,025[a]/8[c]
2.5	3.73752/2.29 × 10^{-5}	1.37549/1.28 × 10^{-5}	1.35822/2.45 × 10^{-5}	2,187,525/7
5	3.81508/2.20 × 10^{-5}	1.37426/2.32 × 10^{-5}	1.38683/4.52 × 10^{-5}	2,187,525/7
7.5	3.92669/1.29 × 10^{-5}	1.41779/5.82 × 10^{-5}	1.38174/1.24 × 10^{-5}	3,125,025/10
10	4.00524/1.02 × 10^{-5}	1.36734/4.27 × 10^{-5}	1.45682/1.45 × 10^{-5}	2,812,525/9
15	4.18243/1.23 × 10^{-5}	1.46264/3.27 × 10^{-5}	1.41846/2.95 × 10^{-5}	3,437,525/11
20	4.47742/2.38 × 10^{-5}	1.39864/3.52 × 10^{-5}	1.56921/3.42 × 10^{-5}	3,750,025/12

[a]Mean value
[b]Standard deviation
[c]Average value required by IRA for each candidate in each run (N_{IRA})

subject to

$$\begin{cases} P(x_{d1}x_{d2}x_{r2} - \ln(x_{r1}) \le 0) \ge R \\ 0 \le x_{di} \le 15, i = 1, 2 \end{cases} \tag{2.25}$$

The problem contains two design variables (x_{d1} and x_{d2}) and two random variables (x_{r1} and x_{r2}), which are normally distributed with the same mean values and variation coefficients considered in the first test case. The target failure probability is $R = 98.9830\%$ ($\beta = 2.32$). Table 2.3 shows the results obtained by Aoues and Chateauneuf [2] considering different strategies. In this case, the initial condition is given by (x_{d1}, x_{d2}) = (12, 12). Note that no convergence was achieved considering the Reliability Index and Karush–Kuhn–Tucker approaches.

Table 2.4 shows the results obtained by using the MEC + IRA + DE approach considering $R = 98.9830\%$ ($\beta = 2.32$). Note that the results obtained by using the proposed strategy are similar to the ones determined by the classical approaches when robustness is disregarded ($\beta = 0$ in Table 2.3). Considering $\delta > 0$, it can be observed that the value of the objective function increases according to δ. The Reliability Index and the Karush–Kuhn–Tucker approaches failed in the convergence for (x_{d1}, x_{d2}) = (12, 12) (see Table 2.3). Differently, the MEC + IRA + DE approach always converged to the optimal solution. Additionally, the total number of evaluations n_{eval} required by the proposed methodology is higher than the one used by the classical approaches.

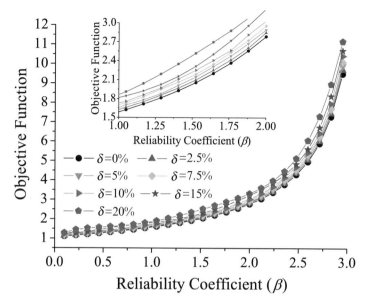

Fig. 2.4 Trade-off frontier between the optimal solution, the robustness parameter and the reliability coefficient for the highly nonlinear limit state problem

Figure 2.4 presents the influence of $\beta(0.1$ (53.9828%) $\leq \beta(R) \leq 2.95$ (99.8411%)) and $\delta \in [0\ 2.5\ 5\ 7.5\ 10\ 15\ 20]$ (%) by using the MEC + IRA + DE algorithm to solve the presented mathematical problem. As observed in the last test case, increasing reliability and robustness implies to increase the value of the objective function. Thus, more reliable and robust solutions increase the value of the objective function and the total number of objective function evaluations.

2.6.3   Short Column Design

This classical engineering test case considers a short column with rectangular cross-section and dimensions x_{r5} and x_{r6}, subjected to the yield stress x_{r1}, the biaxial bending moments x_{r2} and x_{r3}, and the normal force x_{r4} [2]. This model is based on the elastic-plastic constitutive law of the material and contains six random variables $(x_{ri}, i = 1, \ldots, 6)$. The RBO problem is formulated to minimize the cross-sectional area of the column under the reliability coefficient $\beta = 3$ ($R = 99.8651\%$), as given by Eq. (2.26). The statistical data associated with the random variables are presented in Table 2.5.

$$\min_{\mu_{x_r}} f(\mu_{x_{r5}}, \mu_{x_{r6}}) = \mu_{x_{r5}}\mu_{x_{r6}} \tag{2.26}$$

Table 2.5 Statistical data of the random variables used in the short column design problem [2]

	Mean value	Variation coefficient
x_{r1} (MPa)	40	0.10
x_{r2} (kN m)	250	0.30
x_{r3} (kN m)	125	0.30
x_{r4} (kN)	2500	0.2
x_{r5} (m)	$\mu_{x_{r5}}$	0.15
x_{r6} (m)	$\mu_{x_{r6}}$	0.15

Table 2.6 RBO results associated with the short column design problem

Strategy	f (m^2) – Eq. (2.26)	n_{eval}
Reliability index approach	0.3015	945
Performance measure approach	0.3015	684
Karush–Kuhn–Tucker	0.3077	567
Single loop approach	0.2970	46
Sequential optimization and reliability assessment	0.3014	346

subject to

$$\begin{cases} P(G \leq 0) \geq R \\ G = 1 - \dfrac{4x_{r2}}{\mu_{x_{r5}}\mu^2_{x_{r6}}x_{r1}} - \dfrac{4x_{r3}}{\mu^2_{x_{r5}}\mu_{x_{r6}}x_{r1}} - \left(\dfrac{x_{r4}}{\mu_{x_{r5}}\mu_{x_{r6}}x_{r1}}\right)^2 \\ 0 \leq \mu_{x_{r5}}/\mu_{x_{r6}} \leq 2 \end{cases} \qquad (2.27)$$

This problem was solved by Aoues and Chateauneuf [2] in the RBO context considering the initial design $(0.5, 0.5)$ and the same classical optimization strategies presented above (see Table 2.6). As observed by Aoues and Chateauneuf [2], the Karush–Kuhn–Tucker method converges to an optimal solution slightly higher than the minimum values obtained by the other methods. The RBO method converged to $(\mu_{x_{r5}}, \mu_{x_{r6}}) = (0.61, 0.31)$ considering $\mu_{x_{r5}}$ and $\mu_{x_{r6}}$ deterministic (optimal area of 0.1920 m^2). The Karush–Kuhn–Tucker approach converged to $(\mu_{x_{r5}}, \mu_{x_{r6}}) = (0.42, 0.47)$, corresponding to the optimal area of 0.2003 m^2. Different number of objective function evaluations resulted from the application of the mentioned strategies.

Table 2.7 shows the results obtained by using the MEC + IRA + DE approach for $\beta = 3$ ($R = 99.8651\%$). Note that the results obtained by using the MEC + IRA + DE strategy disregarding robustness are similar to the ones determined by the classical approaches (see Table 2.6). Figure 2.5 presents the influence of $\beta(0.1\ (53.9828\%) \leq \beta(R) \leq 4\ (99.9968\%))$ and $\delta \in [0\ 2.5\ 5\ 7.5\ 10\ 15\ 20]\ (\%)$ by using the MEC + IRA + DE algorithm. As observed in the last test case, increasing reliability and robustness implies to increase the value of the objective function.

Table 2.7 Results associated with the short column design problem considering different values of δ

δ (%)	$f(m^2)$ – Eq. (2.26)	$\mu_{x_{r5}}(m)$	$\mu_{x_{r6}}(m)$	n_{eval}
0	0.30151[a]/2.33 × 10⁻⁵[b]	0.38827[a]/3.04 × 10⁻⁵[b]	0.77654[a]/3.44 × 10⁻⁵[b]	3,437,525[a]/11[c]
2.5	0.30855/2.01 × 10⁻⁵	0.43134/2.02 × 10⁻⁵	0.71541/1.37 × 10⁻⁵	3,437,525/11
5	0.31568/4.20 × 10⁻⁵	0.43211/1.32 × 10⁻⁵	0.73093/4.12 × 10⁻⁵	1,875,025/6
7.5	0.32419/1.32 × 10⁻⁵	0.43128/2.44 × 10⁻⁵	0.75220/3.09 × 10⁻⁵	1,875,025/6
10	0.32880/1.43 × 10⁻⁵	0.45666/2.03 × 10⁻⁵	0.72101/2.22 × 10⁻⁵	2,187,525/7
15	0.34929/5.11 × 10⁻⁵	0.48013/3.38 × 10⁻⁵	0.72961/3.30 × 10⁻⁵	2,500,025/8
20	0.36601/4363 × 10⁻⁵	0.51186/1.30 × 10⁻⁵	0.71808/1.53 × 10⁻⁵	2,812,525/9

[a] Mean value
[b] Standard deviation
[c] Average value required by IRA for each candidate in each run (N_{IRA})

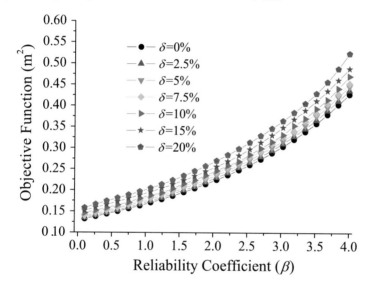

Fig. 2.5 Trade-off frontier between the optimal solution, the robustness parameter and the reliability coefficient for the short column design problem

2.6.4   Cantilever Beam Problem

This test case was previously studied by various authors [17, 18, 23, 24, 38] in the reliability context. The objective of this example is to find the minimum cross-sectional area of the cantilever beam presented in Fig. 2.6, according to Eq. (2.28).

$$\min_{x_d} f(x_{d1}, x_{d2}) = x_{d1} x_{d2} \tag{2.28}$$

Fig. 2.6 Cantilever beam
problem [24]

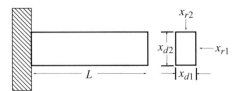

Table 2.8 Statistical data of
the random variables used in
the cantilever beam problem
[17]

	Mean value	Standard deviation
x_{r1} (lb)	500	100
x_{r2} (lb)	1000	100
x_{r3} (psi)	29×10^6	1.45×10^6
x_{r4} (psi)	40,000	2000

Table 2.9 RBO results and performance for the cantilever beam problem

Strategy	$f(\mathrm{cm}^2) -$ Eq. (2.28)	x_{d1}(cm)	x_{d2}(cm)
Reliability index approach [24]	9.5202	2.446	3.8922
Performance measure approach [24]	9.5202	2.446	3.8920
Exact [38]	9.5204	2.4484	3.8884
FORM [18]	9.5202	2.44599	3.89219

where x_{d1} and x_{d2} are the width and the height of the beam, respectively. These
parameters are defined as the design variables (deterministic) with lower and upper
bounds equal to 0 and 10, respectively. The two loads (x_{r1} and x_{r2}) applied at
the free end of the beam, the Young's modulus x_{r3}, and the yield strength x_{r4}
are considered as random variables with normal distribution. The associated mean
values and standard deviations are presented in Table 2.8.

The length L of the cantilever beam is 254 cm and the tip displacement has to
be smaller than the allowable displacement $d_o = 7.724$ cm. The target reliability is
defined as 99.8651% ($\beta = 3$) [17]. The stress (GS) at the fixed end has to be smaller
than the yield strength x_{r4}. Therefore,

$$GS = \frac{600x_{r2}}{x_{d1}x_{d2}^2} + \frac{600x_{r1}}{x_{d1}^2 x_{d2}} - x_{r4} \leq 0 \tag{2.29}$$

The second failure mode GD is defined as the tip displacement exceeding the
allowable displacement d_o, as given by Eq. (2.30).

$$GD = \frac{4L^3}{x_{r3}x_{d1}x_{d2}} \sqrt{\left(\frac{x_{r2}}{x_{d2}^2}\right)^2 + \left(\frac{x_{r1}}{x_{d1}^2}\right)^2} - d_o \leq 0 \tag{2.30}$$

Table 2.9 presents the solution obtained by various authors considering different
strategies.

Table 2.10 Results associated with the cantilever beam problem considering different values of δ

δ (%)	f (cm^2) – Eq. (2.28)	x_{d1} (cm)	x_{d2} (cm)	n_{eval}
0	9.52025[a]/1.11 × 10^{-5} [b]	2.44838[a]/2.37 × 10^{-5} [b]	3.88838[a]/3.64 × 10^{-5} [b]	2,500,025[a]/8[b]
2.5	9.74023/2.24 × 10^{-5}	2.55312/1.16 × 10^{-5}	3.81515/2.99 × 10^{-5}	3,750,025/12
5	9.90947/1.21 × 10^{-5}	2.53019/2.34 × 10^{-5}	3.94261/3.85 × 10^{-5}	2,187,525/7
7.5	10.10964/3.09 × 10^{-5}	2.59731/1.65 × 10^{-5}	3.94082/2.44 × 10^{-5}	3,125,025/10
10	10.26731/2.03 × 10^{-5}	2.66355/3.88 × 10^{-5}	3.91947/2.32 × 10^{-5}	3,437,525/11
15	10.68800/3.99 × 10^{-5}	2.83320/2.02 × 10^{-5}	3.87016/3.29 × 10^{-5}	3,437,525/11
20	11.02776/3.91 × 10^{-5}	2.98477/3.95 × 10^{-5}	3.82826/2.01 × 10^{-5}	4,062,525/13

[a] Mean value
[b] Standard deviation
[c] Average value required by IRA for each candidate in each run (N_{IRA})

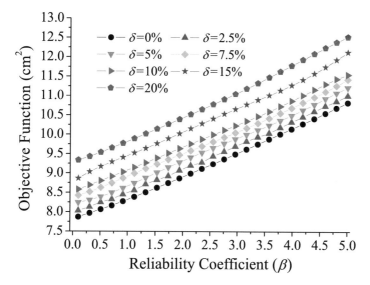

Fig. 2.7 Trade-off frontier between the optimal solution, the robustness parameter and the reliability coefficient for the cantilever beam problem

Table 2.10 shows the results obtained by using the MEC + IRA + DE approach considering $\beta = 3$ ($R = 99.8651\%$). Note that the results obtained by using the MEC + IRA + DE strategy disregarding robustness are similar to the ones determined by the classical approaches, as observed in Table 2.9. Additionally, the insertion of robustness implies an increase of the objective function to find a more robust solution. The total number of evaluations n_{eval} required by the proposed methodology is higher than the one required by the classical approaches. Figure 2.7 presents the influence of $\beta(0.1\ (53.9828\%) \leq \beta(R) \leq 5\ (99.9999\%))$ and $\delta \in [0\ 2.5\ 5\ 7.5\ 10\ 15\ 20]$ (%) by using the MEC + IRA + DE algorithm. Note that increasing reliability and robustness implies the increase of the objective function.

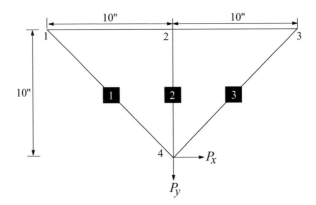

Fig. 2.8 Three bar truss problem [18]

2.6.5 Three Bar Truss Problem

The three bar truss problem was studied by Thanedar and Kodiyalam [34] and Liao and Ivan [18]. The goal of this example is to find the minimum weight of the structure subjected to two deterministic forces through the minimization of the cross-sectional area of the truss members (see Fig. 2.8). Mathematically, this problem is formulated in the RBO context as given by Eqs. (2.31)–(2.37).

$$\min_{x_d} f(x_{d1}, x_{d2}) = 2\sqrt{2}x_{d1} + x_{d2} \tag{2.31}$$

$$P\left(\frac{\sqrt{2}P_x l}{x_{d1}x_{r1}} - x_{r2} \le 0\right) \ge R_1 \tag{2.32}$$

$$P\left(\frac{\sqrt{2}P_y l}{\left(x_{d1} + \sqrt{2}x_{d2}\right)x_{r1}} - x_{r2} \le 0\right) \ge R_2 \tag{2.33}$$

$$\frac{\sqrt{2}}{2}\left|\frac{P_x}{x_{d1}} + \frac{P_y}{\left(x_{d1} + \sqrt{2}x_{d2}\right)}\right| - 5000 \le 0 \tag{2.34}$$

$$\sqrt{2}\left|\frac{P_y}{\left(x_{d1} + \sqrt{2}x_{d2}\right)}\right| - 20000 \le 0 \tag{2.35}$$

Table 2.11 RBO results and performance for the three bar truss problem [18]

Strategy	f (in^2) – Eq. (2.31)	x_{d1} (in^2)	x_{d2} (in^2)	n_{eval}
SQP+PSO+FORM	21.128	6.31	3.265	–

$$\frac{\sqrt{2}}{2} \left| \frac{P_y}{\left(x_{d1} + \sqrt{2}x_{d2}\right)} - \frac{P_x}{x_{d1}} \right| - 5000 \leq 0 \qquad (2.36)$$

$$x_{di} \geq 0.1, i = 1, 2 \qquad (2.37)$$

where x_{d1} is the cross-sectional area of the members 1 and 3 (in^2), x_{d2} is the cross-sectional area of the member 2 (in^2), P_x (10,000 lbs) and P_y (20,000 lbs) are the forces acting on the point 4 along the x and y directions, respectively, l is equal to 10 in, x_{r1} (psi) is the elastic modulus with mean value equal to 10^7 and variation coefficient equal to 0.1, and x_{r2} (in) is the allowable displacement with mean value equal to 0.005 and variation coefficient equal to 0.1. In Liao and Ivan [18], two probabilistic displacement constraints were considered (Eqs. (2.32) and (2.33)). The constraints were formulated to ensure that the resulting displacements at point 4 along the x and y directions meet the predefined reliability coefficients $\beta_1 = 2.85$ (or 99.77%) and $\beta_2 = 2$ (or 97.72%). In addition, three deterministic stress constraints are formulated to ensure that the resulting compressive/tensile stresses of the members 1, 2 and 3, are smaller than 5000 psi, 20,000 psi, and 5000 psi, respectively.

This problem was solved by Liao and Ivan [18] considering the Sequential Quadratic Programming (SQP), Particle Swarm Optimization (PSO), FORM, and the polynomial coefficient method for the reliability analysis. In this strategy, the calculated reliability and its derivative are incorporated into the problem through the SQP optimizer. The PSO algorithm is used to conduct the RBO problem. For this aim, the authors considered 81 different initial estimations for the design variables x_{d1} and x_{d2}. Since the probability density function is not provided in the literature, Thanedar and Kodiyalam [34] and Liao and Ivan [18] considered the random variables with normal distribution. Table 2.11 presents the results obtained by Liao and Ivan [18] for $\beta_1 = 2.85$ (or 99.77%) and $\beta_2 = 2$ (or 97.72%).

Table 2.12 shows the results obtained by using the MEC + IRA + DE approach considering $\beta_1 = 2.85$ (or 99.77%) and $\beta_2 = 2$ (or 97.72%). Note that the result obtained by using the MEC + IRA + DE strategy for the test case without robustness is similar to the one determined by Liao and Ivan [18]. Additionally, the insertion of robustness implies an increase of the objective function to find a more robust solution.

Figure 2.9 presents the influence of δ [0 2.5 5 7.5 10 15 20] (%) and $\beta_1 = \beta_2 = \beta$ (0.1 (53.9828%)$\leq \beta(R) \leq$5 (99.9999%)) by using the MEC + IRA + DE algorithm. In this case, as observed in the previous test cases, increasing reliability and robustness implies an increase of the objective function. Additionally, the objective function increases significantly for β higher than 4.5.

Table 2.12 Results associated with the three bar truss problem considering different values of δ

δ (%)	f (in^2) – Eq. (2.31)	x_{d1} (in^2)	x_{d2} (in^2)	n_{eval}
0	20.98242[a]/2.33 × 10^{-5} [b]	6.27362[a]/2.11 × 10^{-5} [b]	3.23757[a]/3.46 × 10^{-5} [b]	3,125,025[a]/10[c]
2.5	21.37350/1.34 × 10^{-5}	6.42000/1.34 × 10^{-5}	3.20679/2.36 × 10^{-5}	3,125,025/10
5	21.81682/1.37 × 10^{-5}	6.57700/1.22 × 10^{-5}	3.19492/1.76 × 10^{-5}	3,750,025/12
7.5	22.25208/2.29 × 10^{-5}	6.74800/2.08 × 10^{-5}	3.15413/1.99 × 10^{-5}	3,125,025/10
10	22.71876/3.93 × 10^{-5}	6.92543/1.98 × 10^{-5}	3.12245/3.34 × 10^{-5}	3,437,525/11
15	23.72117/2.01 × 10^{-5}	7.31306/1.77 × 10^{-5}	3.03496/3.21 × 10^{-5}	3,125,025/10
20	24.89383/3.09 × 10^{-5}	7.73189/1.65 × 10^{-5}	2.99991/2.04 × 10^{-5}	4,375,025/14

[a] Mean value
[b] Standard deviation
[c] Average value required by IRA for each candidate in each run (N_{IRA})

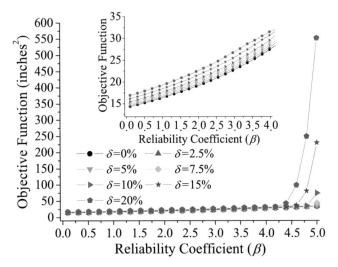

Fig. 2.9 Trade-off frontier between the optimal solution, the robustness parameter and the reliability coefficient for the three bar truss problem

2.7 Conclusion

This chapter investigated the combination of the Mean Effective Concept (MEC) (robustness) and the Inverse Reliability Analysis (IRA) (reliability-based design) with the Differential Evolution (DE) algorithm to solve reliability-based robust optimization problems. The proposed methodology (MEC + IRA + DE) was based on a three loop method (external loop—DE, intermediary loop—MEC, and internal loop—IRA). Five test cases with different complexities were studied to evaluate the performance of the algorithm. The obtained results demonstrated the efficiency of the MEC + IRA + DE methodology. It is important to point out that the number of function evaluations required by MEC + IRA + DE is larger than those required by the approaches restricted either to robustness or reliability. A heuristic optimization strategy was considered by increasing n_{eval}. As observed for the second example where some methods did not converge properly, the proposed methodology was able to successfully determine the solution of the problem. It is important to emphasize that all the presented test cases consider only normal distribution for the variables. However, the proposed methodology can deal with different probability distributions. In this sense, the Rackwitz-Fiessler's Two-parameter Equivalent Normal method should be used.

Acknowledgements The authors acknowledge the Brazilian research agencies CNPq, FAPEMIG (APQ-02284-15), and CAPES for the financial support of this research work through the National Institute of Science and Technology on Smart Structures in Engineering (INCT-EIE).

References

1. Agarwal, H.: Reliability based design optimization: formulations and methodologies. PhD. thesis, University of Notre Dame (2004)
2. Aoues, Y., Chateauneuf, A.: Benchmark study of numerical methods for reliability-based design. Struct. Multidiscip. Optim. **41**, 277–294 (2009)
3. Carter, A.D.S.: Mechanical Reliability and Design. Wiley, New York (1997)
4. Castro, R.E.: Optimization of structures with multi-objective through genetic algorithms. D.Sc. thesis (in Portuguese), COPPE/UFRJ, Rio de Janeiro (2001)
5. Deb, K., Gupta, H.: Introducing robustness in multiobjective optimization. Evol. Comput. **14**(4), 463–494 (2006)
6. Deb, K., Padmanabhan, D., Gupta, S., Mall, A.K.: Handling uncertainties through reliability-based optimization using evolutionary algorithms. IEEE Trans. Evol. Comput. **13**(5), 1054–1074 (2009)
7. Der-Kiureghian, A., De Stefano, M.: Efficient algorithm for second-order reliability analysis. J. Mech. Eng. ASCE1 **17**(12), 2904–2923 (1991)
8. Du, X.: Probabilistic engineering design – first order and second reliability methods. University of Missouri, Rolla (2005)
9. Du, X., Chen, W.: Sequential optimization and reliability assessment method for efficient probabilistic design. J. Mech. Des. **126**, 225–233 (2004)
10. Fiessler, B., Neumann, H.-J., Rackwitz, R.: Quadratic limit states in structural reliability. J. Mech. Eng. ASCE **105**(4), 661–676 (1979)
11. Gholaminezhad, I., Jamali, A., Assimi, H.: Multi-objective reliability-based robust design optimization of robot gripper mechanism with probabilistically uncertain parameters. Neural Comput. Appl. **1**, 1–12 (2016)
12. Jeong, S.B., Park, G.J.: Reliability-based robust design optimization using the probabilistic robustness index and the enhanced single loop single vector approach. In: 10th World Congress on Structural and Multidisciplinary Optimization, Orlando (2013)
13. Keshtegar, B., Chakraborty, S.: An efficient-robust structural reliability method by adaptive finite-step length based on Armijo line search. Reliab. Eng. Syst. Saf. **172**, 195–206 (2018)
14. Lagaros, N.D., Plevris, V., Papadrakakis, M.: Reliability based robust design optimization of steel structures. Int. J. Simul. Multidiscip. Des. Optim. **1**, 19–29 (2007)
15. Lee, M.C.W., Mikulik, Z., Kelly, D.W., Thomson, R.S., Degenhardt, R.: Robust design – a concept for imperfection insensitive composite structures. Compos. Struct. **92**(6), 1469–1477 (2010)
16. Leidemer, M.N.: Proposal of evolutionary robust optimization methodology using the unscented transform applicable to circuits for RF circuits/microwave. MSc. Thesis (in Portuguese), University of Brasilia, Brasilia (2009)

17. Liao, K.W., Ha, C.: Application of reliability-based optimization to earth-moving machine: hydraulic cylinder components design process. Struct. Multidiscip. Optim. **36**(5), 523–536 (2008)
18. Liao, K.W., Ivan, G.: A single loop reliability-based design optimization using EPM and MPP-based PSO. Lat. Am. J. Solids Struct. **11**, 826–847 (2014)
19. Melchers, R.E.: Structural Reliability Analysis and Prediction. Wiley, Chichester (1999)
20. Moreira, F.R., Lobato, F.S., Cavalini, A.A. Jr., Steffen, V. Jr.: Robust multi-objective optimization applied to engineering systems design. Lat. Am. J. Solids Struct. **13**, 1802–1822 (2016)
21. Paenk, I., Branke, J., Jin, Y.: Efficient search for robust solutions by means of evolutionary algorithms and fitness approximation. IEEE Trans. Evol. Comput. **10**, 405–420 (2006)
22. Phadke, M.S.: Quality Engineering Using Robust Design. Prentice Hall, Englewood Cliffs (1989)
23. Qu, X., Haftka, R.T.: Reliability-based design optimization using probabilistic sufficiency factor. Struct. Multidiscip. Optim. **27**(5), 314–325 (2004)
24. Ramu, P., Qu, X., Youn, B.D., Haftka, R.T., Choi, K.K.: Safety factor and inverse reliability measures. In: Proceeding of 45th AIAA/ASME/ASCE/AHS/ASC Structures, Structural Dynamics and Materials Conference, Palm Springs (2004)
25. Ravichandran, G.: Integrated reliable and robust design. Missouri University of Science and Technology (2011)
26. Ritto, T.G., Sampaio, R., Cataldo, E.: Timoshenko beam with uncertainty on the boundary conditions. J. Braz. Soc. Mech. Sci. Eng. **30**(4), 295–303 (2008)
27. Rosenblatt, M.: Remarks on a multivariate transformation. Ann. Math. Stat. **23**, 470–472 (1952)
28. Sampaio, R., Soize, C.: On measures of nonlinearity effects for uncertain dynamical systems application to a Vibro-impact system. J. Sound Vib. **303**, 659–674 (2007)
29. Shahraki, A.F., Noorossana, R.: A combined algorithm for solving reliability-based robust design optimization problems. J. Math. Comput. Sci. **7**, 54–62 (2013)
30. Soize, C.: A comprehensive overview of a non-parametric probabilistic approach of model uncertainties for predictive models in structural dynamics. J. Sound Vib. **288**(3), 623–652 (2005)
31. Souza, D.L., Lobato, F.S., Gedraite, R.: Robust multiobjective optimization applied to optimal control problems using differential evolution. Chem. Eng. Technol. **1**, 1–8 (2015)
32. Storn, R., Price, K.: Differential evolution – a simple and efficient adaptive scheme for global optimization over continuous spaces. Int. Comput. Sci. Inst. **12**, 1–16 (1995)
33. Taguchi, G.: Taguchi on Robust Technology Development – Bringing Quality Engineering Upstream. ASME Press, New York (1993)
34. Thanedar, P.B., Kodiyalam, S.: Structural optimization using probabilistic constraints. Struct. Multidiscip. Optim. **4**, 236–240 (1992)
35. Tichy, M.: First-order third-moment reliability method. Struct. Saf. **16**(2), 189–200 (1994)
36. Wang, S., Li, Q., Savage, G.J.: Reliability-based robust design optimization of structures considering uncertainty in design variables. Math. Probl. Eng. **2015**, 1–8 (2015)
37. Wang, Y., Tuo, Y., Yang, S.X., Biglarbegian, M., Fu, M.: Reliability-based robust dynamic positioning for a turret-moored floating production storage and offloading vessel with unknown time-varying disturbances and input saturation (2018). https://doi.org/10.1016/j.isatra.2017.12.023
38. Wu, Y.T., Shin, Y., Sues, R., Cesare, M.: Safety-factor based approach for probability-based design optimization. In: Proceedings of the 42rd AIAA/ASME/ASCE/AHS/ASC Structures, Structural Dynamics, and Materials Conference (2001)
39. Zhao, Y.G., Ono, T.: A general procedure for first/second-order reliability method (FORM/SORM). Struct. Saf. **21**(1), 95–112 (1999)
40. Zhao, Y.G., Ono, T.: New approximations for SORM: part 2. J. Mech. Eng. ASCE1 **25**(1), 86–93 (1999)

Chapter 3
On Initial Populations of Differential Evolution for Practical Optimization Problems

Wagner Figueiredo Sacco and Ana Carolina Rios-Coelho

3.1 Introduction

In this chapter, we analyze an approach that has been rather overlooked in populational algorithms like the genetic algorithm (GA) [14] and the differential evolution (DE) [39]: the selection of an initial population of solutions [20].

In the literature, there have been some studies on initialization methods applied to evolutionary algorithms, as, for example, Refs. [2, 15, 20]. But, to the best of our knowledge, in all the past contributions the experiments were made with test-functions from the literature, except for an article where a nuclear engineering problem was used [36]. In this work, we test some initialization methods on three challenging practical problems: a non-linear system, a potential energy function, and a parameter estimation problem.

The traditional way of generating an initial population in these algorithms is to use pseudorandom numbers [20]. We test two different initialization schemes against the conventional random generation method: random generation followed by the application of opposition-based learning [41], and the Sobol quasi-random generator [38], which, in spite of the name, is deterministic. We implement these methods under the differential evolution algorithm, which outperformed the more popular genetic algorithms and particle swarm optimization [16] in extensive experiments [43]. For examples of the application of opposition-based learning to optimization (not exclusively for initialization, though), see, for example, Refs. [30, 32]. Regarding the use of Sobol sequence in optimization, see, for instance, Refs. [27, 29, 40].

W. F. Sacco (✉) · A. C. Rios-Coelho
Federal University of Western Pará, Institute of Engineering and Geosciences, Santarém, PA, Brazil
e-mail: wagner.sacco@ufopa.edu.br; ana.coelho@ufopa.edu.br

© Springer Nature Switzerland AG 2019
G. Mendes Platt et al. (eds.), *Computational Intelligence, Optimization and Inverse Problems with Applications in Engineering*,
https://doi.org/10.1007/978-3-319-96433-1_3

The remainder of the chapter is described as follows. The initialization schemes are presented in Sect. 3.2. The practical problems, implementation aspects, and results are presented in Sect. 3.3. Finally, the conclusions are made in Sect. 3.4.

3.2 The Initialization Schemes

3.2.1 The Mersenne Twister Pseudorandom Generator

A pseudorandom number generator is a deterministic algorithm which produces a sequence $(X_i)_{i \geq 0}$ of numbers in the interval $[0, 1)$ which, for virtually all generators used for computer simulation, is purely periodic [18]. Confined to the period, these numbers should behave like independent events of a random variable uniformly distributed in $[0, 1)$ [18]. In this work, we use the Mersenne Twister generator [24].

3.2.2 Mersenne Twister and Opposition-Based Learning

In our second initialization scheme, we generate the solutions combining Mersenne Twister with the opposition-based learning paradigm (OBL). Let us first describe OBL, and then the scheme.

Opposition-Based Learning [41] is a framework for machine intelligence with application to optimization, reinforcement learning, and artificial neural networks [5]. As an example of application of this paradigm to global search algorithms, consider a random initialization of a search procedure. The estimation of candidate solutions provided by random guesses is likely to have poor evaluation, and probably it is not worth intensifying the search in the correspondent region at a first moment. Therefore, since one deals with the absence of a priori knowledge, it is reasonable to explore an opposite region of that search space.

According to Ref. [41], for a multidimensional search space, let $\mathbf{P} = (x_1, x_2, \ldots, x_n)$ be a point in an n-dimensional coordinate system with $x_1, x_2, \ldots, x_n \in \mathbb{R}$ and $x_i \in [low_i, up_i]$. The coordinates i of the opposite point $\mathbf{\breve{P}} = (\tilde{x}_1, \tilde{x}_2, \ldots, \tilde{x}_n)$ are given by

$$\tilde{x}_i = low_i + up_i - x_i \tag{3.1}$$

The authors of Ref. [32] applied OBL to DE in order to accelerate the searching process. Their algorithm employs OBL not only in the initialization, but also to generate, with a certain probability, fitter (i.e., with lower objective function values in a minimization problem) members of the new population. Differently from this previous study, in this work we use the OBL paradigm only to generate the initial population, following the same initialization scheme proposed in Ref. [32]:

1. A population of NP individuals is generated at random. In this work, we used the Mersenne Twister generator;
2. An opposite population of NP individuals is created from the population generated in the previous step;
3. The NP fittest individuals are selected from those generated in steps (1) and (2), becoming the initial population.

3.2.3   The Sobol Low-Discrepancy Sequence

The points in a quasi-random or low-discrepancy sequence are designed to maximally avoid each other. As Ref. [19] says

> While the points generated using pseudorandom numbers try to imitate random points, the points generated using quasi-random sequences try to imitate points with a perfect uniform distribution.

Discrepancy is a measure of deviation from uniformity of a given sequence of points, distributed within another set of points. Here, it is sufficient to consider the discrepancy of $S = \{\mathbf{q}_1, \mathbf{q}_2, \dots, \mathbf{q}_\eta\} \subset [0, 1]^N$, an arbitrary N-dimensional finite sequence with η points generated in $[0, 1]^N$, the unit hypercube of $\mathbb{R}^N$. Thus, we say that $D_\eta(S)$ is the discrepancy of S in $[0, 1]^N$, if

$$D_\eta(S) = \mathrm{Sup}_{\jmath} |F_\eta(\jmath) - M(\jmath)| \tag{3.2}$$

where the supremum is taken over all subintervals $\jmath$ of the unit hypercube, $F_{\eta(\jmath)}$ is the number of points in $S \cap \jmath$ divided by η, and $M(\jmath)$ is the measure of $\jmath$ [9]. Note that the magnitude $|F_\eta(\jmath) - M(\jmath)|$ represents the relative size of the points fraction of S in $\jmath$.

$S = \{\mathbf{q}_1, \mathbf{q}_2, \dots, \mathbf{q}_\eta\}$ is considered a sequence with low-discrepancy in $[0, 1]^N$, if $D_\eta(S)$ becomes small, for η sufficiently large. The Sobol sequence is a deterministic sequence with low discrepancy, which has been used for numerical integration of multidimensional functions via quasi-Monte Carlo methods [28]. For further details, please see Ref. [8].

3.3   Numerical Comparisons

3.3.1   The Practical Problems

3.3.1.1   Chemical Equilibrium Problem

This non-linear system, introduced by Meintjes and Morgan [25], has been widely employed in the literature, see Refs. [10, 12, 22, 34, 42], among others.

It concerns the combustion of propane (C_3H_8) in air (O_2 and N_2) to form ten products. This chemical reaction generates a system of ten equations in ten unknowns, which can be reduced to a system of five equations in five unknowns [25]. We solve this system formulating it as an optimization problem.

The system is given by

$$
\begin{cases}
f_1 = x_1x_2 + x_1 - 3x_5 \\
f_2 = 2x_1x_2 + x_1 + x_2x_3^2 + R_8x_2 - Rx_5 + 2R_{10}x_2^2 + R_7x_2x_3 + R_9x_2x_4 \\
f_3 = 2x_2x_3^2 + 2R_5x_3^2 - 8x_5 + R_6x_3 + R_7x_2x_3 \\
f_4 = R_9x_2x_4 + 2x_4^2 - 4Rx_5 \\
f_5 = x_1(x_2 + 1) + R_{10}x_2^2 + x_2x_3^2 + R_8x_2 + R_5x_3^2 + x_4^2 - 1 + R_6x_3 \\
\quad\ + R_7x_2x_3 + R_9x_2x_4
\end{cases}
$$

$$(3.3)$$

where

$$
\begin{cases}
R = 10 \\
R_5 = 0.193 \\
R_6 = 0.002597/\sqrt{40} \\
R_7 = 0.003448/\sqrt{40} \\
R_8 = 0.00001799/40 \\
R_9 = 0.0002155/\sqrt{40} \\
R_{10} = 0.00003846/40
\end{cases}
$$

Variables x_i are surrogates for atomic combinations, which means that only positive values make physical sense. Among the four real solutions reported in Ref. [25], only one has all-positive components: (3.114102×10^{-3}, 3.459792×10^1, 6.504177×10^{-2}, 8.593780×10^{-1}, 3.695185×10^{-2}). Hence, if the search domain is taken from the positive side, as we did using the interval $[0, 100]^5$, this will be the only solution, with $f(\mathbf{x}^*) = 0$ as global minimum.

3.3.1.2 Lennard-Jones Potential

The optimization of Lennard-Jones clusters is a two-body problem used to simulate heavy atom rare gas clusters such as argon, xenon, and krypton [7]. The Lennard-Jones (LJ) potential energy between two atoms with distance r is given by [26]:

$$
LJ(r) = \frac{1}{r^{12}} - \frac{2}{r^6}
$$

$$(3.4)$$

The total potential energy V of a cluster of N atoms is given by

$$
V = \sum_{i<j} LJ(r_{ij})
$$

$$(3.5)$$

where r_{ij} is the Euclidean distance between atoms i and j, whose coordinates are given, respectively, by $\mathbf{X}_i = (x_i, y_i, z_i)$ and $\mathbf{X}_j = (x_j, y_j, z_j)$.

The Lennard-Jones potential energy minimization is an NP-hard problem, where the number of local minima of an N-atom microcluster grows as $\exp(N^2)$ [21]. Here, as in Ref. [4], we $N = 10$, which has $V = -28.422532$ as minimum [7].

3.3.1.3 Fertilizer Experiment Parameter Estimation

This parameter estimation problem was originally presented in Ref. [11]. Table 3.1 shows the results of an experiment where the variable y represents the yields of wheat corresponding to six rates of application of fertilizer, x, on a coded scale.

The idea is to fit these data to the exponential law of "diminishing returns" [11], given by [6]:

$$y = k_1 + k_2 \exp(k_3 x) \tag{3.6}$$

where k_1, k_2, and k_3 are the parameters to be estimated by a least-squares objective function. The optimal value of this function is given by $f(\mathbf{k}^*) = 13390.23$, where $\mathbf{k}^* = (523.3, -156.9, -0.1997)$ [11]. Such a high functional value is due to the low number of data points [6]. In this work, we define the search region for k_1, k_2, and k_3 as $[-10^5, 10^5] \times [-10^5, 10^5] \times [-10^2, 10^2]$.

3.3.2 Implementation and Setup

Our tests were performed on an Intel® Core™ i7 PC with 12 Gb RAM running Ubuntu 14.04 LTS. DE was implemented in C++ and compiled with GNU g++ version 4.6.3.

The Mersenne Twister and Sobol routines used in our code were developed by other researchers. The Mersenne Twister generator routine was taken from one of its creators' website [23], where it is freely available. The Sobol sequence generator was downloaded from Dr. Frances Kuo's website [17].

Table 3.1 Data for the fertilizer experiment parameter estimation problem [11]

x	y
-5	127
-3	151
-1	379
1	421
3	460
5	426

We used in our tests three population sizes for DE: $NP = 100, \ 500,$ and $1000.$ The other parameters, crossover rate CR and scaling factor F, were set as $CR = 0.9$ and $F = 0.5$, which are values that have been widely employed in the literature [1, 31, 35, 43]. The same one-hundred random seeds (one per execution) were used for the three DEs with the different initialization schemes.

As all optimization problems attacked in this work have known global minima, DE was run using the same termination criterion as in [3, 12, 13, 33, 37], which is ideal for an algorithm's performance assessment:

$$|f(\mathbf{x}^*) - f(\mathbf{x})| \leq \varepsilon_1 |f(\mathbf{x}^*)| + \varepsilon_2 \qquad (3.7)$$

where $f(\mathbf{x}^*)$ is the global optimum, $f(\mathbf{x})$ is the current best, coefficient $\varepsilon_1 = 10^{-4}$ corresponds to the relative error, and $\varepsilon_2 = 10^{-6}$ corresponds to the absolute error [37].

We set a maximum number of generations equal to 100,000 for all population sizes as a stopping criterion, in case the condition given by Eq. (3.7) is not achieved.

3.3.3 Computational Results

Table 3.2 shows the results obtained by each initialization scheme for the three problems and three population sizes. The success rate is the ratio between the number of successful runs (i.e., those where the criterion given by Eq. (3.7) was met) and the total number of runs (one hundred). The terms "Min. NFE," "Max. NFE," and "Avg. NFE" refer, respectively, to the minimum, maximum, and average values of fitness-function evaluations of the successful runs.

Some remarks on the results:

- In Problem 3.3.1.1, the initialization with Sobol had a poor performance with 100 individuals.
- The ten-atom instance of the Lennard-Jones potential (Problem 3.3.1.2) is difficult to general-purpose algorithms. The Sobol-initialized variant with 1000 individuals had the best performance, proving that low-discrepancy sequences with a high number of points can be quite effective.
- In Problem 3.3.1.3, MT+OBL had the worst performance.
- In terms of average success rate, the best were Mersenne Twister, for $Pop = 100$, and Sobol, for $Pop = 500$ and $Pop = 1000$. Based on these results, we can say that these methods are the best for, respectively, small and large populations. Mersenne Twister associated with Opposition-Based Learning, by its turn, was always the second best, proving to be quite robust in all situations.
- In terms of computational cost (expressed by NFE), in the average Sobol was the most successful for $Pop = 100$ and $Pop = 500$ and the second for $Pop = 1000$. Mersenne Twister was the best with $Pop = 1000$ and the worst with the other two population sizes.

Table 3.2 Results for each initialization scheme

Mersenne Twister initialization scheme				Sobol initialization scheme			
Population	100	500	1000	Population	100	500	1000
Problem 3.3.1.1				*Problem 3.3.1.1*			
Success rate	0.02	1.00	1.00	Success rate	0.01	1.00	1.00
Min. NFE	15,197	396,801	721,771	Min. NFE	212,665	375,383	638,120
Max. NFE	81,186	514,717	867,292	Max. NFE	212,665	513,215	873,291
Avg. NFE	48,191	455,152	800,690	Avg. NFE	212,665	454,642	797,079
Problem 3.3.1.2				*Problem 3.3.1.2*			
Success rate	0.06	0.10	0.13	Success rate	0.06	0.10	0.16
Min. NFE	262,536	4,535,147	10,000,000	Min. NFE	450,351	4,474,187	10,000,000
Max. NFE	6,668,278	45,280,948	90,331,810	Max. NFE	1,746,705	17,653,500	92,035,756
Avg. NFE	1,518,918	26,705,408	46,863,274	Avg. NFE	899,377	10,436,043	50,378,399
Problem 3.3.1.3				*Problem 3.3.1.3*			
Success rate	0.33	1.00	1.00	Success rate	0.32	1.00	1.00
Min. NFE	9836	55,086	114,377	Min. NFE	10,614	57,223	105,987
Max. NFE	45,308	141,505	208,199	Max. NFE	103,009	154,232	212,391
Avg. NFE	17,767	84,530	148,709	Avg. NFE	23,603	82,837	153,050

MT + OBL initialization scheme			
Population	100	500	1000
Problem 3.3.1.1			
Success rate	0.04	1.00	1.00
Min. NFE	13,374	355,028	709,025
Max. NFE	186,385	513,606	888,898
Avg. NFE	57,529	454,553	800,064
Problem 3.3.1.2			
Success rate	0.05	0.12	0.11
Min. NFE	277,518	4,152,111	10,000,000
Max. NFE	958,393	42,934,572	99,803,321
Avg. NFE	503,548	18,186,145	66,535,936
Problem 3.3.1.3			
Success rate	0.22	1.00	1.00
Min. NFE	11,005	50,624	113,532
Max. NFE	34,863	148,790	221,279
Avg. NFE	18,159	82,170	154,183

3.4 Conclusions

In this work, besides pseudorandom generation, we test two other population initialization schemes applied to practical problems: random generation followed by the application of opposition-based learning and the use of a quasi-random sequence, namely the Sobol sequence.

In terms of effectiveness, we recommend Mersenne Twister for small populations and the Sobol sequence for large population sizes. This corroborates Ref. [8], which says that quasi-random sequences work better with a large sample size (the initial solutions in our case).

Costwise, our experiments show that the best option is the initialization by Sobol, except for $Pop = 1000$, when Mersenne Twister prevails. Still in terms of computational cost, we advise the reader to increase the population size only if necessary.

As mentioned in the Introduction, the experiments in the previous studies on initialization methods applied to evolutionary algorithms, to the best of our knowledge, were made with test-functions from the literature, except for a recent effort [36]. The main contribution of this work is the use of practical problems to test the initialization methods. Real-world problems like those attacked here often have more complex search spaces than tailor-made optimization test problems, being thus more challenging for the optimization methods [33].

As further development, we plan to test other initialization schemes using practical problems, as Ref. [15] did for test-functions from the literature.

Acknowledgements The authors acknowledge the financial support provided by CNPq (Conselho Nacional de Desenvolvimento Científico e Tecnológico, Ministry of Science & Technology, Brazil).

References

1. Ali, M.M., Törn, A.: Population set-based global optimization algorithms: some modifications and numerical studies. Comput. Oper. Res. **31**(10), 1703–1725 (2004). https://doi.org/10.1016/S0305-0548(03)00116-3
2. Ali, M., Pant, M., Abraham, A.: Unconventional initialization methods for differential evolution. Appl. Math. Comput. **219**(9), 4474–4494 (2013). https://doi.org/10.1016/j.amc.2012.10.053
3. Csendes, T., Pál, L., Sendín, J.O.H., Banga, J.R.: The GLOBAL optimization method revisited. Optim. Lett. **2**(4), 445–454 (2008). https://doi.org/10.1007/s11590-007-0072-3
4. Deep, K., Arya, M.: Minimization of Lennard-Jones potential using parallel particle swarm optimization algorithm. In: Ranka, S., Banerjee, A., Biswas, K.K., Dua, S., Mishra, P., Moona, R., Poon, S.H., Wang, C.L. (eds.) Contemporary Computing, pp. 131–140. Springer, Berlin (2010)
5. Engelbrecht, A.P.: Computational Intelligence: An Introduction. Wiley, Chichester (2007)
6. Englezos, P., Kalogerakis, N.: Applied Parameter Estimation for Chemical Engineers. Chemical Industries. CRC Press, New York (2000)
7. Floudas, C.A., Pardalos, P.M.: Handbook of Test Problems in Local and Global Optimization. Nonconvex Optimization and Its Applications. Kluwer Academic Publishers, Dordecht (1999)
8. Galanti, S., Jung, A.: Low-discrepancy sequences. J. Deriv. **5**(1), 63–83 (1997). https://doi.org/10.3905/jod.1997.407985
9. Gentle, J.E.: Random Number Generation and Monte Carlo Methods. Statistics and Computing. Springer, New York (2003)
10. Grosan, C., Abraham, A.: A new approach for solving nonlinear equations systems. IEEE Trans. Syst. Man Cybern. A Syst. Hum. **38**(3), 698–714 (2008). https://doi.org/10.1109/TSMCA.2008.918599

11. Hartley, H.O.: The modified Gauss-Newton method for the fitting of nonlinear regression functions by least squares. Technometrics **3**(2), 269–280 (1961)
12. Hirsch, M.J., Meneses, C.N., Pardalos, P.M., Resende, M.G.C.: Global optimization by continuous GRASP. Optim. Lett. **1**(2), 201–212 (2007). https://doi.org/10.1007/s11590-006-0021-6
13. Hirsch, M.J., Pardalos, P.M., Resende, M.G.C.: Speeding up continuous GRASP. Eur. J. Oper. Res. **205**(3), 507–521 (2010). https://doi.org/10.1016/j.ejor.2010.02.009
14. Holland, J.H.: Adaptation in Natural and Artificial Systems. MIT Press, Cambridge (1992)
15. Kazimipour, B., Li, X., Qin, A.K.: Initialization methods for large scale global optimization. In: 2013 IEEE Congress on Evolutionary Computation, pp. 2750–2757 (2013). https://doi.org/10.1109/CEC.2013.6557902
16. Kennedy, J., Eberhart, R.: Particle swarm optimization. In: Proceedings of the IEEE International Conference on Neural Networks, Piscataway, vol. 4, pp. 1942–1948 (1995)
17. Kuo, F.: Sobol sequence generator (2010). http://web.maths.unsw.edu.au/~fkuo/sobol/index.html. Accessed 23 Feb 2017
18. Leeb, H., Wegenkittl, S.: Inversive and linear congruential pseudorandom number generators in empirical tests. ACM Trans. Model. Comput. Simul. **7**(2), 272–286 (1997). https://doi.org/10.1145/249204.249208
19. Maaranen, H., Miettinen, K., Mäkelä, M.: Quasi-random initial population for genetic algorithms. Comput. Math. Appl. **47**(12), 1885–1895 (2004). https://doi.org/10.1016/j.camwa.2003.07.011
20. Maaranen, H., Miettinen, K., Penttinen, A.: On initial populations of a genetic algorithm for continuous optimization problems. J. Glob. Optim. **37**(3), 405–436 (2006). https://doi.org/10.1007/s10898-006-9056-6
21. Maranas, C.D., Floudas, C.A.: A global optimization approach for Lennard-Jones microclusters. J. Chem. Phys. **97**(10), 7667–7678 (1992). https://doi.org/10.1063/1.463486
22. Maranas, C.D., Floudas, C.A.: Finding all solutions of nonlinearly constrained systems of equations. J. Glob. Optim. **7**(2), 143–182 (1995). https://doi.org/10.1007/BF01097059
23. Matsumoto, M.: Mersenne twister home page (2011). http://www.math.sci.hiroshima-u.ac.jp/~m-mat/MT/emt.html. Accessed 23 Feb 2017
24. Matsumoto, M., Nishimura, T.: Mersenne twister: a 623-dimensionally equidistributed uniform pseudo-random number generator. ACM Trans. Model. Comput. Simul. **8**(1), 3–30 (1998). https://doi.org/10.1145/272991.272995
25. Meintjes, K., Morgan, A.P.: Chemical equilibrium systems as numerical test problems. ACM Trans. Math. Softw. **16**(2), 143–151 (1990). https://doi.org/10.1145/78928.78930
26. Moloi, N.P., Ali, M.M.: An iterative global optimization algorithm for potential energy minimization. Comput. Optim. Appl. **30**(2), 119–132 (2005). https://doi.org/10.1007/s10589-005-4555-9
27. Nakib, A., Daachi, B., Siarry, P.: Hybrid differential evolution using low-discrepancy sequences for image segmentation. In: IEEE 26th International Parallel and Distributed Processing Symposium Workshops & PhD Forum (IPDPSW), Piscataway, pp. 634–640 (2012). https://doi.org/10.1109/IPDPSW.2012.79
28. Niederreiter, H.: Random number generation and Quasi-Monte Carlo methods. J. Soc. Ind. Appl. Math. (1992). https://doi.org/10.1137/1.9781611970081
29. Peng, L., Wang, Y.: Differential evolution using Uniform-Quasi-Opposition for initializing the population. Inf. Technol. J. **9**(8), 1629–1634 (2010)
30. Rahnamayan, S., Tizhoosh, H.R., Salama, M.M.A.: A novel population initialization method for accelerating evolutionary algorithms. Comput. Math. Appl. **53**(10), 1605–1614 (2007). https://doi.org/10.1016/j.camwa.2006.07.013
31. Rahnamayan, S., Tizhoosh, H.R., Salama, M.M.: Opposition versus randomness in soft computing techniques. Appl. Soft Comput. **8**(2), 906–918 (2008). https://doi.org/10.1016/j.asoc.2007.07.010
32. Rahnamayan, S., Tizhoosh, H.R., Salama, M.M.A.: Opposition-based differential evolution. IEEE Trans. Evol. Comput. **12**(1), 64–79 (2008). https://doi.org/10.1109/TEVC.2007.894200

33. Rios-Coelho, A.C., Sacco, W.F., Henderson, N.: A metropolis algorithm combined with Hooke–Jeeves local search method applied to global optimization. Appl. Math. Comput. **217**(2), 843–853 (2010). https://doi.org/10.1016/j.amc.2010.06.027

34. Sacco, W.F., Henderson, N.: Finding all solutions of nonlinear systems using a hybrid metaheuristic with fuzzy clustering means. Appl. Soft Comput. **11**(8), 5424–5432 (2011). https://doi.org/10.1016/j.asoc.2011.05.016

35. Sacco, W.F., Henderson, N.: Differential evolution with topographical mutation applied to nuclear reactor core design. Prog. Nucl. Energy **70**, 140–148 (2014). https://doi.org/10.1016/j.pnucene.2013.09.012

36. Sacco, W.F., Rios-Coelho, A.C., Henderson, N.: Testing population initialisation schemes for differential evolution applied to a nuclear reactor core design. Int. J. Nucl. Energy Sci. Technol. **8**(3), 192–212 (2014). https://doi.org/10.1504/IJNEST.2014.063008

37. Siarry, P., Berthiau, G., Durdin, F., Haussy, J.: Enhanced simulated annealing for globally minimizing functions of many-continuous variables. ACM Trans. Math. Softw. **23**(2), 209–228 (1997). https://doi.org/10.1145/264029.264043

38. Sobol', I.M.: On the distribution of points in a cube and the approximate evaluation of integrals. USSR Comput. Math. Math. Phys. **7**(4), 86–112 (1967). https://doi.org/10.1016/0041-5553(67)90144-9

39. Storn, R., Price, K.: Differential evolution – a simple and efficient heuristic for global optimization over continuous spaces. J. Glob. Optim. **11**(4), 341–359 (1997). https://doi.org/10.1023/A:1008202821328

40. Thangaraj, R., Pant, M., Abraham, A., Badr, Y.: Hybrid evolutionary algorithm for solving global optimization problems. In: Corchado, E., Wu, X., Oja, E., Herrero, Á., Baruque, B. (eds.) Hybrid Artificial Intelligence Systems, pp. 310–318. Springer, Berlin (2009)

41. Tizhoosh, H.R.: Opposition-based learning: a new scheme for machine intelligence. In: International Conference on Computational Intelligence for Modelling, Control and Automation and International Conference on Intelligent Agents, Web Technologies and Internet Commerce (CIMCA- IAWTIC'06), vol. 1, pp. 695–701 (2005). https://doi.org/10.1109/CIMCA.2005.1631345

42. Van Hentenryck, P., McAllester, D., Kapur, D.: Solving polynomial systems using a branch and prune approach. SIAM J. Numer. Anal. **34**(2), 797–827 (1997). https://doi.org/10.1137/S0036142995281504

43. Vesterstrom, J., Thomsen, R.: A comparative study of differential evolution, particle swarm optimization, and evolutionary algorithms on numerical benchmark problems. In: Proceedings of the 2004 Congress on Evolutionary Computation, vol. 2, pp. 1980–1987 (2004). https://doi.org/10.1109/CEC.2004.1331139

Chapter 4
Application of Enhanced Particle Swarm Optimization in Euclidean Steiner Tree Problem Solving in R^N

Wilson Wolf Costa, Marcelo Lisboa Rocha, David Nadler Prata, and Patrick Letouzé Moreira

4.1 Introduction

Humanity, in the daily routines of people or an the great projects developed by organizations or even entire societies, has always been faced with the question of the efficiency of the processes associated with these activities. With the advent of computational systems, there was the development of computational mathematics and, within this area of knowledge, combinatorial optimization. It has been used to solve a wide range of problems, both practical and theoretical, in the areas of telecommunications, transport, energy, production of goods and services, among others [11].

Among the problems of combinatorial optimization, the Steiner Problem, or Steiner's Tree Problem, consists of connecting points in space, through segments of lines, in order to minimize the total distance, that is, the sum of the length of line segments is minimal. It differs from the Minimum Generating Tree Problem by the possibility of adding points other than the proposed points, known as Steiner Points.

The Steiner Tree Problem (STP) is said to be Euclidean when points are considered in a metric space where the rules of Euclidean geometry apply, and in particular the definition of distance. Historically the problem dates back to the seventeenth century, when Fermat proposed in a letter: "Given three points on a plane, find a fourth point such that the sum of its distances to the three points is minimal." This problem became known as the Fermat Problem and is a particular case of Steiner's Problem [9]. Brazil et al. [2] made a historical review of the problem, since its origin, giving prominence to the pre-computational phase.

W. W. Costa · M. L. Rocha (✉) · D. N. Prata · P. L. Moreira
Postgraduate Program in Computational Modelling of Systems, Federal University of Tocantins, Palmas, Brazil

© Springer Nature Switzerland AG 2019
G. Mendes Platt et al. (eds.), *Computational Intelligence, Optimization and Inverse Problems with Applications in Engineering*,
https://doi.org/10.1007/978-3-319-96433-1_4

There are several exact algorithms, but limited to problems with few points, limited to 17 or 18, using branch-and-bound techniques, as proposed by Smith [14] or whole programming, as proposed by Fampa and Anstreicher [5]. For a greater number of points, approximation algorithms are used to find solutions close to optimal in a reasonable execution time. Heuristics such as GRASP with Path-Relinking [13], Micro-Canonic Optimization [10], and local search [16] were successfully exploited.

This work seeks to demonstrate the use of another heuristic for the solution of Steiner's Problem, that is, Particle Swarm Optimization (PSO), which was described by Kennedy and Eberhart [8] as a Bio-Inspired Metaheuristic, with artificial life ties, such as swarm theory. The scope of this work goes beyond the use of the standard model proposed in [8], by using the Prim algorithm [12] as an optimization guide, together with a Steiner point repositioning process that does not conform geometrically with a solution to the problem.

4.2 Definition of Euclidean Steiner Tree Problem (ESTP)

The Euclidean Steiner Tree Problem (ESTP) is as follows: Given P points (also called obligatory) in R^N with Euclidean metric, find a minimum spanning tree (MST) that connects the given points, using, if necessary, extra points known as Steiner points (see Fig. 4.1). The remainder of this section presents the main ESTP characteristics.

Suppose given P points $x_i \in R^N, i = 1, 2, 3, \ldots, P$ in N dimensional space. Then, a solution of ESTP, called Steiner Minimum Tree (SMT) must present the following properties [13, 14]:

- The maximum number of Steiner points (K) is $P - 2$;
- A Steiner point must have valence (or degree) equal to 3;
- Edges emanating from a Steiner point lie in the same plane and have mutual angles of $120°$.

If a tree (minimum or not) satisfies such properties, then we call it a Steiner tree. We call as Steiner topology the graph that represents a Steiner tree. The total number of different topologies with K Steiner points is $C_{P,K+2} = \frac{(P+K-2)!}{(K!2^K)}$. When $K = P - 2$, we have a Full Steiner Tree (FST) and the number of

Fig. 4.1 Steiner Tree with four obligatory points and two Steiner points (S_1 and S_2)

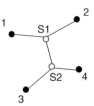

different topologies is $f(P) = \frac{(2P-4)!}{2^{P-2}(P-2)!}$. Considering, for example, $P = 10$, the total number of full topologies is $f(10) = 2,027,025$. This is the number of full topologies to be minimized by a brute force method. In Garey at al. [7] and Garey and Johnson [6], the ESTP is shown to be NP-Hard, fundamentally due to this combinatorial explosion, and is not possible to find an exact polynomial time algorithm for this problem. One minimization method is proposed by Smith [14], and will be approached in Sect. 4.2.1. Another existing exact method is the one proposed by Fampa and Anstreicher [5], named Smith+, which is an improved version of Smith's algorithm [14]. However, Fampa and Anstreicher [5] method was not considered in this work, because both source-code and used test instances are not available. Furthermore, this method has been applied only to small instances ($P \leq 18$). It is also worth noting that several areas of Science and Technology, such as Biotechnology, research in phylogenetic inference [10], Engineering, design and operation of networks of mines [1], and telecommunications [3], among many other problems of practical interest, can be modeled and solved based on ESTP in the R^N.

4.2.1   Smith's Algorithm

Smith [14] proposed an algorithm for the solution of the ESTP in the space with dimension $N \geq 3$ based on the enumeration of the filled topologies, followed by the minimization and calculation of the Steiner points for each topology found. To avoid bad solutions, a branch-and-bound algorithm was used.

This algorithm can be summarized considering that P points are given as follows:

1. Find all topologies filled to the P points.
2. For each filled topology, find the lowest cost tree by optimizing the spatial positions of Steiner's points.

The optimization of the positions of the Steiner points of Step 2 is done by an iterative method based on the solution of linear systems with sparse matrices. The optimum condition used in the algorithm is that all angles between three edges are greater than or equal to 120°.

4.3   Particle Swarm Optimization

Particle Swarm Optimization (PSO) has been described by Kennedy and Eberhart as a bioinspired metaheuristic, with artificial life links in general, and with flocks of birds and swarms of fishes, in particular [8].

In the standard model for Particle Swarm Optimization, a particle i, which can be understood as a bird, for example, occupies a position in space and moves in this space at a certain speed. Thus, this ith particle is described by two vectors in the space of D dimensions: the position vector $\mathbf{X_i}$ and the velocity vector $\mathbf{V_i}$. The index i refers to the ith particle of a population of size N. Each of the particles moves in this vector space at each iteration, or time, t of the algorithm, based on its current position $\mathbf{X_i}$, and its velocity $\mathbf{V_i}$, as follows:

$$\mathbf{X_i(t+1)} = \mathbf{X_i(t)} + \mathbf{V_i(t+1)} \tag{4.1}$$

That is, the current position vector of the particle is the sum of the position vector in the previous time and the current velocity vector. The contribution of the standard model lies in the way the velocity vector is constructed. It is formed by two main components in the form of two vectors: $\mathbf{Pbest_i}$, the cognitive component, which represents the best position occupied by particle i itself, and by $\mathbf{Gbest}$, the social component, which is the best position occupied by any particle of the population, that is, the best value of any $\mathbf{Pbest_i}$. This best position is calculated based on a cost function $f(\cdot)$ and the best position is, in general, the one that occupies the lowest value of this function, when dealing with minimization problems. Using the idea that the particle is a bird and that the cost function is the distance from a food source, the birds fly in space at a speed, and this velocity is modified by the perception that it is moving away from the source of food, trying to return to a position that has already been occupied and is closer to the food source and also by the best position that a bird, throughout the population, has occupied up to the current time. This "social information" or "social component" is communicated to the entire population at each iteration.

In this way, it can be formulated that:

$$\mathbf{V_i(t+1)} = w \cdot \mathbf{V_i(t)} + c_1 \cdot r_1 \cdot (\mathbf{Pbest_i(t)} - \mathbf{X_i(t)}) + c_2 \cdot r_2 \cdot (\mathbf{Gbest(t)} - \mathbf{X_i(t)}) \tag{4.2}$$

where:

- w is a constant of inertial weight;
- c_1 and c_2 are acceleration coefficients;
- r_1 and r_2 are random values generated in the interval $[0, 1]$.

Eberhart and Shi [4] estimated empirical values for $w = 0.7298$ and $c_1 = c_2 = 1.49618$ that produce good results. van den Bergh and Engelbrecht [15] made studies on the ranges and limits for these coefficients.

As we are interested in a minimization problem:

$$\mathbf{Gbest(t)} = \min_{i=1,\dots,n} \mathbf{Pbest_i(t)} \tag{4.3}$$

The default model has many variants, as described in the original article of Kennedy and Eberhart [8], several other authors used their ideas and concepts. An imposition of the standard model is that Eqs. (4.1) and (4.2) are defined in a linear vector space. When using PSO as metaheuristic, and the particles cannot be defined linearly, the model must be adapted by modifying Eq. (4.2) to account for the non-linearity.

4.4   Resolution of the Euclidean Tree Problem in R^N with the Particle Swarm Optimization

The proposed resolution is an adaptation of the PSO metaheuristic to the ESTP solution. The exposition of this resolution involves a correct definition of the particle, the discussion about the problem approach, geometric considerations, the adaptation of traditional algorithms, and the conjugation of these in a resolution method.

4.4.1   Definition of a Particle

The particle used in the PSO algorithm is a PASE solution. In other words, a particle in the PSO is a Steiner Tree, with its mandatory data points, its Steiner points and the edges connecting these points, forming a topology. Considering complete topologies, a problem proposed with P data points, has $S = P - 2$ Steiner points, which results in a total of $2P - 3$ edges. Generating n initial particles, we have n trees with all these components.

Due to the combinatorial character of the topology, it is not possible to establish a vector resolution for the topology that obeys the properties necessary for a vector space, such as the sum of vectors or product to be scaled. That is, Eq. (4.2), which produces the velocity vector, can not be used in this situation simply because the operations described therein are not definable.

Thus, it is necessary to redefine the two characteristics of the particle: their position and velocity. The position can easily be identified with the Steiner tree, which is ultimately a graph $G = (V, A)$. For speed definition, something more must to be considered, as will be presented in Sect. 4.4.4.1.

4.4.2   Overview of the Proposed Resolution

As discussed previously, in Sect. 4.2.1, Smith [14] asserted that the resolution for ESTP would be restricted to the combinatorial part of the problem, since it proposes a fast algorithm for obtaining the location of Steiner's points since a topology is provided. The solutions derived from it act in the form of the enumeration of topologies using several approaches to solve combinatorial problems, such as branch-and-bound [14] and GRASP with path-relinking [13], among others.

The resolution proposed in this work is based on using the geometry to surround the different topologies, in order to avoid solutions that are too far away from the optimal solutions. In order to do so, we sought to observe the characteristics of the optimal global solution. This solution has the points in the best positions, and the Steiner Tree, being minimal, is the Minimum Spanning Tree (MST) considering all points, both the required data and the Steiner points calculated. In this way, a Minimum Generating Tree algorithm is used as a guideline for better solutions.

4.4.3 Use of the Minimum Spanning Tree and Geometry to Obtain Topology

Initially, we have the required points and the Steiner points generated (randomly in the initial solution or through the execution of the algorithm throughout the process). Since, a priori, the topology is not defined, all points can bind to any others, which generates a fully connected graph, where the weight of each edge is the Euclidean distance between the vertices connected to this edge. One can then apply a MST algorithm to transform these graphs into a minimal tree. The Prim algorithm is commonly used to solve such problem [12]. The Prim algorithm is presented in Algorithm 4.1.

Algorithm 4.1 Simplified Prim's algorithm to ESTP

1: Choose, arbitrarily, an initial vertex for the tree;
2: **while** there are vertices that do not belong to the tree **do**
3:    Add to the tree the vertex that does not belong to the tree and has the smallest distance to any other vertex of the tree.
4: **end while**

Figure 4.2 shows an ESTP with five compulsory points: P_1 to P_5, and three Steiner points: S_1 to S_3, with the situation prior to the application of the Prim algorithm, where the graph is considered to be completely connected (on the left), and after the application of the algorithm (on the right). In this example Steiner's points are not yet well positioned. The application of the Prim algorithm generates a minimal tree, but this is not a Steiner tree because Steiner's points are not in the ideal positions. This generates several nonconformities that need to be addressed.

4.4.3.1 Vectorial Calculus of a Steiner Point from Three Given Points

The calculation of Steiner's point from three data points deserves to be emphasized due to the fact that since the problem is in $\mathbb{R}^n$, the dimension of the problem space is not known in advance. Three-dimensional solutions for the computation of this

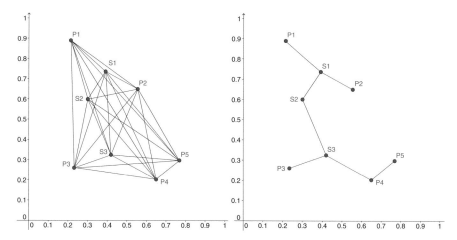

Fig. 4.2 Before (left) and after (right) the application of the Prim algorithm

point have difficulty in being extended, computationally, to an n-dimensional space. A vector solution was then adopted.

It is understood that three points define an N-dimensional Euclidean plane, given the points P_1, $P_2 \wedge P_3$ in this space, be $\mathbf{A}$, $\mathbf{B} \wedge \mathbf{C}$ their associated vectors, respectively. Given the scalars λ_1 and λ_2, the vector $\mathbf{V}$ belongs to the plane defined by these vectors if:

$$\mathbf{V} = \mathbf{A} + \lambda_1(\mathbf{B} - \mathbf{A}) + \lambda_2(\mathbf{C} - \mathbf{A})$$

To facilitate the visualization of the equations, consider the substitution of variables. Let $\mathbf{D} = (\mathbf{B} - \mathbf{A})$ and $\mathbf{E} = (\mathbf{C} - \mathbf{A})$. The dot product can be written as: $\mathbf{D} \cdot \mathbf{E} = ||\mathbf{D}|| \cdot ||\mathbf{E}|| \cdot \cos \theta$. One can define $\mathbf{X} = ||\mathbf{E}|| \cdot \cos \theta = \frac{\mathbf{D} \cdot \mathbf{E}}{||\mathbf{D}||}$. Being the unit vector in the direction of $\mathbf{u} = \frac{\mathbf{D}}{||\mathbf{D}||}$, we have $\mathbf{X} = \frac{\mathbf{D} \cdot \mathbf{E}}{\mathbf{D} \cdot \mathbf{D}} \cdot \mathbf{D}$. The vector $\mathbf{Y}$, orthogonal to $\mathbf{X}$ and belonging to the same plane, is defined as: $\mathbf{Y} = \mathbf{E} - \mathbf{X}$, and the unit vector in direction of the vector $\mathbf{Y}$ is $\mathbf{v} = \frac{\mathbf{Y}}{||\mathbf{Y}||}$. From the derivation of the unit vectors $\mathbf{u}$ and $\mathbf{v}$, it is possible to determine the vector $\mathbf{R}$, whose end is the point P_4, third vertex of the equilateral triangle formed by P_1, $P_2 \wedge P_3$, P_4. Thus $\mathbf{R} = \mathbf{A} + \frac{1}{2}||\mathbf{D}|| \cdot \mathbf{u} + \frac{\sqrt{3}}{2}||\mathbf{D}|| \cdot \mathbf{v}$.

In such way it is possible to determine the point P_4 and, consequently, the segment $P_3 P_4$. In a symmetrical way, changing $\mathbf{D}$ by $\mathbf{E}$, it is possible to determine the point P_5 and the segment $P_2 P_5$. The intersection between the segment $P_3 P_4$ and the segment $P_2 P_5$ is the Steiner point. Figure 4.3 shows, on the left, the Steiner's point determination, given P_1, P_2, and P_3, and on the right, the determination of Steiner point S_2, given points S_2, P_4, and P_5 on the tree of Fig. 4.2.

An important observation to be emphasized is that this vectorial calculation, allows to find an exact solution, assuming that three points are given. But for any case where the number of given points is greater than 3, at least one point will be

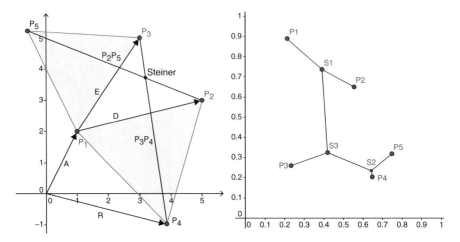

Fig. 4.3 Vectorial determination of the Steiner point (left) and determination of the point S2 from S3, P4 and P5 (right)

a Steiner point, and thus subject to the variation of its position. Consequently, this implies reworking the vectorial calculus.

4.4.4 A Modified Particle Swarm Optimization for ESTP Resolution

As discussed in Sect. 4.3, Particle Swarm Optimization requires that particles, which are solutions for ESTP, undergo changes based on their current "position", their cognitive component and the social component. The cognitive component of the particle is the best "position" it has occupied throughout the process, and the social component is the best "position" occupied by any particle in the swarm. "Position", in the case of ESTP, means the set of spatial coordinates of the points and the topology, represented by the set of edges of the tree.

ESTP has a hybrid nature since it has a continuous component in Euclidean space, and a combinational component due to the problem tree structure. Thus, it was necessary to adapt the original PSO to a combinational space, expressed by the topology tree.

Given this dichotomy, three approaches can be made:

1. Neglect the combinatorial part and work with the spatial positions of Steiner's points, and their velocity vectors, or;
2. Focus on the topology and despise the positions and spatial velocities of the points, or still;
3. Consider both aspects simultaneously.

The first approach was considered in the preliminary versions of this work, but it was not adopted, because it presents a very slow convergence, even in small problems. The second approach is used by most works on ESTP with the application of a heuristic for topology definition and positioning using the Smith algorithm [13, 14]. The third approach was then considered in the work described in this chapter.

4.4.4.1 Particle Velocity Based on the Spatial Characteristics of Steiner's Points

This approach is based on the standard Particle Swarm Optimization model where the concepts of cognitive and social components have been modified and a guide to indicate the optimization path has been modified. The guide is the Prim algorithm. This was chosen by a very obvious observation: The tree representing the optimal global solution of ESTP is also a Minimum Spanning Tree considering the previously given and Steiner points. In this way, the Prim algorithm can be used as a guide to combine the particles. Algorithm 4.2 describes the "Particle Step."

In reference to the algorithm "Generate Topology", the only difference in relation to the "Particle Speed" algorithm is that the first one has a set of given points and Steiner points provided, and the second one has as input the sets of given points and Steiner points of the current particle as well as the set of Steiner points of another particle to be combined. Both algorithms have as output the generation of a minimum distance topology from a set of obligatory and Steiner points provided.

The general algorithm of the proposed solution is composed considering "particle step" algorithm (Algorithm 4.2) in the general structure of the Particle Swarm Optimization Metaheuristic considering geometric improvements to the ESTP, denominated "PSO-Steiner."

The process of combining the particle with its best previous version becomes the cognitive component of the PSO. Similarly, its social component becomes the act of combining the particle with the best particle of all until then.

Initially the particles with the required points are created and the Steiner points are randomly generated. The algorithm "Generate Topology" is responsible for producing a feasible topology for each particle. In this process the best overall particle is selected using as the comparison metrics the lower value of the cost function f which calculates the total particle distance.

At each iteration two combinations of each particle are performed X_i with its best previous version, $Pbest_i$, its cognitive component, and with the best particle of all, $Gbest$, the social component. Then the verification occurs if the modified particle has a cost lower than its previous version. If so, this particle is stored in $Pbest_i$ and if this is better than $Gbest$, it is stored in the latter variable. After this process, the result is the particle $Gbest$. Algorithm 4.3 describes the methodology at its highest level.

Algorithm 4.2 General proposed algorithm: PSO-Steiner—particle step

---

Input:
    Positions of obligatory points of current particle;
    Positions of Steiner points of current particle;
    Positions of Steiner points of another particle;
Output:
    A topology resulting from the combination of input particles.

1: Execute Prim algorithm, considering initial graph fully connected formed by all input points;
2: **while** there are Steiner points with a degree equal to 1 **do**
3:    Remove from the graph the edge incident on the Steiner point with a degree equal to 1;
4:    Store the Steiner point removed in a set Available;
5: **end while**
6: **while** there are Steiner points with a degree equal to 2 **do**
7:    Identify the points adjacent to the Steiner point with a degree equal to 2;
8:    Create an edge by connecting adjacent identified points;
9:    Remove from the graph the edges incident on the Steiner point with a degree equal to 2;
10:    Store the Steiner point removed in a set Available;
11: **end while**
12: **while** there are given points that have a degree greater than 1 **do**
13:    **if** degree of given point is greater than 2 **then**
14:        Select the pair of adjacent points that form the smallest angle with the given point
15:    **else**
16:        Select the single pair of adjacent points;
17:    **end if**
18:    **if** formed by the selected adjacent pair of points is greater than 120° **then**
19:        Position a Steiner point of the set Available over the obligatory point;
20:    **else**
21:        Position a Steiner point of the set Available through the vector calculation of the position among the obligatory point and the adjacent pair selected;
22:    **end if**
23:    Include 3 edges between the repositioned Steiner point and the obligatory point and points of the selected adjacent pair;
24: **end while**
25: **while** there are Steiner points that have a degree greater than 3 **do**
26:    Select the pair of adjacent points that form the smallest angle with the Steiner point observed;
27:    **if** angle formed by the selected adjacent pair of points is greater than 120° **then**
28:        Position a Steiner point of the set Available on the Steiner point observed;
29:    **else**
30:        Position a Steiner point of the set Available through the vector calculation of the position, given the Steiner point observed and the pair of adjacent selected;
31:    **end if**
32:    Include 3 edges between the re-positioned Steiner point, the observed Steiner point and the selected adjacent pair points;
33: **end while**
34: Neglect Steiner points that were not reinserted in topology;

Algorithm 4.3 Algorithm PSO-Steiner

Input:
    Quantity and positions of obligatory points of the problem;
    Quantity p of particles to be generated;
    Number N of iterations;
Output:
    The best solution found and its minimum cost.
Let:
    $Pbest_i$ the best version of particle p_i with the lower cost.
    $Gbest$ the best version among all particles with lower global cost.
    $f(X_i)$ is the function that calculates the total cost (length of Steiner tree) of the particle

1: **for** each particle X_i, $i = 1, ..., p$ **do**
2:     assign the mandatory points of entry to X_i obligatory;
3:     generate the Steiner points randomly and assign the $X_i.Stainer$;
4:     generate the topology by the algorithm "Generate Topology";
5:     **if** $f(X_i) < f(Gbest)$ **then**
6:       $Gbest \leftarrow X_i$;
7:     **end if**
8: **end for**
9: **for** each iteration $n = 1, ..., N$ **do**
10:    **for** each particle X_i, $i = 1, ..., p$ **do**
11:      merge particle X_i with $Pbest_i$ using "Particle Step"İ algorithm;
12:      merge particle X_i with $Gbest$ using "Particle Step"İ algorithm;
13:      **if** $f(X_i) < f(Pbest_i)$ **then**
14:        $Pbest_i \leftarrow X_i$;
15:        **if** $f(Pbest_i) < f(Gbest)$ **then**
16:          $Gbest \leftarrow Pbest_i$;
17:        **end if**
18:      **end if**
19:    **end for**
20: **end for**
21: **return** $Gbest$ and $f(Gbest)$;

4.5   Computational Results

The previously presented algorithms that are parts of PSO-Steiner were implemented in the C++ language and executed using multiple instance sets of 1000 and
10,000 random points generated in the range [0; 1] in a three-dimensional space.
A GRASP heuristic with Path-relinking from literature was also used. The results
were collected for a comparative analysis.

4.5.1   *Computational Test Environment*

For the computational tests the following computational elements were used:

- Computer consisting of a virtual machine in a VMware ESXi 5.5 virtualization environment with 4 non-clocked VCPUs, 1 GByte of RAM, configured so as not to share these VCPUs. Host computers are Dell PowerEdge R710 Rev 2 servers, featuring 2 Intel®Xeon^{TM}X5690 processors, 6 cores, 12 MByte L2 cache, 3466 MHz normal clock, maximum clock 3600 MHz, with 96 GBytes of 1333 DDR RAM MHz;
- Linux distribution system Fedora release 23 for 64-bit (x86_64);
- Compiler g++ (GCC) version 5.1.1-4.

4.5.2 Test Methodology

The methodology adopted was the comparative analysis of the test battery execution in two distinct implementations: the software produced from the description presented in Sect. 4.4.4, hereinafter referreds to as "PSO-Steiner" and another solution based on the GRASP algorithm with Path-relinking provided by the authors of Ref. [13], hereinafter referred as "AGMHeur4".

The test battery had the following characteristics:

1. Compilation of the two implementations in the same environment described in Sect. 4.5.1.
2. The same stopping criterion was adopted for the two methods: the execution of 100 (one hundred) iterations.
3. For the input data of the problem, two sets of test files with randomly generated points were used in a three-dimensional space in the range [0; 1], both containing 10 files each. The first set contains files with 1000 obligatory points, while the second set contains files with 10,000 obligatory points. Although other sets of smaller number of points were used in the implementation phase tests and the initial comparisons, the 1000 and 10,000 obligatory points sets were chosen because they were significantly larger than those used by other solutions presented in the literature.
4. Each of the two programs (PSO-Steiner and AGMHeur4) was run 10 times for each test instance.

The computational result of the two heuristics considered in this work (PSO-Steiner and AGMHeur4) obtained over each instance of the two set (with 1000 and 10,000 obligatory points) was compared considering the mean value of the execution time and the mean value of minimum solution cost will be presented in Sections 4.5.2.1 to 4.5.2.4. For the sake of comparison of mean values of execution time and minimum solution cost between the heuristics considered, for statistical analysis, Student's T-test was performed, with level of significance of 5%. For the Student's T-test, the null hypothesis (H_0) chosen was that the mean values of the results of the two heuristics are the same.

The statistical analysis using Student's T-test to compare the performance between PSO-Steiner and AGMHeur4 heuristics is presented in Tables 4.2, 4.5, 4.7, and 4.10. Each of these four tables presents the columns as follows: Instance file: name of the instance considered. Variances: indicate if the two samples have the same or different variance. df: number of degrees of freedom considered. T critic two-tailed: critic value for the Student's T distribution with df degrees of freedom. Stat t: calculated value of T-test statistic. Conclusion: indicates if H_0 is accepted or rejected.

Another question to be considered in the rest of the work are related to the performance analysis of PSO-Steiner against AGMHeur4. First question is related to the improvement of mean time factor between PSO-Steiner and AGMHeur4, identified as Δ_{time}, being calculated as Δ_{time} = [(PSO-Steiner Mean Time − AGMHeur4 Mean Time)/(AGMHeur4 Mean Time)] × 100%. Second question is related to the improvement of mean cost between PSO-Steiner and AGMHeur4, identified as Δ_{cost}, being calculated as Δ_{cost} = [(PSO-Steiner Mean Cost − AGMHeur4 Mean Cost)/(AGMHeur4 Mean Cost)] × 100%.

4.5.2.1 Results: Set of 1000 Points—Execution Time

The results obtained with the set of 1000 points with respect to the execution time are summarized in Table 4.1. The minimum and average times in seconds and the time variance for the proposed solution, PSO-Steiner, and the program used as the base AGMHeur4, are shown for each instance file used, considering the 10 executions to which they were submitted.

The results of the application of the Student's T test in the data obtained by the execution of the programs in the 1000 points instance files are described in Table 4.2. It is observed that the values of the statistical calculations are, in module, very different from the values of two-tailed T_{critic}. Therefore, we can conclude by the rejection of the null hypothesis in all cases.

Table 4.1 Results for the set of 1000 points with respect to the execution time

| Instance file | PSO-Steiner | | | AGMHeur4 | | | Δ_{time} (%) |
	Minimum time (s)	Mean time (s)	Time variance	Minimum time (s)	Mean time (s)	Time variance	
n1000d3_1	202.51	219.46	98.86	363.80	370.62	44.36	−40.8
n1000d3_2	216.67	227.31	71.28	382.43	389.83	39.88	−41.7
n1000d3_3	219.96	227.62	24.30	365.36	378.05	46.99	−39.8
n1000d3_4	208.91	225.80	69.48	388.12	398.38	35.84	−43.3
n1000d3_5	219.29	228.54	26.84	365.58	371.33	22.36	−38.5
n1000d3_6	211.54	221.66	34.52	390.08	395.22	13.70	−43.9
n1000d3_7	201.29	211.57	26.89	371.16	383.56	31.23	−44.8
n1000d3_8	210.76	220.78	35.44	378.31	387.86	56.70	−43.1
n1000d3_9	216.87	229.47	53.11	379.76	389.01	52.60	−41.0
n1000d3_10	216.48	222.53	8.25	370.11	379.13	37.96	−41.3

Table 4.2 Results of the T test to 1000 points instances with respect to the execution time

T-test: two samples

Instance file	Variances	df	T_{critic} two-tailed	Stat t	Conclusion
n1000d3_1	Equivalents	18	2.10092204	-39.93994851	$|$Stat $t| > T_{critic} =>$ reject H_0
n1000d3_2	Equivalents	18	2.10092204	-48.74346366	$|$Stat $t| > T_{critic} =>$ reject H_0
n1000d3_3	Equivalents	18	2.10092204	-56.33333528	$|$Stat $t| > T_{critic} =>$ reject H_0
n1000d3_4	Equivalents	18	2.10092204	-53.17318260	$|$Stat $t| > T_{critic} =>$ reject H_0
n1000d3_5	Equivalents	18	2.10092204	-64.36826847	$|$Stat $t| > T_{critic} =>$ reject H_0
n1000d3_6	Equivalents	18	2.10092204	-79.03215373	$|$Stat $t| > T_{critic} =>$ reject H_0
n1000d3_7	Equivalents	18	2.10092204	-71.33251858	$|$Stat $t| > T_{critic} =>$ reject H_0
n1000d3_8	Equivalents	18	2.10092204	-55.04116660	$|$Stat $t| > T_{critic} =>$ reject H_0
n1000d3_9	Equivalents	18	2.10092204	-49.06602672	$|$Stat $t| > T_{critic} =>$ reject H_0
n1000d3_10	Differents	13	2.16036866	-72.84079802	$|$Stat $t| > T_{critic} =>$ reject H_0

Table 4.3 Total execution time for the experiment with 1000 points instances

Algorithm	Total time (in seconds)	Total time (in hours, minutes and seconds)
PSO-Steiner	22,347.89	6 h 12 min 27 s
AGMHeur4	38,429.98	10 h 40 min 29 s

Still in Table 4.1, we demonstrate the improvement of the mean time factor (Δ_{time}) with the PSO-Steiner application in regard to the AGMHeur4 solution. It is observed that there is a considerable reduction of time, in the range of 38.5% and 44.8% lower, for the cases studied.

Another important result for the 1000 points set experiment is the computational effort required which is linked to the computational time. To obtain the PSO-Steiner results, it took 6.2 h of batch processing, and for the solution based on AGMHeur4, it took 10.6 h. Table 4.3 presents these results in more detail.

4.5.2.2 Results: Set of 1000 Points—Cost Function

The results with the set of 1000 points with respect to the cost function (total length of the Steiner tree) are summarized in Table 4.4. The minimum and mean costs and cost variance for the proposed solution, PSO-Steiner, and for the program used as the basis of comparison, AGMHeur4, are shown for each instance file used, considering the 10 executions to which they were submitted.

The results of the application of the Student's T-test in the data obtained by the execution of the programs in the 1000 points instance files are described in Table 4.5. It is observed that the values of the statistical calculations are, in module, very different from the values of two-tailed T_{critic}. In this way, we can conclude by rejecting the null hypothesis in all cases.

Table 4.4 Results for the set of 1000 points with respect to the cost (total distance)

Instance file	PSO-Steiner			AGMHeur4			Δ_{cost} (%)
	Minimum cost	Mean cost	Cost variance	Minimum cost	Mean cost	Cost variance	
n1000d3_1	62.759918274	62.810810394	0.001560241	63.702473973	63.728955940	0.000234334	−1.40
n1000d3_2	63.397453554	63.457294073	0.000965602	64.347102889	64.370892728	0.000407446	−1.38
n1000d3_3	64.105215913	64.433770403	0.088429996	64.895233403	64.907417009	0.000063343	−0.71
n1000d3_4	63.173992988	63.266763533	0.002499324	64.143524578	64.171127600	0.000303961	−1.37
n1000d3_5	64.548917462	64.623302856	0.002045067	65.410065215	65.448493907	0.000816644	−1.20
n1000d3_6	62.992508360	63.391539052	0.118144878	63.918698996	63.933935627	0.000239202	−0.82
n1000d3_7	64.105174282	64.343156533	0.098883829	64.942768817	64.971697017	0.000340709	−0.92
n1000d3_8	63.653864571	63.708487378	0.000937240	64.720957101	64.753902150	0.000256406	−1.56
n1000d3_9	64.262254371	64.699545573	0.056118387	64.911205479	64.980107869	0.001185187	−0.34
n1000d3_10	63.950600180	64.374749386	0.095632742	64.982742884	65.020751619	0.000546033	−0.94

Table 4.5 Results of the T-Test for 1000 points instances related with respect to the cost (total distance)

T-test: two samples					
Instance file	Variances	df	T_{critic} two-tailed	Stat t	Conclusion
n1000d3_1	Differents	12	2.17881283	-68.53789530	$\lvert$Stat $t\rvert > T_{critic} =>$ reject H_0
n1000d3_2	Equivalents	18	2.10092204	-77.96731926	$\lvert$Stat $t\rvert > T_{critic} =>$ reject H_0
n1000d3_3	Differents	9	2.26215716	-5.03499611	$\lvert$Stat $t\rvert > T_{critic} =>$ reject H_0
n1000d3_4	Differents	11	2.20098516	-54.01440831	$\lvert$Stat $t\rvert > T_{critic} =>$ reject H_0
n1000d3_5	Equivalents	18	2.10092204	-48.77997702	$\lvert$Stat $t\rvert > T_{critic} =>$ reject H_0
n1000d3_6	Differents	9	2.26215716	-4.98505880	$\lvert$Stat $t\rvert > T_{critic} =>$ reject H_0
n1000d3_7	Differents	9	2.26215716	-6.30991796	$\lvert$Stat $t\rvert > T_{critic} =>$ reject H_0
n1000d3_8	Differents	14	2.14478669	-95.68656194	$\lvert$Stat $t\rvert > T_{critic} =>$ reject H_0
n1000d3_9	Differents	9	2.26215716	-3.70628193	$\lvert$Stat $t\rvert > T_{critic} =>$ reject H_0
n1000d3_10	Differents	9	2.26215716	-6.58710218	$\lvert$Stat $t\rvert > T_{critic} =>$ reject H_0

Table 4.4 shows the improvement of the mean cost (Δ_{cost}) between the PSO-Steiner application and the AGMHeur4 solution. It is observed that there is a reduction of the mean cost, from 0.34% to 1.56%, for the cases studied.

4.5.2.3   Results: Set of 10,000 Points—Execution Time

The results with the set of 10,000 points related to the execution time are summarized in Table 4.6. The minimum and mean times in seconds and the time variance for the proposed solution, PSO-Steiner, and for the program used as the base AGMHeur4, are shown for each instance file used, considering the 10 executions to which they were submitted.

The results of the application of the Student's T-test (with level of significance of 5%) in the data obtained by the execution of the programs in the 10,000 points instance files are described in Table 4.7. It is observed that the values of the statistical calculations are, in module, very different from the values of two-tailed T_{critic}. In this way, we can conclude by rejecting the null hypothesis in all cases.

Table 4.6 shows the improvement of the mean time factor (Δ_{time}) between the PSO-Steiner application and the AGMHeur4 solution. It is observed that there is a considerable reduction of time, from 78.2% to 79.6%, for the cases studied.

The computational effort undertaken is another important result of the experiment with the 10,000 points sets and is also a function of time. To obtain the PSO-Steiner results it took 2 days and 11 h of batch processing, and for the solution based on AGMHeur4, it took 11 days and 14 h. Table 4.8 presents these results in detail.

Table 4.6 Results for the set of 10,000 points with respect to the execution time

Instance file	PSO-Steiner			AGMHeur4			
	Minimum time (s)	Mean time (s)	Time variance	Minimum time (s)	Mean time (s)	Time variance	Δ_{time} (%)
n10000d3_1	2054.33	2089.33	675.17	9698.94	9918.85	26,286.44	−78.9
n10000d3_2	2051.77	2127.27	2089.09	9618.72	9972.15	21,166.86	−78.7
n10000d3_3	2165.98	2198.33	559.25	9831.03	10,075.89	34,095.95	−78.2
n10000d3_4	2084.96	2139.63	829.01	9690.37	10,027.64	31,292.62	−78.7
n10000d3_5	2132.23	2163.78	1020.90	9808.87	10,105.47	56,409.64	−78.6
n10000d3_6	2074.28	2098.32	462.22	9728.9	9929.24	19,696.01	−78.9
n10000d3_7	2054.25	2096.68	767.54	10,072.61	10,256.03	33,819.96	−79.6
n10000d3_8	2067.69	2126.16	1083.06	9667.21	9971.25	24,313.07	−78.7
n10000d3_9	2098.86	2127.70	288.08	9803.55	9960.34	21,297.79	−78.6
n10000d3_10	2080.64	2097.97	207.89	9735.25	9987.86	24,609.18	−79.0

Table 4.7 Results of the T-test to 10,000 points instances with respect to the execution times

T-test: two samples					
Instance file	Variances	df	T_{critic} two-tailed	Stat t	Conclusion
n10000d3_1	Differents	9	2.26215716	−150.78716473	$\|$Stat $t\| > T_{critic}$ => reject H_0
n10000d3_2	Differents	11	2.20098516	−162.67445318	$\|$Stat $t\| > T_{critic}$ => reject H_0
n10000d3_3	Differents	9	2.26215716	−133.81587286	$\|$Stat $t\| > T_{critic}$ => reject H_0
n10000d3_4	Differents	9	2.26215716	−139.17727304	$\|$Stat $t\| > T_{critic}$ => reject H_0
n10000d3_5	Differents	9	2.26215716	−104.79516292	$\|$Stat $t\| > T_{critic}$ => reject H_0
n10000d3_6	Differents	9	2.26215716	−174.41589228	$\|$Stat $t\| > T_{critic}$ => reject H_0
n10000d3_7	Differents	9	2.26215716	−138.73819130	$\|$Stat $t\| > T_{critic}$ => reject H_0
n10000d3_8	Differents	10	2.22813885	−155.67323431	$\|$Stat $t\| > T_{critic}$ => reject H_0
n10000d3_9	Differents	9	2.26215716	−168.58666177	$\|$Stat $t\| > T_{critic}$ => reject H_0
n10000d3_10	Differents	9	2.26215716	−158.37835944	$\|$Stat $t\| > T_{critic}$ => reject H_0

Table 4.8 Total execution time for the experiment with 10,000 points instances

Algorithm	Total time (in seconds)	Total time (in days, hours, minutes and seconds)
PSO-Steiner	212,652.15	2 day 11 h 4 min 12 s
AGMHeur4	1,002,047.97	11 day 14 h 20 min 48 s

4.5.2.4 Results: Set of 10,000 Points—Cost Function

The results with the set of 10,000 points, considering 10 executions, for the proposed solution (PSO-Steiner) and the competitor of the literature (AGMHeur4), related to the minimum cost, average cost and cost variance in relation to the total tree length. Steiner (cost function), for each instance used, are presented in Table 4.10.

The results of the application of the Student's T-test in the data obtained by the execution of the programs (PSO-Steiner and AGMHeur4) in the 10,000 points instance files are described in Table 4.9. It is observed that the values of the statistical

Table 4.9 Results of the T-test for 10,000 points instances with respect to the cost (total distance)

T-test: two samples					
Instance file	Variances	df	T_{critic} two-tailed	Stat t	Conclusion
n10000d3_1	Differents	9	2.26215716	-17.83491166	|Stat t| $> T_{\text{critic}}$ => reject H_0
n10000d3_2	Differents	9	2.26215716	-11.69775829	|Stat t| $> T_{\text{critic}}$ => reject H_0
n10000d3_3	Differents	10	2.22813885	-37.19830389	|Stat t| $> T_{\text{critic}}$ => reject H_0
n10000d3_4	Differents	12	2.17881283	-48.23252467	|Stat t| $> T_{\text{critic}}$ => reject H_0
n10000d3_5	Differents	9	2.26215716	-12.07896339	|Stat t| $> T_{\text{critic}}$ => reject H_0
n10000d3_6	Differents	9	2.26215716	-13.27842414	|Stat t| $> T_{\text{critic}}$ => reject H_0
n10000d3_7	Differents	9	2.26215716	-7.21239109	|Stat t| $> T_{\text{critic}}$ => reject H_0
n10000d3_8	Differents	9	2.26215716	-25.34607963	|Stat t| $> T_{\text{critic}}$ => reject H_0
n10000d3_9	Differents	9	2.26215716	-12.33368861	|Stat t| $> T_{\text{critic}}$ => reject H_0
n10000d3_10	Differents	13	2.16036866	-51.08106681	|Stat t| $> T_{\text{critic}}$ => reject H_0

calculations are, in module, very different from the values of two-tailed T_{critic}. In this way, we can conclude by rejecting the null hypothesis in all cases.

Table 4.10 shows the improvement of the mean cost (Δ_{cost}) between the PSO-Steiner application and the AGMHeur4 solution. It is observed that there is a reduction of the mean cost, between 0.53% and 0.76%, for the cases studied.

4.6   Discussion on the Experimental Results

When discussing the experimental results, we must consider some very important criteria: the quality of the solution and the computational effort. They will be commented in the following subsections.

4.6.1   Solution Quality

The experimental results regarding execution time and cost function were presented in Sect. 4.5 considering 1000 and 10,000 points instance sets. As can be seen, in all cases there was rejection of the null hypothesis (H_0) in the T-test at 5% level of significance. This rejection leads to the conclusion that the alternative hypothesis can be accepted in which the mean values are different. Therefore, it is concluded that the methods considered present different results. Hence, as both in the time criterion and in the cost criterion, smaller values means better performance, it can be stated that, given the conditions of the experiment, the PSO-Steiner is superior to the solution based on the AGMHeur4.

Considering the cost criteria, that is, the total length of the tree, there was also a reduction, though with less expressive values at first sight. For sets of 1000 points,

Table 4.10 Results for the set of 10,000 points with respect to the cost (total distance)

Instance file	PSO-Steiner			AGMHeur4			Δ_{cost} (%)
	Minimum cost	Mean cost	Cost variance	Minimum cost	Mean cost	Cost variance	
n10000d3_1	292.840701907	293.125321621	0.085255356	294.701323744	294.791305328	0.002001502	−0.53
n10000d3_2	290.975586967	292.235669499	0.382446617	294.488122156	294.524657049	0.000450040	−0.76
n10000d3_3	292.522022503	292.734709951	0.025343519	294.588518330	294.634809222	0.000748417	−0.63
n10000d3_4	293.331933547	293.459279739	0.012110651	295.205214123	295.297054754	0.002407279	−0.59
n10000d3_5	291.160399010	292.612184429	0.306750978	294.664776234	294.735347256	0.002212591	−0.70
n10000d3_6	291.669639276	292.917686675	0.254889211	294.993596651	295.042053722	0.001067216	−0.70
n10000d3_7	291.534536628	293.108076462	0.692618209	294.981673717	295.006940328	0.000534819	−0.64
n10000d3_8	293.122874635	293.270416533	0.045488076	294.927451818	294.999694875	0.001060695	−0.56
n10000d3_9	291.427399338	292.882044335	0.273065837	294.885857015	294.921248738	0.000294538	−0.68
n10000d3_10	293.969393388	294.197171033	0.010184114	295.889044261	296.001455129	0.002292304	−0.57

this reduction was between 0.34% and 1.56%, and for sets of 10,000 points it was between 0.53% and 0.76%. The following should be observed for these values:

1. Statistical tests show that it is possible to reject the null hypothesis, that is, the methods are distinct with a confidence of 95% will present different solutions;
2. Percentage gains in total distance should be analyzed in the light of Steiner's ratio, which shows that there is only 27.76% freedom between the extreme conditions of the distribution of the required points. In conditions where the distribution of these points is random, this interval would be even smaller and the results obtained more significant;
3. The data files for the computational tests were generated on the condition that they were confined in a cube of unit dimensions. With the increase in the number of points from 1000 to 10,000 the points became closer to each other, decreasing the mean distance and the absolute gains by applying the ESTP resolution algorithms.

4.6.2 Computational Effort

The required computational effort is also depicted in Sect. 4.5. While the mean times and statistics involved with these results have been considered in the previous section, the total times also deserve to be examined.

Regarding the time criterion, in which the stopping criterion was conditioned to the execution of 1000 iterations over the set of Steiner points, the PSO-Steiner presents mean times much smaller than that of the AGMHeur4. For the sets with 1000 points, this reduction was between 38.5% and 44.8%, and for sets of 10,000 points it was between 78.2% and 79.6%, being quite expressive.

The total execution time of the experiment to run the two heuristics (PSO-Steiner and AGMHeur4), considering the test methodology of Sect. 4.5.2, was 1,275,477.99 s, equivalent to 14 days, 18 h, and 18 min.

The individual computational effort is another important result of the experiment for both 1000 points and 10,000 points sets. For the 1000 points sets, it took 6.2 h of processing for the PSO-Steiner, while it took 10.6 h for AGMHeur4. For the 10,000 points sets, the PSO-Steiner results consumed 2 days and 11 h of processing, while the AGMHeur4-based solution consumed 11 days and 14 h.

Looking to this computational effort is possible to observe that PSO-Steiner has much better performance than AGMHeur4 (mainly in bigger instances), being almost two times faster in 1000 points instances and almost six times faster in 10,000 points instances.

4.7 Conclusion and Future Work

The present work aimed to demonstrate the possibility of applying the Particle Swarm Optimization Metaheuristics, Particle Swarm Optimization—PSO, to the Euclidean Steiner Tree Problem—ESTP, in a space R^N, in the search for a problem solving method that achieves a good computational performance compared to known methods.

The original Particle Swarm Optimization Metaheuristics was developed for problems in a continuous space of solutions, a vector space where the cognitive and social components of the particles should obey the conditions of linearity and, therefore, could be summed up producing a vector resulting from the displacement of the particle. It should be noted that the particle in this case is a Steiner tree that solves the problem of connecting the points provided, although not the least cost, which is the final objective of solving the problem.

Due to the hybrid feature of ESTP, which has a continuous component of Euclidean space and a combinational component due to the problem tree structure, it was necessary to adapt the original PSO to a combinational space, expressed by the topology tree. The proposed method uses the Prim algorithm to combine the particles. In doing so, various nonconformities are produced with the model of the Complete Steiner Tree. The application of a set of elimination and re-positioning procedures yields a particle, tree, whose cost is equal to or less than that of the original particles. The process of combining the particle with its best previous version becomes the cognitive component of the PSO, while combining the particle with the best particle of all, until then, becomes its social component.

Based on these concepts, an algorithm was implemented and used in a computational experiment, where the implementation of a GRASP based solution with Path-relinking, the AGMHeur4, was used as a basis for comparison. For the tests, two sets of 10 problems each were used, one with 1000 obligatory points and other with 10,000 obligatory points. The experiment required a total run time of 11 days and 14 h. Note that the two implementations were monoprocessed.

The results obtained were statistically analyzed and presented, in all cases, the rejection of the null hypothesis, that is, showing that there is a distinction between the two methods, and with better results for the method presented in this chapter (PSO-Steiner). In terms of execution time, a reduction of more than 38.5% was achieved for problems of 1000 points and greater than 78.2% for those of 10,000 points. In regard to the minimum cost obtained, reductions were achieved between 0.34% and 1.56% for the problems of 1000 points sets and between 0.53% and 0.76% for the 10,000 points sets.

In view of the above, it can be concluded that the methodology for solving the Problem of the Euclidean Steiner Tree in $\mathbb{R}^N$ presented opens a new line of investigation for the problem. The results are promising and prove the robustness and efficiency of the proposed approach, obtaining results in problems of unusual size in the literature.

The results obtained allow to foresee a series of unfolding strategies that can circumvent the initial constraints of the problem. The first proposal for future work is in the implementation of a multiprocessed version of the algorithm. This is expected to have a significant impact on execution times, since in the experiments performed the computational time was very high, particularly in the problems of 10,000 points. Another research line is to apply the algorithms in sets of points not confined to the cube of unit dimensions, since in all test files used in the experiment, the points were randomly generated within this limit. It is expected to verify the impact on the gains obtained in the cost function, that is, in the sum of the distances between the points of the tree.

Yet another proposal consists of changing the stopping criteria of the software implementations. In this work, it was fixed in both algorithms. It is possible to investigate a criterion that establishes a maximum time previously set. We believe that this would lead PSO-Steiner to perform even better than AGMHeur4, which could result in more significant differences on the values obtained from the cost function. Also, an investigation of problems with a larger spatial dimension ($N > 3$) and the impact of the dimension on the achievement of results in both the time and cost domain must be performed.

Acknowledgements The authors acknowledge the reviewers for important and helpful contributions to this work. The development of this research benefited from the UFT Institutional Productivity Research Program (PROPESQ/UFT).

References

1. Alford, C., Brazil, M., Lee, D.H.: Optimisation in Underground Mining, pp. 561–577. Springer, Boston (2007). https://doi.org/10.1007/978-0-387-71815-6_30
2. Brazil, M., Graham, R.L., Thomas, D.A., Zachariasen, M.: On the history of the Euclidean Steiner tree problem. Arch. Hist. Exact Sci. **68**(3), 327–354 (2014). https://doi.org/10.1007/s00407-013-0127-z
3. Du, D., Hu, X.: Steiner Tree Problems in Computer Communication Networks. World Scientific Publishing, River Edge (2008)
4. Eberhart, R.C., Shi, Y.: Comparing inertia weights and constriction factors in particle swarm optimization. In: Proceedings of the 2000 Congress on Evolutionary Computation. CEC00 (Cat. No.00TH8512), vol. 1, pp. 84–88 (2000). https://doi.org/10.1109/CEC.2000.870279
5. Fampa, M., Anstreicher, K.M.: An improved algorithm for computing Steiner minimal trees in Euclidean d-space. Discret. Optim. **5**(2), 530–540 (2008). https://doi.org/10.1016/j.disopt.2007.08.006
6. Garey, M.R., Johnson, D.S.: Computers and Intractability; a Guide to the Theory of NP-Completeness. W.H. Freeman, San Francisco (1979)
7. Garey, M.R., Graham, R.L., Johnson, D.S.: The complexity of computing Steiner minimal trees. SIAM J. Appl. Math. **32** (1977). https://doi.org/10.1137/0132072
8. Kennedy, J., Eberhart, R.: Particle swarm optimization. In: Proceedings of IEEE International Conference on Neural Networks, vol. 4, pp. 1942–1948 (1995). https://doi.org/10.1109/ICNN.1995.488968
9. Kuhn, H.W.: "Steiner's" Problem Revisited, pp. 52–70. Mathematical Association of America, Washington (1974)

10. Montenegro, F., Torreão, J.R.A., Maculan, N.: Microcanonical optimization algorithm for the Euclidean Steiner problem in Rn with application to phylogenetic inference. Phys. Rev. E **68** (2003). https://doi.org/10.1103/PhysRevE.68.056702
11. Nemhauser, G.L., Wolsey, L.A.: Integer and Combinatorial Optimization. Wiley-Interscience, New York (1988)
12. Prim, R.C.: Shortest connection networks and some generalizations. Bell Syst. Tech. J. **36**(6), 1389–1401 (1957). https://doi.org/10.1002/j.1538-7305.1957.tb01515.x
13. Rocha, M.L.: An hybrid metaheuristic approach to solve the Euclidean Steiner tree problem in Rn. In: Proceedings of XLV Brazilian Symposium on Operational Research, vol. 1, pp. 1881–1892 (2013)
14. Smith, W.D.: How to find Steiner minimal trees in Euclidean d-space. Algorithmica **7**(1), 137–177 (1992). https://doi.org/10.1007/BF01758756
15. van den Bergh, F., Engelbrecht, A.: A study of particle swarm optimization particle trajectories. Inf. Sci. **176**(8), 937–971 (2006). https://doi.org/10.1016/j.ins.2005.02.003
16. Zachariasen, M.: Local search for the Steiner tree problem in the Euclidean plane. Eur. J. Oper. Res. **119**(2), 282–300 (1999). https://doi.org/10.1016/S0377-2217(99)00131-9

Chapter 5
Rotation-Based Multi-Particle Collision Algorithm with Hooke–Jeeves Approach Applied to the Structural Damage Identification

Reynier Hernández Torres, Haroldo Fraga de Campos Velho, and Leonardo Dagnino Chiwiacowsky

5.1 Overview

Structural health monitoring (SHM) is an application in System Identification that becomes a high trend topic of research for many areas, such as civil, mechanical, and aerospace engineering. SHM combines the sensing technology with methods for Structural Damage Identification (SDI) that detect, localize, and quantify damage in structures. Detecting damage in an early time allows repairing structures before failures occur. Therefore, it is important to perform a regular monitoring on the structures.

For critical systems, like buildings, aircraft, and aerospace structures, it is essential to detect damages in an accurate and safe way. In this case, a structural failure could cause catastrophic consequences, such as human and material losses.

Damage is defined as changes introduced into a system, affecting its current or future performance. A damage becomes significant if the structural response differs from the expected or designed response. The damage identification is carried out by comparing the current structural state with the designed structure. Changes in the system may be due alterations on:

R. H. Torres · H. F. de Campos Velho (✉)
Associated Laboratory for Computing and Applied Mathematics (LAC), National Institute for Space Research (INPE), São José dos Campos, SP, Brazil
e-mail: haroldo.camposvelho@inpe.br

L. D. Chiwiacowsky
Graduate Program in Industrial Engineering (PPGEP), University of Caxias do Sul (UCS), Bento Gonçalves, RS, Brazil
e-mail: ldchiwiacowsky@ucs.br

© Springer Nature Switzerland AG 2019
G. Mendes Platt et al. (eds.), *Computational Intelligence, Optimization and Inverse Problems with Applications in Engineering*,
https://doi.org/10.1007/978-3-319-96433-1_5

- material and geometric properties,
- boundary conditions, or
- system connectivity.

In mechanical vibration, the dynamics modeling is characterized by knowing the structures geometry, material properties, initial and boundary conditions, and the forcing terms. The output of the model such as displacement, accelerations, mode shapes, etc. depends on the physical properties of the structures. Changes in the physical properties that could be caused by cracks, loosening of connections, or any other damage should be detectable through the modal properties.

The damage identification process can be classified based on the performance, attending five levels [29, 39]:

1. *Detection*: the presence of any damage is verified.
2. *Localization*: determines where the damages are located.
3. *Classification*: determines the kind of damage.
4. *Extension*: gives an estimation of the extent and severity of the damage.
5. *Prognosis*: tries to predict the remaining service life of the structure.

The literature dealing with SDI is vast and classifies the methods depending on their approaches, such as frequency changes, mode shape changes, dynamically measured flexibility, matrix update methods, nonlinear methods, Neural Networks-based methods, and other methods.

Most of the works attend the first two levels of performance (detecting and locating the damage). Moreover, the literature is limited in the last three levels, comprising the classification, extension, and prognosis [26].

There is an important number of papers and theses handling the SDI as optimization problems, and solving them using Computational Intelligence. The SDI problem formulated as an optimization problem is considered one of the most effective strategies [43]. Different metaheuristic algorithms are used for solving these problems. For example, Table 5.1 shows a list of some metaheuristics and hybrid algorithms found in the literature. Hybrid Metaheuristics are the most promising used methods for this purpose.

5.2 Hybrid Algorithm for SDI

Hybrid metaheuristics are methods that combine a metaheuristic with other optimization approaches, such as algorithms from mathematical programming, constraint programming, machine learning, or artificial intelligence.

Hybridizing different algorithmic concepts allows obtaining a better performance, exploiting and combining the advantages of single strategies [2].

Table 5.1 Literature overview of computational intelligence methods used for SDI

Metaheuristics	
Genetic algorithm (GA)	[3, 4, 9, 13, 22]
Improved PSO (IPSO)	[42]
Tabu search (TS)	[1]
Continuous ACO (CnACO)	[43]
Ant colony optimization (ACO)	[21, 43]
Swarm intelligence (SI)	[44]
Particle swarm optimization (PSO)	[23, 32]
Big bang-big crunch (BB-BC)	[35]
Simulated annealing (SA)	[19]
Differential evolution (DE)	[12, 33, 34]
Global artificial fish swarm algorithm (GAFSA)	[41]
Rank-based ant system (RAS)	[5]
Ant system with heuristic information (ASH)	[5]
Elitist ant system (EAS)	[5]
Hybrid algorithms	
Genetic algorithm + Conjugated gradient method (GA-CGM)	[6, 8]
Genetic algorithm + Levenberg–Marquardt (GA-LM)	[14]
Genetic fuzzy system (GFS)	[27]
Simulated annealing genetic algorithm (SAGA)	[18]
Genetic algorithm and particle swarm optimization (GA-PSO)	[31]
Ant System with heuristic information + Hooke–Jeeves (ASH-HJ)	[5]
PSO with Nelder–Mead (PSO-NM)	[7]
Multi-particle collision algorithm with Hooke–Jeeves (MPCA-HJ)	[16]
q-Gradient with Hooke–Jeeves (qG-HJ)	[15]

Hybrid algorithms can be classified into two groups according to their taxonomy [36]:

- Collaborative hybrids combine two or more algorithms that could work in three ways:

  - Multi-stage, combining two stages: a global search followed by a local search;
  - Sequential, running both algorithms alternatively until a stopping criterion is met; or,
  - Parallel, where the algorithms run simultaneously over the same population.

- Integrative hybrids, where a master algorithm has other algorithm embedded working in two possible ways:

  - with a full manipulation of the population at every iteration, or
  - with the manipulation of a portion of the population.

5.2.1 *Rotation-Based Multi-Particle Collision Algorithm with Hooke-Jeeves*

Rotation-Based Multi-Particle Collision Algorithm with Hooke–Jeeves (RBMPCA-HJ) is a hybrid algorithm that uses an integration scheme of the Multi-Particle Collision Algorithm (MPCA) and the Rotation-Based Sampling (RBS), becoming an improved global search stage, followed by an intensification stage performed by the Hooke–Jeeves method (HJ), as shown in Fig. 5.1.

In the next subsections, each stage of the RBMPCA-HJ will be described.

5.2.2 *Multi-Particle Collision Algorithm*

MPCA is a metaheuristic inspired by the physics of nuclear particle collision interactions [30]. It is based on the scattering and absorption phenomena that occur inside a nuclear reactor. Scattering is when an incident particle is scattered by a target nucleus, while absorption is when an incident particle is absorbed by the target nucleus, as depicted in Fig. 5.2.

In MPCA, a set of candidate solutions, called particles, travel through the search space. There are three main functions: perturbation, exploitation, and scattering. New solutions are created perturbing the particles. If a new particle is better than the previous particle, then an intensification is made in its neighborhood. If an

Fig. 5.1 Operating flow of the hybrid algorithm rotation-based multi-particle collision algorithm with Hooke–Jeeves

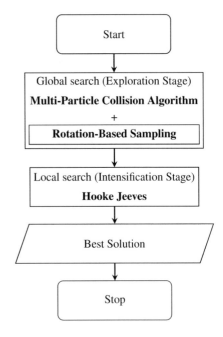

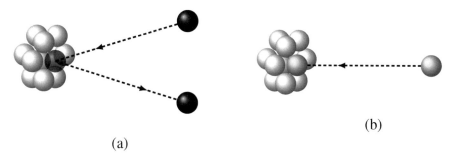

Fig. 5.2 Phenomena inside a nuclear reactor that inspire MPCA. (**a**) Scattering. (**b**) Absorption

improvement is not achieved, with a calculated probability, a new random particle is generated.

The initial set of n_p particles of MPCA is generated randomly in the search space. A mechanism called Blackboard allows sending the best solution overall reached in certain moments of the algorithm to the other particles. That moment is determined by the number of function evaluations. At every $\text{NFE}_{\text{blackboard}}$ evaluations, an update of the best particle is done.

As for stopping criterion, a maximum number of function evaluations (NFE_{mpca}) is defined. The MPCA algorithm flow is shown in Fig. 5.3.

5.2.2.1 Perturbation Function

This function performs a random variation of a particle within a defined range. A perturbed particle $\mathbf{P}^\star = (P_1, P_2, \ldots, P_n)$ is completely defined by its coordinates:

$$P_i^\star = P_i + [(UB_i - P_i) \times R] - [(P_i - LB_i) \times (1 - R)] \tag{5.1}$$

where $\mathbf{P}$ is the particle to be perturbed, $\mathbf{UB}$ and $\mathbf{LB}$ are the upper and the lower limits of the defined search space, respectively, and $R = \text{rand}(0, 1)$.

An objective function $f(\cdot)$ is used to evaluate a solution for an optimization problem. If $f(\mathbf{P}^\star) < f(\mathbf{P})$ (for minimization), then the particle $\mathbf{P}$ is replaced by $\mathbf{P}^\star$, and the EXPLOITATION function is activated, performing an exploitation in the neighborhood of the particle. If the new perturbed particle $\mathbf{P}^\star$ is worse than the current particle $\mathbf{P}$, the SCATTERING function is launched.

5.2.2.2 Exploitation Function

In this stage, the algorithm performs a series of small perturbations, computing a new particle $\mathbf{P}^\S$ each time, using the following equation:

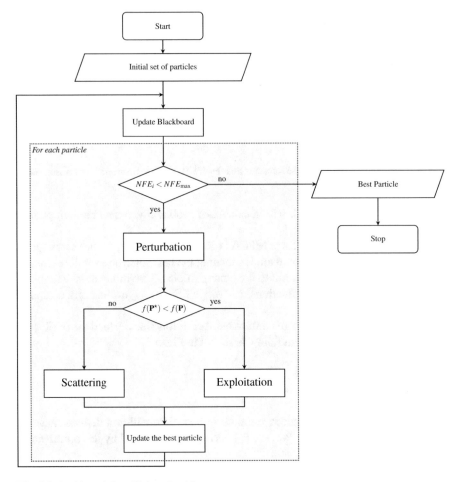

Fig. 5.3 Multi-particle collision algorithm

$$P_i^\S = P_i + [(u - P_i) \times R] - [(P_i - l) \times (1 - R)] \tag{5.2}$$

where $u = P_i \times \text{rand}(1, SL)$ is the small upper limit, and $l = P_i \times \text{rand}(IL, 1)$ is the small lower limit, with a superior value SL and an inferior value IL for the generation of the respective random number.

5.2.2.3 Scattering Function

The SCATTERING function works as a Metropolis scheme. The particle **P** is replaced by a totally new random solution **P**, or the exploitation is made, with a probability found by a Cauchy distribution:

$$p^s = \frac{1}{\pi \gamma \left(1 + \left(\frac{f(\mathbf{P}) - f(\mathbf{P}^{\text{best}})}{\gamma}\right)^2\right)}, \quad \gamma = 1 \tag{5.3}$$

where $\mathbf{P}^{\text{best}}$ is the best particle.

5.2.2.4 Blackboard Updating Function

At every $\text{NFE}_{\text{blackboard}}$ function evaluations, the best particle overall is elected and sent to all the particles, updating the reference P^{best}.

5.2.3 Opposition-Based Learning and Some Derived Mechanisms

The OBL mechanism was created by Tizhoosh [37] in 2005. The idea of OBL is to consider the opposite of a candidate solution, which has a certain probability of being closer to the global optimum.

Some mechanisms derived from OBL have been developed, such as *Quasi-opposition* that reflects a point to a random point between the center of the domain and the opposite point; *Quasi-reflection*, that projects the point to a random point between the center of the domain and itself; and *Center-based* sampling, that creates a point between itself and its opposite [10, 28, 38].

In a short period of time, these mechanisms have been utilized in different soft computing areas, improving the performance of various techniques of Computational Intelligence, such as metaheuristics, artificial neural networks, fuzzy logic, and other applications [40].

For better understanding of the mechanism, it is necessary to define the concept of some specific numbers used in such mechanism.

Definition 5.1 Let $z \in [a, b]$ be a real number, and $c = (a + b)/2$. The opposite number z_o, the quasi-opposite number z_{qo}, the quasi-reflected number z_{qr}, and the center-based sampled number z_{cb} are defined as:

$$z_o = a + b - z \tag{5.4}$$

$$z_{qo} = \text{rand}(z_o, c) \tag{5.5}$$

$$z_{qr} = \text{rand}(c, z) \tag{5.6}$$

$$z_{cb} = \text{rand}(z_o, z) \tag{5.7}$$

Figure 5.4 shows a graphical representation of these numbers.

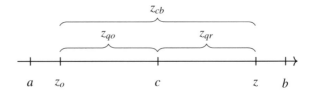

Fig. 5.4 Graphical representation of the opposition number (z_o), quasi-opposite number (z_{qo}), quasi-reflected number (z_{qr}), and center-based sampled number (z_{cb}) from the original number z

Definition 5.2 Let $\mathbf{Z} = (z_1, z_2, \ldots, z_n)$ be a point in an n-dimensional space, where $z_1, z_2, \ldots, z_n \in \mathbb{R}$, and $z_i \in [a_i, b_i]$, $\forall i \in \{1, \ldots, n\}$. The opposite point $\mathbf{Z}_o$, the quasi-opposite point $\mathbf{Z}_{qo}$, the quasi-reflected point $\mathbf{Z}_{qr}$, and the center-based sampled point $\mathbf{Z}_{cb}$ are completely defined by their coordinates:

$$z_{o_i} = a_i + b_i - z_i \tag{5.8}$$

$$z_{qo_i} = \mathrm{rand}(z_{o_i}, c_i) \tag{5.9}$$

$$z_{qr_i} = \mathrm{rand}(c_i, z_i) \tag{5.10}$$

$$z_{cb_i} = \mathrm{rand}(z_{o_i}, z_i) \tag{5.11}$$

respectively.

Definition 5.3 (Opposition-Based Optimization) Let $\mathbf{Z}$ be a point in n-dimensional space (i.e., candidate solution), and $\mathbf{Z}_o$ an opposite point in the same space (i.e., opposite candidate solution). If $f(\mathbf{Z}_o) \leq f(\mathbf{Z})$, then the point $\mathbf{Z}$ can be replaced by $\mathbf{Z}_o$, which is better, otherwise it will maintain its current value.

The solution and the opposite solution are evaluated simultaneously, and the optimization process will continue with the better one.

The same idea of the Opposition-based Optimization is applicable for the other points.

5.2.3.1 Rotation-Based Learning and Rotation-Based Sampling

The RBL mechanism is another extension of the OBL [20], and the Rotation-Based Sampling (RBS) mechanism is a combination of the Center-Based Sampling and RBL mechanisms.

Definition 5.4 Let $z \in [a, b]$ be a real number, and $c = (a + b)/2$ be the center. Draw a circle with center c and radius $c - a$. The point $(z, 0)$ is projected on the circle. The quantity u is defined as u, and the random deflection angle β is found using the normal distribution $\beta \sim \mathcal{N}(\beta_0, \delta)$, with mean β_0, and standard deviation δ. The rotation number z_r, and the rotation-based sampling number z_{rbs} are defined as:

Fig. 5.5 Geometric
interpretation of the Rotation
(z_r) and the Rotation-based
Sampling (z_{rbs}) numbers in
2D space

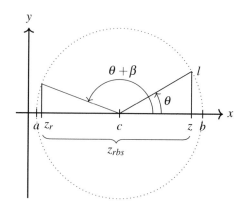

$$z_r = c + u \times \cos \beta - v \times \sin \beta \qquad (5.12)$$

$$z_{rbs} = \text{rand}(z, z_r) \qquad (5.13)$$

Similarly to Definition 5.2, the concepts of Rotation and Rotation-based sampling Points are enunciated:

Definition 5.5 Let $\mathbf{Z} = (z_1, z_2, \ldots, z_n)$ be a point in an n-dimensional space, where $z_1, z_2, \ldots, z_n \in \mathbb{R}$, and $z_i \in [a_i, b_i]$, $\forall i \in \{1, \ldots, n\}$. The rotation point $\mathbf{Z}_r$, and the rotation-based sampling number $\mathbf{Z}_{rbs}$ are completely defined by their coordinates:

$$z_{r_i} = c_i + u_i \times \cos \beta - v_i \times \sin \beta \qquad (5.14)$$

$$z_{rbs_i} = \text{rand}(z_{r_i}, z_i) \qquad (5.15)$$

respectively.

The geometric representation in a 2D-space of the Rotation and the Rotation-Based Sampling numbers is shown in Fig. 5.5.

5.2.3.2 Integration of MPCA and RBS

The RBS mechanism is integrated to the MPCA in four different moments. In the Population Initialization, each particle is compared with its opposite particle found within the search space $[LB_i, UB_i]$. The best particle between them will continue the travel. In the same way, after creating a new particle in the PERTURBATION and the SCATTERING functions, the opposite particle is calculated within the search space $[LB_i, UB_i]$. The best particle between them will continue the travel. In the EXPLOITATION function, after a small perturbation is performed on the particle, with probability J_r (jumping rate), the opposite particle is found within the small lower and small upper values $[l, u]$. Again, the best particle between them will continue the travel.

5.2.4 Hooke–Jeeves Pattern Search Method

The direct search method proposed by Hooke and Jeeves (HJ) [17] consists of the repeated application of exploratory moves around a base point **s** defined in an n-dimensional space. If these moves are successful, they are followed by pattern moves. Figures 5.6 and 5.7 show the flowchart of the algorithm.

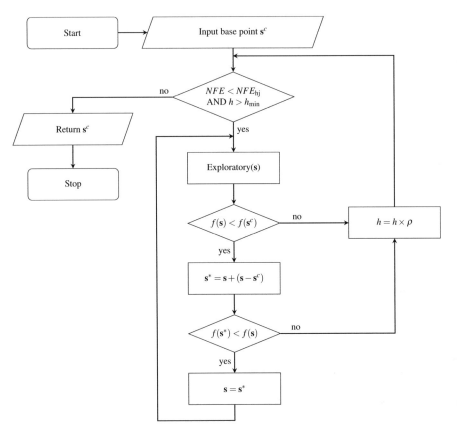

Fig. 5.6 Main function—Hooke–Jeeves direct search

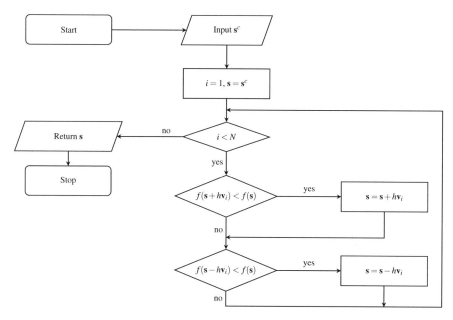

Fig. 5.7 Exploratory function—Hooke–Jeeves direct search

5.2.4.1 Exploratory Move

In the exploratory move, the solution is changed adding and subtracting one column of the search direction matrix $\mathbf{V}$ scaled by a step size h, denoted as $\mathbf{v}_i$.

If $\mathbf{V}$ is the identity matrix, then the modification is performed on the ith element of the solution $\mathbf{s}$ each time. This process is done for all the dimensions of the problem. The original solution is compared with the added solution and with the subtracted solution, and the best among them will be accepted for the next iteration.

The exploratory move returns the best solution found.

5.2.4.2 Pattern Move

A new pattern point $\mathbf{s}^*$ is calculated from the solution $\mathbf{s}$ obtained from the exploratory move, and the current solution $\mathbf{s}^c$ as follows:

$$\mathbf{s}^* = \mathbf{s} + \left(\mathbf{s} - \mathbf{s}^c\right) \tag{5.16}$$

If the pattern point $\mathbf{s}^*$ is better than $\mathbf{s}^c$, the latter can be replaced by $\mathbf{s}^*$. If there is no improvement, the step size h is reduced ρ times.

A minimum step size $h_{\min}$ and a maximum number of function evaluations NFE_{hj} are defined as the stopping criteria.

5.3 Vibration-Based Damage Identification Problem as an Optimization Problem

The dynamic response of motion under kinematic excitations is shown in Eq. (5.17):

$$\mathbf{M\ddot{u}}(t) + \mathbf{C\dot{u}}(t) + \mathbf{Ku}(t) = \mathbf{F}(t) \tag{5.17}$$

In the equation, $\mathbf{M}$, $\mathbf{C}$, and $\mathbf{K}$ represent the $d \times d$ mass, damping, and stiffness matrices, respectively; d is the number of degrees of freedom of the structure. $\mathbf{F}$ and $\mathbf{u}$ are the external force and the displacement vectors, respectively. The initial conditions for the model are given by Eqs. (5.18) and (5.19).

$$\mathbf{u}(0) = \mathbf{u}_0 \tag{5.18}$$

$$\mathbf{\dot{u}}(0) = \mathbf{\dot{u}}_0 \tag{5.19}$$

The numerical solution for this model is obtained using the Newmark method since no analytical solution exists for any arbitrary functions of $\mathbf{M}$, $\mathbf{C}$, $\mathbf{K}$, and $\mathbf{F}$ [24].

In this work, the inverse problem is formulated by localizing and quantifying damages on the structure as an optimization problem. An optimization algorithm will minimize the squared difference between the computed displacements $\mathbf{u}^{\text{mod}}$ (obtained after running the structural model with a stiffness vector $\mathbf{k}$) and the measured displacements $\mathbf{u}^{\text{obs}}$ (acquired from the sensors in vibration experiments), as follows:

$$J(\mathbf{k}) = \sum_{i=0}^{d_m} \left[\mathbf{u}_i^{\text{obs}}(t) - \mathbf{u}_i^{\text{mod}}(\mathbf{k}, t) \right]^2 \tag{5.20}$$

where t represents the time, d_m is the number of measured displacements, and $\mathbf{k} = (k_1, k_2, \ldots, k_n)$ contains the values of the stiffness for each element, with n elements in total.

Figure 5.8 shows a graphical representation of the inverse solution for a generic problem. Parameter $\boldsymbol{\Theta}^{\text{e}}$ represents the influence of environmental and operational conditions (temperature, humidity, etc.) [11].

Estimated damages $\boldsymbol{\Theta}^{\text{d}}$ (in percent) are represented by the loss of stiffness:

$$\boldsymbol{\Theta}^{\text{d}} = \left(1 - \frac{\mathbf{k}^{\text{d}}}{\mathbf{k}^{\text{u}}} \right) \times 100\% \tag{5.21}$$

where $\mathbf{k}^{\text{d}} = \left(k_1^{\text{d}}, k_2^{\text{d}}, \ldots, k_n^{\text{d}} \right)$ and $\mathbf{k}^{\text{u}} = \left(k_1^{\text{u}}, k_2^{\text{u}}, \ldots, k_n^{\text{u}} \right)$ are the estimated stiffness vector for the damaged system, and the stiffness vector of the undamaged system, respectively.

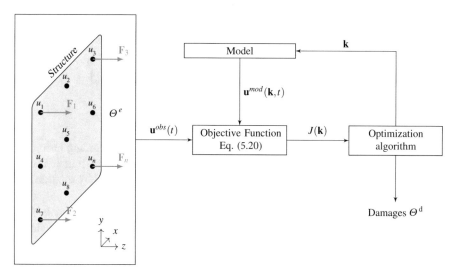

Fig. 5.8 Vibration-based damage identification as optimization problem

5.4   Damage Identification in a Cantilevered Beam

Beams are a simple but common type of structural component. They are used mostly in Civil and Mechanical Engineering. These structural members have the main function of supporting transverse loading and carrying it to the supports. Their shape is bar-like, being one dimension larger than the others, and they deform only in the directions perpendicular to the x-axis.

The experiments will be performed on a plane beam. This type of beam resists primarily transverse loading on a preferred longitudinal plane.

5.4.1   Cantilevered Beam

The cantilevered beam shown in Fig. 5.9 is modeled with ten beam finite elements. It is clamped at the left end, and each aluminum beam element, with $\rho = 2700 \, \text{kg/m}^3$ and $E = 70 \, \text{GPa}$, has a constant rectangular cross section area with $b = 15 \times 10^{-3} \, \text{m}$ and $h = 6 \times 10^{-3} \, \text{m}$, a total length $l = 0.43 \, \text{m}$, and an inertial moment $I = 3.375 \times 10^{-11} \, \text{m}^4$. The damping matrix is assumed proportional to the undamaged stiffness matrix $\mathbf{C} = 10^{-3} \, \mathbf{K}$. An external varying force $F(t) = 5.0 \times 2.0 \sin(\pi t) \, \text{N}$ is applied to the tenth element, in the free extreme of the beam.

Initial conditions are null, $\mathbf{u}(0) = 0$ and $\dot{\mathbf{u}}(0) = 0$. For the experiments, the numerical simulation was performed assuming $t_f = 2 \, \text{s}$, with a time step $\Delta t = 4 \times 10^{-3} \, \text{s}$.

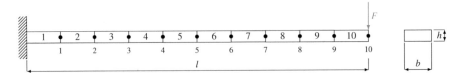

Fig. 5.9 20-DOF beam structure

Synthetic data were obtained from the structure response to a forcing term. The forward model is executed representing the beam structure. The displacements could be measured by strain-gages, while the rotations could be measured by rotation rate sensors or gyroscopes [45].

The measurements are commonly corrupted by noise in sensor signals. The solution of an inverse problem can be affected by small perturbations or noise, introducing spurious oscillations.

For testing the robustness of the algorithm in the solution of the inverse problem, some noise is added to the synthetic measured data [25]:

$$\hat{\mathbf{u}}_i(t) = \mathbf{u}_i(t) + \sigma\mu \qquad (5.22)$$

where $\hat{\mathbf{u}}_i(t)$ is the noisy data, $\mu \sim \mathcal{N}(0, \sigma)$ is a random value, and $\mathcal{N}$ the Gaussian distribution with zero mean and standard deviation σ. Three cases are tested: noiseless data ($\sigma = 0.00$), noisy data with $\sigma = 0.02$, and noisy data with $\sigma = 0.05$.

The next configuration was used for the algorithms: a set of 20 particles was used in MPCA, the blackboard updating occurs at every $NFE_{blackboard} = 100,000$. The limits for the exploitation were set as $IL = 0.7$ and $SL = 1.1$. The stopping criterion for MPCA was assumed $NFE_{mpca} = 200,000$. For the RBS mechanism, $\beta_0 = 3.14$ rad, and $\delta = 0.25$. For HJ, $\rho = 0.8$, $h_{min} = 1 \times 10^{-11}$, and $NFE_{hj} = 100,000$. The search space was set as $\mathbf{S} = [0.5, 1.05]\mathbf{k}^u$.

5.4.2 Experimental Results: Damage Identification from a Full Dataset

In the first set of experiments, observed data were taken from all nodes, with a total of 20 time series with 500 points each one. For the analysis in each case, it was calculated the mean of 15 runs of the inverse solution.

5.4.2.1   Case 1: Single Damage

In case 1, a single damage of 10% was simulated on the first element of the structure, maintaining the other elements without changes.

Figure 5.10 shows the identification process results. The damage was well estimated in all experiments. In the experiments with noisy data, MPCA-HJ identified a little damage of less than 1% of 10th element, while RBMPCA-HJ identified a negative value of damage of this element. This negative value could represent an increase of the stiffness value, which is not allowed in the scope of this work.

5.4.2.2   Case 2: Mixed Multi-Damage

In case 2, a damage configuration of 10% on the 2nd element; 20% on the 4th; 30% on the 6th, 5% on the 9th element, and 10% for the 10th element was set. The remaining elements are assumed as undamaged.

Figure 5.11 shows the results. All the damages were well identified in all experiments. For the 10th element, some error was detected. It is important to highlight that in two runs, with $\sigma = 0.02$, the MPCA-HJ totally missed the damage in the 10th element.

5.4.3   Experimental Results: Damage Identification from a Reduced Dataset

In the second set of experiments, observed data were taken from some degrees of freedom: displacements from node 2 and node 10, and rotation from node 5 and node 10. In total, in this case there are four time-series with 500 points each one. For the analysis at each case, it was calculated the average from 15 runs of the inverse solution.

5.4.3.1   Case 1: Single Damage

Similar to the experiments with a full dataset, a single damage of 10% was simulated on the first element, maintaining the other elements undamaged.

Figure 5.12 shows the mean values (bars) of the damage parameters for the damage detection process. For the noiseless data, the damage was well identified and none false damage appeared.

For the experiments with $\sigma = 0.02$, MPCA-HJ launched a false alarm for element 10. This is caused because this algorithm found a damage of about 10% in three runs, and of 20% in one run. RBMPCA-HJ launched a false alarm for the 10th element just in two single runs. In the other elements, some errors were detected, most of them negative.

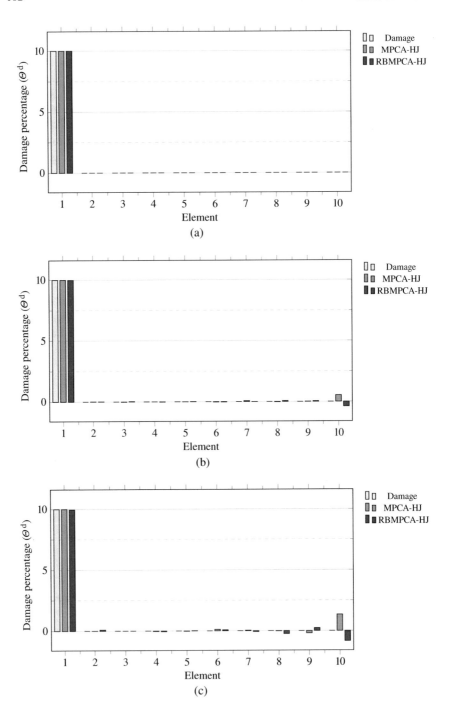

Fig. 5.10 Results for the damages identification in a beam structure using a full dataset. A single damage is located in the fixed element. (**a**) Noiseless data. (**b**) Noisy data with $\sigma = 0.02$. (**c**) Noisy data with $\sigma = 0.05$

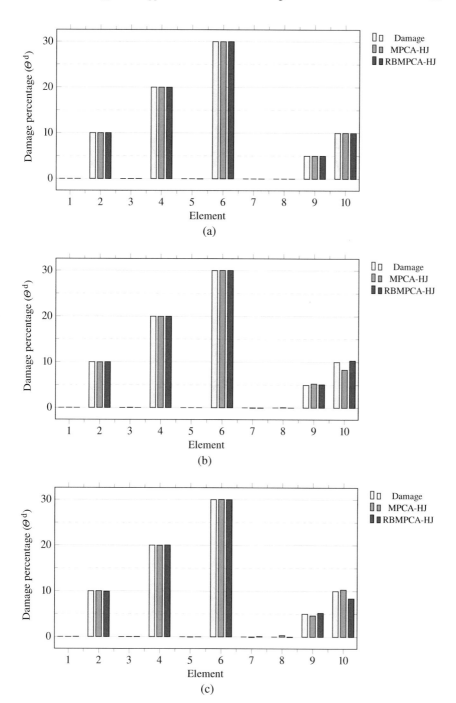

Fig. 5.11 Results for the damages identification in a beam structure using a full dataset. A damage configuration with mixed damages was set. (**a**) Noiseless data. (**b**) Noisy data with $\sigma = 0.02$. (**c**) Noisy data with $\sigma = 0.05$

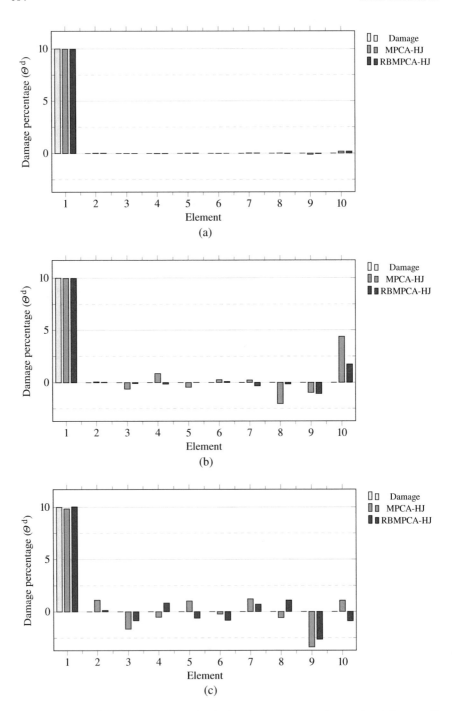

Fig. 5.12 Results for the damages identification in a beam structure using a reduced dataset with a few time series. A single damage is located in the fixed element. (**a**) Noiseless data. (**b**) Noisy data with $\sigma = 0.02$. (**c**) Noisy data with $\sigma = 0.05$

For the experiments with $\sigma = 0.05$, it is noticeable that at the 9th element a negative damage was identified for both algorithms. Both algorithms detected negligible damages for the other elements. It is important to highlight the increasing level of variation for the results from the noise level.

5.4.3.2 Case 2: Mixed Multi-Damage

In this experiment, the same damage configuration of the Sect. 5.4.2.2 was used.

Figure 5.13 shows the results for the damage identification. All the damages were detected. For the 6th and the 9th elements, the damage values were well estimated, with a small error. The estimated damage on the 2nd, the 4th, and the 10th elements have an error about 5% around the true value. A false alarm appeared for the 3rd element with a damage about 5%.

For the experiments with $\sigma = 0.02$, the 5% damage at the 9th element and the 10% at the 2nd element were not well identified—high level of errors appear caused by the presence of the noise. The damage for the 4th element was detected, but with an error greater than 5%. On the other hand, the damages at the 6th and the 10th elements were detected, and the values have an error less than 5%. Again, a false alarm appeared at the 3rd element.

For the experiments with $\sigma = 0.05$, the results are equivalent to those achieved with $\sigma = 0.02$. The damage at the 9th element was overshadowed by the effects of the noise. For the 3rd element, which is undamaged, a damage equal to the estimated at the 4th element was estimated, which have a damage of 20%. Also, false alarms appeared for the 5th and the 7th elements.

5.5 Final Remarks

The Structural Damage Identification was solved as an optimization problem, using the hybrid metaheuristic RBMPCA-HJ. The results were compared with those obtained using MPCA-HJ. Two damage configurations were tested: one with a single damage in the first element, and the other with a mixed configuration, with five damages of different intensity. Estimations were made using two data sets: a full dataset obtained from all the DOF from the structure, and a reduced dataset obtained from four DOF (two displacement variables and two rotation variables). The use of these hybrid algorithms allows to obtain good estimations using a full set of data, or containing low level of noise.

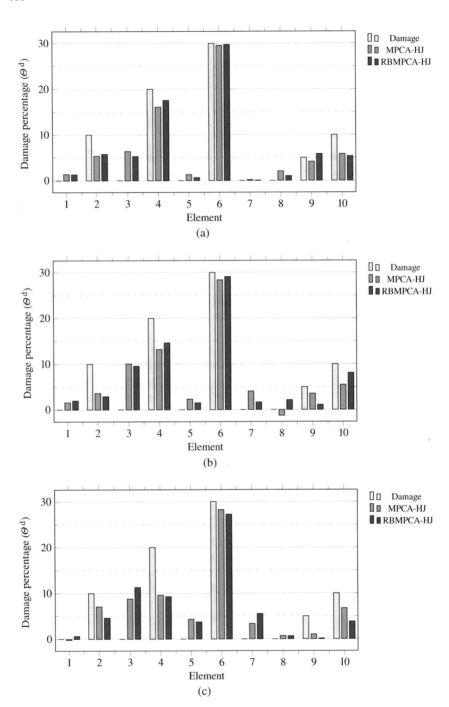

Fig. 5.13 Results for the damages identification in a beam structure using a reduced dataset with a few time series. A damage configuration with mixed damages was set. (**a**) Noiseless data. (**b**) Noisy data with $\sigma = 0.02$. (**c**) Noisy data with $\sigma = 0.05$

Acknowledgements The authors acknowledge the support from the National Council for Research and Development (CNPq) under grants numbers 159547/2013-0 and 312924/2017-8.

References

1. Arafa, M., Youssef, A., Nassef, A.: A modified continuous reactive tabu search for damage detection in beams. In: 36th Design Automation Conference, Parts A and B, vol. 1, pp. 1161–1169. ASME (2010). https://doi.org/10.1115/DETC2010-28389
2. Blum, C., Puchinger, J., Raidl, G., Roli, A.: A brief survey on hybrid metaheuristics. In: 4th International Conference on Bioinspired Optimization Methods and Their Applications, pp. 3–16 (2010)
3. Boonlong, K.: Vibration-based damage detection in beams by cooperative coevolutionary genetic algorithm. Adv. Mech. Eng. **6**, 1–13 (2014). https://doi.org/10.1155/2014/624949
4. Borges, C.C.H., Barbosa, H.J.C., Lemonge, A.C.C.: A structural damage identification method based on genetic algorithm and vibrational data. Int. J. Numer. Methods Eng. **69**(13), 2663–2686 (2007). https://doi.org/10.1002/nme
5. Braun, C.E., Chiwiacowsky, L.D., Gomez, A.T.: Variations of ant colony optimization for the solution of the structural damage identification problem. Proc. Comput. Sci. **51**, 875–884 (2015). https://doi.org/10.1016/j.procs.2015.05.218
6. Campos Velho, H.F., Chiwiacowsky, L.D., Sambatti, S.B.: Structural damage identification by a hybrid approach: variational method associated with parallel epidemic genetic algorithm. Scientia Interdiscip. Stud. Comput. Sci. **17**(1), 10–18 (2006)
7. Chen, Z., Yu, L.: An improved PSO-NM algorithm for structural damage detection. In: Advances in Swarm and Computational Intelligence, pp. 124–132. Springer (2015). https://doi.org/10.1007/978-3-319-20466-6$_$14
8. Chiwiacowsky, L.D., Campos Velho, H.F., Gasbarri, P.: A variational approach for solving an inverse vibration problem. Inverse Prob. Sci. Eng. **14**(5), 557–577 (2006). https://doi.org/10.1080/17415970600574237
9. Chou, J.H., Ghaboussi, J.: Genetic algorithm in structural damage detection. Comput. Struct. **79**(14), 1335–1353 (2001). https://doi.org/10.1016/S0045-7949(01)00027-X
10. Ergezer, M., Simon, D., Du, D.: Oppositional biogeography-based optimization. In: IEEE International Conference on Systems, Man and Cybernetics, 2009, pp. 1009–1014. IEEE, New York (2009)
11. Fritzen, C.P., Kraemer, P.: Self-diagnosis of smart structures based on dynamical properties. Mech. Syst. Signal Process. **23**(6), 1830–1845 (2009). https://doi.org/10.1016/j.ymssp.2009.01.006
12. Fu, Y.M., Yu, L.: A DE-based algorithm for structural damage detection. Adv. Mater. Res. **919**, 303–307 (2014). https://doi.org/10.4028/www.scientific.net/AMR.919-921.303
13. Gomes, H.M., Silva, N.R.S.: Some comparisons for damage detection on structures using genetic algorithms and modal sensitivity method. Appl. Math. Model. **32**(11), 2216–2232 (2008). https://doi.org/10.1016/j.apm.2007.07.002
14. He, R.S., Hwang, S.F.: Damage detection by an adaptive real-parameter simulated annealing genetic algorithm. Comput. Struct. **84**(31–32), 2231–2243 (2006). https://doi.org/10.1016/j.compstruc.2006.08.031
15. Hernández, R., Scarabello, M.C., Campos Velho, H.F., Chiwiacowsky, L.D., Soterroni, A.C., Ramos, F.M.: A hybrid method using q-gradient to identify structural damages. In: Dumont, N.A. (ed.) Proceedings of the XXXVI Iberian Latin-American Congress on Computational Methods in Engineering, Rio de Janeiro (2015)
16. Hernández, R., Chiwiacowsky, L.D., Campos Velho, H.F.: Multi-particle collision algorithm with Hooke-Jeeves for solving a structural damage detection problem. In: Araújo, A.L., Correia, J.R., Soares, C.M.M. (eds.) 10th International Conference on Composite Science and Technology, Lisbon (2015)

17. Hooke, R., Jeeves, T.A.: "Direct Search" solution of numerical and statistical problems. J. ACM **8**(2), 212–229 (1961)
18. Kokot, S., Zembaty, Z.: Damage reconstruction of 3d frames using genetic algorithms with Levenberg–Marquardt local search. Soil Dyn. Earthq. Eng. **29**(2), 311–323 (2009). https://doi.org/10.1016/j.soildyn.2008.03.001
19. Kourehli, S.S., Bagheri, A., Amiri, G.G., Ghafory-Ashtiany, M.: Structural damage detection using incomplete modal data and incomplete static response. KSCE J. Civil Eng. **17**(1), 216–223 (2013). https://doi.org/10.1007/s12205-012-1864-2
20. Liu, H., Wu, Z., Li, H., Wang, H., Rahnamayan, S., Deng, C.: PRICAI2014: trends in artificial intelligence. In: 13th Pacific Rim International Conference on Artificial Intelligence. Rotation-Based Learning: A Novel Extension of Opposition-Based Learning, pp. 511–522. Springer International Publishing (2014). https://doi.org/10.1007/978-3-319-13560-1$_$41
21. Majumdar, A., Maiti, D.K., Maity, D.: Damage assessment of truss structures from changes in natural frequencies using ant colony optimization. Appl. Math. Comput. **218**(19), 9759–9772 (2012). https://doi.org/10.1016/j.amc.2012.03.031
22. Mares, C., Surace, C.: An application of genetic algorithms to identify damage in elastic structures. J. Sound Vib. **195**(2), 195–215 (1996). https://doi.org/10.1006/jsvi.1996.0416
23. Mohan, S., Maiti, D., Maity, D.: Structural damage assessment using FRF employing particle swarm optimization. Appl. Math. Comput. **219**(20), 10387–10400 (2013). https://doi.org/10.1016/j.amc.2013.04.016
24. Newmark, N.M.: A method of computation for structural dynamics. J. Eng. Mech. Div. **85**(3), 67–94 (1959)
25. Nichols, J.M., Murphy, K.D.: Modeling and Estimation of Structural Damage. Wiley, New York (2016)
26. Ooijevaar, T.H.: Vibration based structural health monitoring of composite skin-stiffener structures. Ph.D. thesis, University of Twente (2014)
27. Pawar, P.M., Ganguli, R.: Genetic fuzzy system for online structural health monitoring of composite helicopter rotor blades. Mech. Syst. Signal Process. **21**(5), 2212–2236 (2007). https://doi.org/10.1016/j.ymssp.2006.09.006
28. Rahnamayan, S., Tizhoosh, H.R., Salama, M.: Quasi-oppositional differential evolution. In: IEEE Congress on Evolutionary Computation, 2007. CEC 2007, pp. 2229–2236. IEEE, New York (2007)
29. Rytter, A.: Vibrational based inspection of civil engineering structures. Ph.D. thesis, Aalborg University (1993)
30. Sacco, W.F., Oliveira, C.R.E.: A new stochastic optimization algorithm based on a particle collision metaheuristic. In: Proceedings of 6th WCSMO (2005)
31. Sandesh, S., Shankar, K.: Application of a hybrid of particle swarm and genetic algorithm for structural damage detection. Inverse Prob. Sci. Eng. **18**(7), 997–1021 (2010). https://doi.org/10.1080/17415977.2010.500381
32. Seyedpoor, S.: A two stage method for structural damage detection using a modal strain energy based index and particle swarm optimization. Int. J. Non-Linear Mech. **47**(1), 1–8 (2012). https://doi.org/10.1016/j.ijnonlinmec.2011.07.011
33. Seyedpoor, S.M., Yazdanpanah, O.: Structural damage detection by differential evolution as a global optimization algorithm. Iran. J. Struct. Eng. **1**(1), 52–62 (2014)
34. Seyedpoor, S.M., Shahbandeh, S., Yazdanpanah, O.: An efficient method for structural damage detection using a differential evolution algorithm-based optimisation approach. Civ. Eng. Environ. Syst. 1–21 (2015). https://doi.org/10.1080/10286608.2015.1046051
35. Tabrizian, Z., Afshari, E., Amiri, G.G., Ali Beigy, M.H., Nejad, S.M.P.: A new damage detection method: big bang-big crunch (BB-BC) algorithm. Shock Vib. **20**(4), 633–648 (2013). https://doi.org/10.3233/SAV-130773
36. Ting, T.O., Yang, X.S., Cheng, S., Huang, K.: Hybrid Metaheuristic algorithms: past, present, and future. In: Yang, X.-S. (ed.) Recent Advances in Swarm Intelligence and Evolutionary Computation. Studies in Computational Intelligence, vol. 585, pp. 71–83. Springer International Publishing (2015). https://doi.org/10.1007/978-3-319-13826-8$_$4

37. Tizhoosh, H.R.: Opposition-based learning: a new scheme for machine intelligence. In: International Conference on Computational Intelligence for Modelling, Control and Automation and International Conference on Intelligent Agents. Web Technologies and Internet Commerce (CIMCA-IAWTIC'06), vol. 1, pp. 695–701. IEEE, New York (2005). https://doi.org/10.1109/CIMCA.2005.1631345
38. Tizhoosh, H.R., Ventresca, M.: Oppositional Concepts in Computational Intelligence. Studies in Computational Intelligence. Springer, Berlin (2008)
39. Worden, K., Dulieu-Barton, J.M.: An overview of intelligent fault detection in systems and structures. Struct. Health Monit. **3**(1), 85–98 (2004). https://doi.org/10.1177/1475921704041866
40. Xu, Q., Wang, L., Wang, N., Hei, X., Zhao, L.: A review of opposition-based learning from 2005 to 2012. Eng. Appl. Artif. Intell. **29**, 1–12 (2014)
41. Yu, L., Li, C.: A global artificial fish swarm algorithm for structural damage detection. Adv. Struct. Eng. **17**(3), 331–346 (2014). https://doi.org/10.1260/1369-4332.17.3.331
42. Yu, L., Wan, Z.: An improved PSO algorithm and its application to structural damage detection. In: 2008 Fourth International Conference on Natural Computation, vol. 1, pp. 423–427. IEEE (2008). https://doi.org/10.1109/ICNC.2008.224
43. Yu, L., Xu, P.: Structural health monitoring based on continuous ACO method. Microelectron. Reliab. **51**(2), 270–278 (2011). https://doi.org/10.1016/j.microrel.2010.09.011
44. Yu, L., Xu, P., Chen, X.: A SI-based algorithm for structural damage detection. In: Advances in Swarm Intelligence, pp. 21–28. Springer (2012). https://doi.org/doi:10.1007/978-3-642-30976-2_3
45. Zembaty, Z., Kokot, S., Bobra, P.: Application of rotation rate sensors in measuring beam flexure and structural health monitoring, pp. 65–76. Springer International Publishing (2016). https://doi.org/10.1007/978-3-319-14246-3_6

Chapter 6
Optimization in Civil Engineering and Metaheuristic Algorithms: A Review of State-of-the-Art Developments

Gebrail Bekdaş, Sinan Melih Nigdeli, Aylin Ece Kayabekir, and Xin-She Yang

6.1 Introduction

Civil engineering concerns many applications, including transportation and storage of water resources, superstructure and infrastructure constructions, mass transportation and resources, traffic in transportation, stabilization of soil backfills and improvement of the soil. In fact, all applications providing a living environment essentially fall into the area of civil engineering. The world has different geographic areas with different resources and differential risks of natural disasters. Due to differences in resources, different regions may have different material and labor costs. The most important factor in the design of civil engineering applications is safety. Due to natural hazards and earth conditions, the defined safety standards are various. Consequently, several regulations exist in different regions and countries. Another important factor is the requirements and demands of people and social authorities. For that reason, an optimum design of civil engineering in one region may not be the best and acceptable design in a different place. Hence, the optimum designs must be user oriented.

For example, in designing a product of mechanical engineering, the rules of physics are a defining factor. The excitation conditions are well-known and a precise production is possible in a factory. However, these conditions may not be acceptable for a structure construction in practice, and the area conditions

G. Bekdaş · S. M. Nigdeli · A. E. Kayabekir
Department of Civil Engineering, Istanbul University, Avcılar, Istanbul, Turkey
e-mail: bekdas@istanbul.edu.tr; melihnig@istanbul.edu.tr

X.-S. Yang (✉)
School of Science and Technology, Middlesex University, London, UK
e-mail: x.yang@mdx.ac.uk

© Springer Nature Switzerland AG 2019
G. Mendes Platt et al. (eds.), *Computational Intelligence, Optimization and Inverse Problems with Applications in Engineering*,
https://doi.org/10.1007/978-3-319-96433-1_6

are specific. For another example, the area may be in a major earthquake zone. In this case, additional design analyses must be considered since the effects of several factors cannot be neglected. Additionally, structures have huge components and production is usually done in a construction yard. Especially, the hardening of concrete and the positioning of the reinforcement bars are done during the construction of reinforced concrete structures. A precise design is not possible and the design variables must be assigned with discrete variables. Similarly, in construction of steel structure, the profile sizes in the local market are standard and the supplying of a special section is not an economical solution. Due to these reasons, the civil engineering applications have nonlinear constraints which depend on the special conditions. Therefore, designs should be robust to uncertainties in materials properties and manufacturability as well as construction uncertainties. To solve such design optimization problems can be very challenging, and numerical algorithms are more suitable than any mathematical approaches in optimization.

In recent years, metaheuristic algorithms have been employed in optimization methodologies developed for civil engineering applications. Metaheuristic algorithms use some inspiration from nature for problem-solving. The purposes of optimization is to find best or most robust design options so as to design variables and parameters are realistic and the overall objective is achievable in practice.

Therefore, this chapter is organized as follows: Sect. 6.2 provides some brief formulation of optimization and the introduction of several metaheuristic algorithms. Then, Sect. 6.3 provides a detailed review of various applications in civil engineering. Finally, Sect. 6.4 draws conclusions briefly.

In civil engineering, metaheuristic based applications were presented in books for general areas [209], structures and infrastructures [204]. Here, we will focus on the most recent state-of-the-art developments in civil engineering applications.

6.2   Optimization and Metaheuristic Algorithms

An optimum design formulation can be in general written as

$$\text{Minimize } f_i(x), \quad x \in \mathbb{R}^n, \ (i = 1, 2, \ldots, M) \tag{6.1}$$

subject to equality constraints:

$$h_j(x) = 0, \ (j = 1, 2, \ldots, J) \tag{6.2}$$

and inequality constraints:

$$g_k(x) \leq 0, \ (k = 1, 2, \ldots, K) \tag{6.3}$$

where **x** is a vector of design variables defined as

$$\mathbf{x} = (x_1,\ x_2,\ \ldots,\ x_n)^T \tag{6.4}$$

The objective functions $f_i(\mathbf{x})$ are minimized (or maximized) by finding the best sets of design variables $(\mathbf{x})$. However, design problems tend to be multi-objective with M different objectives, but here will focus on the case of single objective optimization (thus $M = 1$). The design constraints are handled and incorporated into the objective using a penalty function. This allows us to emphasize the review of optimization techniques and applications without any concern of constraint-handling techniques. The rest of this section outlines several widely used metaheuristic methods.

6.2.1 Genetic Algorithm

Genetic Algorithm (GA) is the one of the well-known metaheuristic algorithms, developed by John Holland [81]. The basis of this algorithm is Charles Darwin's natural selection and the main operators include the crossover and recombination, mutation and selection. The main steps of this algorithm are as follows:

Step 1. Encode optimization objective.
Step 2. Define a fitness function or criterion for selection of an individual.
Step 3. Initialize a population of individuals.
Step 4. Evaluate the fitness function for all individuals.
Step 5. Generate a new population by using crossover, mutation and proportionate reproduction.
Step 6. Evolve the population until the defined stopping criterion is met.
Step 7. Decode the results.

Goldberg and Samtani [71] used GA for a civil engineering problem where the optimum design of a 10-bar truss structure was done in that study. Since then, there are hundreds of papers in the context of civil engineering applications and we will highlight a few applications of GA in Sect. 6.3.

6.2.2 Simulated Annealing

Simulated Annealing is an algorithm for mimicking the process of metal annealing to increase ductility and strength of the metal by internally re-arranging atomic structure. Kirkpatrick et al. [110] developed this algorithm, and Černý [26] was among the first to use simulated annealing (SA) in several civil engineering and more applications appear recently [36, 87, 189], but the hybridization of SA with GA has also been explored [83, 128].

6.2.3 Particle Swarm Optimization

Particle swarm optimization (PSO) is a swarm-intelligence based metaheuristic algorithm, developed by Kennedy and Eberhart [108], imitating swarming behavior of birds and fish. The particles of a swarm are randomly generated and new solutions are iteratively updated according to two iterative equations which depend on the current locations, velocities, and the current best solution. This is a population-based algorithm and the global optimum is achievable when the design space is well explored by multiple particles in a sufficiently long time.

For example, the application of PSO in structural engineering is presented by Kaveh [88], and more examples will be given in Sect. 6.3.

6.2.4 Harmony Search

Harmony Search (HS) is a music-inspired algorithm, developed by Geem et al. [67]. The design assembles are coded into a set of harmonies as solutions, and the set of design variables may be in a different part of the solution range or the optimum results may be close to existing solutions. This can be effective if the design solutions are in the neighborhood. This method has been applied in several engineering problems, including civil engineering applications [212].

6.2.5 Firefly Algorithm

Firefly algorithm (FA) is a swarm-intelligence based algorithm, developed by Yang [200]. This algorithm uses the characteristics of attraction mechanism and flashing of fireflies. Since the short-distance attraction is stronger than long-distance attraction, the swarm can automatically subdivide into multiple subswarms, thus the firefly algorithm is especially suitable for nonlinear multimodal optimization problems [201].

FA has been employed in the development of methodologies of civil engineering applications such as structural engineering problems [59], tower structures [180], steel slabs casting quality [138], truss structures [140], and path planning [130].

6.2.6 Cuckoo Search

Cuckoo search (CS) developed by Yang and Deb [205] is another population-based algorithm. It was inspired by the brood parasitism of some cuckoo species. The eggs and nests are coded as solutions, and thus the eggs with high quality in the best nest will be carried over the next generation. New solutions/eggs are generated using Lévy flights so as to explore the search space more effectively. In civil engineering, CS has been employed in the optimization of structural engineering problems [61], steel frames [90], and trusses [63].

6.2.7 Bat Algorithm

Bat algorithm developed by Yang [202] has been used in various civil engineering applications. The principle of the algorithm is to idealize the echolocation behavior of microbats with varying frequencies, loudness, and pulse emission rates. Such properties are used to update the equations of the bat algorithm so as to try to balance between exploration and exploitation. The majority of the civil engineering applications using bat algorithm are about the structural engineering [60, 206]. Structural engineering applications include optimization of steel plate shear walls [70], skeletal structures [102], trusses [182], and reinforced concrete beam [13].

6.2.8 Flower Pollination Algorithm

Flower Pollination Algorithm (FPA) is a new metaheuristic algorithm, developed by Yang [204], inspired by the characteristics of the pollination process of flowering plants. The pollination can be done in two ways. The pollen can be transferred by pollinators such as insects, birds, bats, other animals, or wind. Some flower types use self-pollination. The pollination done by the pollinators is called as cross-pollination and it can be considered as a global pollination process. The local pollination process is generated by the self-pollination. The combination of cross-pollination, self-pollination, flower constancy, and switch probability can be used to encode solutions and thus solve optimization problems effectively. The pseudocode of the algorithm is given in Fig. 6.1.

L obeys a Lévy distribution, while ϵ obeys a uniform distribution $U(0, 1)$. Recent studies show that the algorithm is suitable and effective on structural problems [150], including space trusses [14].

6.2.9 Other Metaheuristic Algorithms Used in Civil Engineering

There are many other algorithms that fall into the category of metaheuristics and they are also used in civil engineering applications. Due to the limited length of this book chapter, we will only briefly mention them. Ant Colony Optimization (ACO)

Objective minimize or maximize $f(x)$, $x = (x_1, x_2, \ldots, x_n)$ Initialize a population of n flowers with random solutions Find the best solution (g^*) of the initial population Define a switch probability (p) **while** $t <$ *Number of iterations* **do**

    for $i \leftarrow 1$ **to** n **do**   `/* n is the number of flowers */`

        if *rand* $< p$ **then**

            Global pollination using $x_i^{t+1} = x_i^t + L(x_i^t - g^*)$

        else

            Global pollination using $x_i^{t+1} = x_i^t + \varepsilon(x_i^t - g^*)$

        end if

        Evaluate new solutions Update the better solutions in the population

    end for

    Find the current best solution (g^*)

end while

Fig. 6.1 Pseudocode of the FPA

is a well-known metaheuristic algorithm developed by Dorigo [45], and ACO has been used in structural engineering [169]. The recent civil engineering applications include vehicle routing problem [162], traffic engineering [43], slope stability [65], and structural engineering [6, 169].

The big-bang and big crunch algorithm [49] was also used for structural engineering problems such as trusses [17, 79, 99, 100], steel frames [78], parameter estimation of structures [183], and retaining walls [18].

Electrostatic and Newtonian mechanic laws are the source of inspiration for the development of Charged System Search (CSS). The algorithm was developed by Kaveh and Talatahari [101], and the algorithm has been employed in damage detection in skeletal structures [93], optimization of castellated beams [96], structures [94, 105], and tuned mass dampers [106, 109].

Another algorithm used in structural engineering is the Krill Herd algorithm developed by Gandomi and Alavi [58]. Ray optimization inspired by the reflection of light [91] was employed in applications on structural [92] and transportation engineering [50].

6.3 Applications and Optimization in Civil Engineering

In this section, we highlight some of the most recent studies on optimization in the context of civil engineering applications. The emphasis of applications is on truss structures, frame structures, bridges, reinforced concrete members, construction management, tuned mass dampers, transportation engineering, hydraulics, infrastructures, geotechnical engineering, and other applications.

6.3.1 Truss Structures

Truss structures are widely used in practice for engineering designs. Truss systems form the framework of constructions like bridges, towers, roof supporting structures, etc. Truss structures have been widely investigated for generating and verifying the recently developed optimization algorithms in several types such as:

- Multi-objective optimization
- Discrete optimization
- Mixed-type optimization
- Hybrid optimization

Most of the optimization methods mentioned in the previous section can in general be used for sizing optimization (cross-sectional areas of the members are the design variables), shape optimization (nodal coordinates are the design variables), and topology optimization (the locations of links are the design variables).

In the documented methods, several approaches used metaheuristic based algorithms. Genetic Algorithm (GA) is frequently used for truss structures. For example, Rajeev and Krishnamoorthy [160] used discrete variables and GA with a penalty parameter depending on constraint violation. Koumousis and Georgiou [113] solved the mixed layout and sizing optimization problem of a typical steel roof using a genetic algorithm. Adeli and Kumar [2] used a distributed genetic algorithm for optimization of large structures on a cluster of workstations connected via a local area network (LAN). The nodal locations were treated as continuous design variables using a hybrid natural approach for shape optimal design by Rajan [159].

Coello and Christiansen [35] developed a GA-based multiobjective method using a min-max optimization approach. Erbatur et al. [48] employed GA in the optimum design of space and plane trusses. Krishnamoorthy et al. [114] used an object-oriented approach to implement a flexible genetic algorithm for space trusses. The article [77] reports and investigates the application of evolution strategies to optimize the design of truss bridges. The size, shape, and topology design variables defined in this process should be considered simultaneously for the most effective optimization procedure. The fuzzy formulation was combined with a genetic algorithm for solving fuzzy multi-objective truss optimization problems by Kelesoglu [107]. Šešok and Belevičius [173] developed a modified genetic algorithm for topology optimization of truss systems, where the repair of the genotype is used instead of some constraint. Furthermore, the improved GA was employed by Toğan and Daloğlu [187] for the optimization of truss structures using an initial population strategy and self-adaptive member grouping. For truss-like structures, an approach based on kinematic stability repair was used to improve the GA by Richardson et al. [163]. Li [124] developed a methodology by using improved species-conserving genetic algorithm.

Also, PSO has been employed in the sizing and layout optimization of truss structures [171]. Li et al. [126] developed a methodology based on the particle swarm optimizer with passive congregation and a HS scheme for optimum design of trusses. Also, Perez and Behdinan [155] employed PSO in optimization of truss structures. Kaveh and Talatahari [98] developed a discrete heuristic particle swarm ant colony optimization (DHPSACO) method combining a particle swarm optimizer with passive congregation (PSOPC), ant colony optimization (ACO), and harmony search scheme (HS) for the design of truss structures with discrete variables.

Ant Colony Optimization is another metaheuristic algorithm used in optimization of truss structures [19]. Teaching Learning Based optimization is an education inspired algorithm and it has been employed in the optimum truss design [20, 38, 39]. Artificial Bee Colony combined with adaptive penalty function (ABC-AP) was used to minimize the weight of truss structures [177]. In addition, metaheuristic algorithms such as Firefly algorithm [140], Cuckoo Search [61], Bat algorithm [182] and Big Bang Big Crunch (BB-BC) algorithm [17, 79, 99, 100] were also employed to find the best design of truss structures.

Sadollah et al. [177] used the so-called Mine Blast Algorithm (MBA) for solving truss structure optimization problems with many design variables and constraints. Gandomi et al. [62] investigated the optimum design of trusses by Cuckoo Search Algorithm. Kaveh et al. [104] combined swarm intelligence and chaos theory for optimal design of truss structures. Bekdaş et al. [14] minimized the weight of 3D and 2D truss structures using the Flower Pollination Algorithm. Ho-Huu et al. [80] integrated a sequential optimization and reliability assessment (SORA) with the improved constrained differential evolution algorithm (ICDE) for solving reliability-based design optimization (RBDO) problems of truss structures.

In addition, mathematical optimization techniques have been developed for truss structures. Ben-Tal and Nemirovski [15] developed an approach, where a solution to the stabilized problem should be feasible for all allowed data, using semidefinite programming for truss structures and they demonstrated that the approach can be extended to other mathematical programming problems. Achtziger [1] optimized cross-sectional areas and the positions of joints simultaneously by methods of mathematical programming. Two types of structural optimization formulations have been developed to handle system uncertainties. These formulations are reliability based design optimization (RBDO) and robust design optimization (RDO) [85].

Torii et al. [190] presented an approach for design complexity control in truss optimization. The complexity measures are non-convex and for that reason, a global gradient based optimization algorithm is generated. Two design complexity measures were built and these were continuous differentiable approximations for the total number of nodes and bars. Afterwards, these measures were included in the optimization problem with a penalized objective function.

Furthermore, other optimization techniques have also been used in the analyses of truss structures. Toklu et al. [188] proposed a method using HS algorithm to obtain minimum potential energy of truss structural systems. Geometrically nonlinear analysis of trusses using Particle Swarm Optimization was investigated by Temür et al. [185].

6.3.2   Reinforced Concrete Members

Design of cross-sections and reinforcements of reinforced concrete (RC) members are traditionally determined by the knowledge of engineers. In the process, the variable requirements are checked with the requirements defined in several design codes accepted by local rules. Optimization of cross-sections and reinforcements of RC members may be more economical in construction. For that reason, several approaches have been proposed. Generally, metaheuristic algorithm based approaches were used. Genetic Algorithm (GA) is a widely employed algorithm in the optimization of RC members such as beams [54, 72] and columns [21]. The hybrid form of two metaheuristic algorithms such as GA and Simulated Annealing (SA) was applied in the optimum design methodology for RC continuous beams in [120]. In addition, a hybrid optimization algorithm combining GA and discretized form of the Hook and Jeeves method was proposed to design of RC flat slab buildings by Sahab et al. [168].

Harmony Search Algorithm was also employed in the optimum design approaches for continuous beams [4], T-shaped RC beams [10], and RC columns [12]. Reference [149] developed a novel HS based optimization process for cost optimization of RC biaxially loaded columns. A random search technique was proposed for the detailed design of RC continuous beams [148].

Cost optimization of RC retaining wall was applied to obtain optimal shape, structural stability, bending moment minimization, and optimum location. Ceranic et al. [25] used a simulated annealing algorithm for the minimum cost design of reinforced concrete retaining wall together with material and labor costs associated with concreting, reinforcing, and form working. Castillo et al. [24] developed a method that simultaneously considered safety factors and probabilities of failure. This method was also used for minimum cost optimization for RC retaining walls. Yepes et al. [210] used a simulated annealing algorithm for optimum design of variables of the earth-retaining walls.

Ahmadi-Nedushan and Varaee [3] used PSO for optimal design of reinforced concrete earth-retaining walls. The formulation of the problem includes important design variables such as the geometrical ones; describing the thickness of the kerb and the footing, as well as the toe and the heel lengths and the variables describing the reinforcement set-up. Yepes et al. [211] aimed to minimize the embedded carbon dioxide (CO_2) emissions and the minimization of economic cost. For the hybrid optimization method based on a variable neighborhood search, threshold acceptance strategy is used for the design of reinforced concrete cantilever retaining walls. Kaveh and Abadi [89] utilized Harmony Search Algorithms for optimum design of the RC retaining wall. Smith and Cubrinovski [176] developed a limit analysis method: discontinuity layout optimization (DLO) for the solution of retaining wall problems involving earthquake loading. Yang [203] used analytical solutions for seismic passive pressures under earthquake loads with the nonlinear failure criterion. Ghazavi and Salavati [69] used the Bacterial Foraging Optimization Algorithm (BFOA) for an economic optimization and sensitivity analysis of reinforced concrete cantilever (RCC) retaining walls.

Kaveh et al. [103] used a multi-objective genetic algorithm for optimization of geometry, concrete grades, and reinforcement. Kaveh and Soleimani [97] utilized colliding bodies optimization and democratic particle swarm optimization in the optimum design of RC retaining walls. Temur and Bekdaş [184] employed teaching learning based optimization in the optimum dimension and reinforcement design of RC retaining walls. Also, BB-BC [18] and charged system search [179] have been employed for the optimization of RC retaining walls.

6.3.3   Frame Structures

Another major application in civil engineering is the optimum design of frame structures. These frames can be constructed from steel or reinforced concrete members. The major contributions are as follows.

Mijar et al. [141] used a continuum structural topology optimization formulation with Voigt and Reuss rules to obtain optimal bracing concept designs under prescribed loading (lateral-wind and seismic-type loading), and boundary restraints. Pezeshk et al. [156] used genetic algorithm for the design of 2D, geometrical, nonlinear steel-framed structures. Kameshki and Saka [84] used Genetic Algorithm (GA) for optimum design of nonlinear steel frames with semi-rigid connection.

On the other hand, Li et al. [125] address the design optimization of structures involving a unilateral contact constraint using non-gradient approach, namely evolutionary structural optimization (ESO) method, and they investigated also the difference of optimal frame designs between these two element types. Liu and Ye [128] investigated life-cycle cost for multi-objective design optimization of seismic steel moment-resisting frame (SMRF) structures. Camp et al. [23] solved the optimization problem of steel frame using an ACO method with discrete optimization. Saka [170] demonstrated that the stochastic search techniques are more powerful than the mathematical methods for solving discrete design problems of steel frame.

Perea et al. [154] compared some heuristic methods (random walk and the descent local search) and metaheuristic methods (threshold accepting and the simulated annealing) for the economic optimization of reinforced concrete bridge frames used in road construction. Furthermore, several metaheuristic based methods have been developed for optimum design of reinforced concrete frames for cost optimization [5, 22, 73, 95, 118, 152, 161], and CO_2 emission minimization [21, 153].

Fesanghary et al. [55] combined the sequential quadratic programming and the harmony search algorithm. The developed hybrid algorithm was used for a welded beam design, a 10-bar plane truss and a four-story, two-bay steel frame. Toğan [186] employed a Teaching Learning Based Optimization (TLBO) technique for discrete optimization of planar steel frames design. Kociecki and Adeli [111] used two-phase genetic algorithm for simultaneous sizing and topology optimization of free-form steel space frame roof structures consisting of discrete commercially available rectangular hollow structural sections.

Talatahari et al. [181] developed a method using eagle strategy (ES) with differential evolution. This method was applied to weight minimization problems of steel frames with discrete variables. Kanno [86] developed a mixed-integer second-order cone programming formulation for topology optimization of a planar frame structure with discrete design variables, demonstrating that the formulation can be applied for spatial frames. A gradient-based structural optimization method was improved for tailoring the hardening/softening behavior of nonlinear mechanical systems by Dou and Jensen [46], and this method was applied to plane frames. Aydoğdu et al. [7] developed a novel algorithm for optimum design of real size steel space frames by using artificial bee colony algorithm improved by adding Lévy flight distribution in the search of scout bees. Optimum W-sections were selected and structural weight was minimized by using this algorithm.

6.3.4 Bridges

One of the major construction types is bridges, and the optimization of bridges has been considered in the documented methods. Frangopol et al. [56] developed a method for reliability-based life-cycle cost design of deteriorating concrete structures and the method was used for reinforced concrete T-girders of a highway bridge. Kong and Frangopol [112] developed a method based on reliability index profile superstation method using a computer program for analyzing life-cycle performance of deteriorating bridge structures based on reliability. Robelin and Madanat [164] formulated a realistic history-dependent model of bridge deck deterioration as a Markovian model to determine optimal maintenance and replacement policies for one facility. Lee et al. [119] used a two-step approach based on the unit load method to obtain the optimal tensioning strategy for an atypical asymmetric cable-stayed bridge.

Metaheuristic based methods have also been employed for optimum design of bridges. Sgambi et al. [174] investigated the dependability assurance in the design of suspension bridges with long-span by using GA. An Ant Colony Optimization (ACO) based method developed by Martínez et al. [135] was compared with Genetic Algorithm (GA) and threshold acceptance algorithm for economic optimization of reinforced concrete (RC) bridge piers with hollow rectangular sections. This method was used for the design and analysis of Reinforced Concrete tall road piers of 90 m in height with hollow rectangular sections [136]. Also, Martinez-Martin et al. [137] developed a method using SA and GA for the design of reinforced concrete bridge piers. The aim of that study was cost minimization, the reinforcing steel congestion, and the embedded CO_2 emissions.

Dong et al. [44] developed a probabilistic method for pre-earthquake retrofit optimization of bridge networks to mitigate seismic damage to society, economy, and environment. Three reliability-based design optimization (RBDO) (Reliability Index Approach methods, Performance Measure Approach and Sequential Optimization and Reliability Assessment) were performed on Messina Bridge to obtain the minimum volumes of the main cables and bridge girder by Kusano et al. [117]. Saad et al. [165] used the reliability-based design optimization (RBDO) method to improve the life-cycle cost formulation for better design of concrete bridge structures.

6.3.5 Tuned Mass Damper

A tuned mass damper (TMD) is a device which is used in all types of mechanical systems in order to reduce vibrations. In civil engineering, TMDs are also used in the reduction of structural vibrations of bridges, tower structures, and high buildings. These structures may be subject to unstable vibrations resulting from earthquakes, strong winds, and traffic. The optimization problem of TMDs covers the optimum parameter design of TMDs such as mass, stiffness, and damping. For this purpose, several simple expressions were proposed, but these expressions are not effective on considering multiple vibration modes, excitations with random frequency, inherent damping, stroke limitation of TMD, and user-defined variable ranges [76, 167, 194]. In that case, many metaheuristic methods have been employed for the problem of civil engineering structures. The proposed methods are based on GA [41, 74, 134, 157, 175], PSO [121, 122], bionic optimization [178], HS [9, 11, 146], ACO [52] artificial bee optimization [51], shuffled complex evolution [53], and CSS [106].

In addition, the optimum design of TMDs was done in order to prevent the pounding of adjacent structures [147]. Lievens et al. [127] used a robust optimization approach for the design of a TMD to minimize the mass for a footbridge.

6.3.6 Construction Management

Time and cost are the two most important factors to be considered in every construction project. Construction planners must select appropriate resources, including the team size, equipment, methods, and technologies to perform the tasks of a construction project. The following optimization methods have been proposed.

Liu et al. [129] developed the LP/IP hybrid method by combining linear and integer programming for assisting construction planners in making time-cost trade-off decisions. Chan et al. [28] used GA with time constraint for resource scheduling in construction projects. El-Rayes and Moselhi [47] developed an automated model that utilizes dynamic programming formulation and incorporates a scheduling algorithm, and an interruption algorithm, for optimizing resource utilization for repetitive construction projects. Senouci and Eldin [172] minimized the total construction cost with resource scheduling using genetic algorithm. Zheng et al.

[216] employed GA-based multi-objective approach for optimum total time and total cost simultaneously. Yang [199] used elitist PSO algorithm to solve bi-criterion time-cost trade-off analysis. Xiong and Kuang [198] employed ACO to solve time–cost trade-off problems. Geem [66] employed Harmony Search (HS) algorithm to minimize both project cost and time.

Zhang and Thomas [214] minimized the project duration and cost concurrently by using a model based on the ant colony system techniques. In dynamic multi-cloud scenarios, a method for cost optimization of virtual infrastructure was proposed by Lucas-Simarro et al. [131]. Tran et al. [191] developed a hybrid multiple objective evolutionary algorithm based on hybridization of artificial bee colony and differential evolution to solve time–cost–quality trade-off problem, and Tran et al. [192] simultaneously minimized the project duration, cost, and the utilization of evening and night work shifts by using objective symbiotic organisms search optimization algorithm.

6.3.7 Hydraulics and Infrastructures

Hydraulics is a major civil engineering subject in which optimization is needed. Up to date, several approaches have been proposed. Wurbs [197] used a water allocation module (WAM) for a hypothetical water management scenario, utilizing moisture dependent irrigation. Lund and Ferreira [132] used deterministic optimization to develop strategic operating rules of large-scale water resource systems.

Geem et al. [68] solved standard nonlinear pipe network test problems for optimum solution by employing harmony search algorithm. Maier et al. [133] solved water distribution system optimization problems using ant colony optimization algorithms. Mohammadi et al. [142] calculated Auto Regressive Moving Average (ARMA) model coefficients by goal programming for river flow forecasting. Hajebi et al. [75] used Water Distribution Network Cluster technique to solve this partitioning problem for gravity driven water distribution networks. Sabzkouhi and Haghighi [166] developed a many-objective particle swarm optimization (MO-PSO) model to find extreme values of nodal pressures and pipe velocities efficiently.

PSO was also employed in water distribution system problems [143]. Mugisha [144] singled out effective infrastructure optimization through high-impact change management plans and incorporating strong water loss management strategies. Furthermore, the study includes the investigation of the modes of infrastructure performance monitoring. Middleton and Brandt [139] presented an integrated framework which simultaneously considers economic and engineering decisions, and the developed model optimizes CO_2 management infrastructure at a variety of carbon prices for the oil sands industry.

6.3.8 Transportation Engineering

Transportation is an important application of optimization techniques. Desrochers et al. [40] solved the vehicle routing problem with time windows (VRPTW) by column generations. Bookbinder and Désilets [16] generated a model for optimization of transfers in a transit network and used the model to study the optimization of transfers under random bus travel times and no-hold policies according to various objectives. Chakroborty et al. [27] used genetic algorithms (GAs) with transfer time considerations to optimize transit system schedules. Fwa et al. [57] developed a method based on genetic algorithm to solve multi-objective network level pavement maintenance programming problems.

GA has also been used for urban traffic flow optimization [42], traffic signal coordination problem [158], emergency logistic scheduling [29], and calibration of rail transit assignment models [217]. Chien and Qin [34] developed an analytical model to minimize total cost (i.e., the sum of supplier and user costs) by optimizing the number and locations of stops subject to non-additive users' value of time.

Wiering et al. [195] used adaptive reinforcement learning algorithms for learning to control traffic lights. This algorithm considered the expected waiting times of cars for red and green lights at each intersection, and sets the traffic lights to green for the configuration maximizing individual car gains. Zhao and Zeng [215] developed a mathematical stochastic methodology based on an integrated simulated annealing and genetic algorithm to minimize transfers with reasonable route directness while maximizing service coverage. Kuan et al. [116] used genetic algorithms and ant colony optimization for solving the feeder bus network design problem with minimum operator and user costs. Ant colony optimization was also employed in transportation problems such as the vehicle routing problem [162] and traffic engineering problem [43].

In addition, SA was employed for planning high-speed rail systems by Costa et al. [36]. Kroon et al. [115] generated a stochastic optimization model to minimize the average delay of the train. D'Ariano et al. [37] developed a branch and bound algorithm for the real-time management of a complex rail network. Jin et al. [82] developed a discrete method by particle swarm optimization (PSO) for transmission network expansion planning (TNEP). Yang et al. [207] optimized total energy consumption and the total traversing time using a mathematical model to find optimal train movements under the consideration of operational interactions. Yang et al. [208] used a fuzzy optimization framework to reschedule trains in a double-track railway network when the capacity reduction is caused by a low-probability incident. Nikolić et al. [151] minimized total waiting time, taking care simultaneously of the number of passengers served using based on the bee colony optimization (BCO) technique. Woo and Yeo [196] optimized a flexible inspection schedule by minimizing the life-cycle cost of pavement management systems. Walraven et al. [193] optimized traffic flow using a method based on reinforcement learning. Yu et al. [213] used a method based on a fuzzy programming approach to optimize the signal timing for isolated intersection.

6.3.9 Geotechnics

In geotechnical engineering, the slope stability problem is a major engineering practice. For the problem, several metaheuristic methods such as PSO [64], ACO [65], HS [33], and SA [30] have been employed. Also, Cheng et al. [32] compared six major types of heuristic methods: SA, GA, PSO, simple HS, modified HS, and Tabu search. Also, optimization was used in the following geotechnical engineering practices.

Cheng and Li [31] developed a methodology using a multi-objective genetic algorithm for the robust geotechnical design (RGD) of braced excavations in clayey soils. Basha and Bubu [8] integrated a reliability method and a deterministic design procedure to determine the penetration depth, anchor pull, and section modulus for optimum design of an anchored cantilever sheet pile wall. In the theoretical study on pile length optimization of pile groups and piled rafts, Leung et al. [123] demonstrated that the overall foundation behavior can be enhanced by varying pile lengths across the pile group or piled raft. Nicholson [145] presented optimization concepts and capabilities for design of wind turbine tower and foundation systems.

6.4 Conclusions

This chapter has focused on the review of latest developments concerning optimization and designs in civil engineering. We have briefly outlined some recent metaheuristic algorithms, followed by the detailed literature of various applications, ranging from structures and concrete members to bridges and transportation. The diversity of such applications is higher, which shows the flexibility and effectiveness of metaheuristic algorithms.

Such diversity and effectiveness also necessitate further understanding of metaheuristic algorithms. Therefore, further research can focus on theoretical analysis, parametric studies, and parameter tuning as well as more real-world applications.

Acknowledgements The authors acknowledge their universities for the support and also would like to thank the reviewers for their detailed comments.

References

1. Achtziger, W.: On simultaneous optimization of truss geometry and topology. Struct. Multidiscip. Optim. **33**(4), 285–304 (2007). https://doi.org/0.1007/s00158-006-0092-0
2. Adeli, H., Kumar, S.: Distributed genetic algorithm for structural optimization. J. Aerosp. Eng. **8**(3), 156–163 (1995). https://doi.org/10.1061/(ASCE)0893-1321(1995)8:3(156)
3. Ahmadi-Nedushan, B., Varaee, H.: Optimal design of reinforced concrete retaining walls using a swarm intelligence technique. In: Topping, B.H.V., Tsompanakis, Y. (eds.) Proceedings of the 1st International Conference on Soft Computing Technology in Civil, Structural

and Environmental Engineering, Stirlingshire, pp. 1–12 (2009). https://doi.org/10.4203/ccp. 92.26

4. Akin, A., Saka, M.P.: Optimum detailed design of reinforced concrete continuous beams using the harmony search algorithm. In: Topping, B.H.V., Adam, J.M., Pallarés, F.J., Bru, R., Romero, M.L. (eds.) Proceedings of the Tenth International Conference on Computational Structures Technology, pp. 1–20. Civil-Comp Press, Stirlingshire (2010). https://doi.org/10. 4203/ccp.93.131

5. Akin, A., Saka, M.P.: Harmony search algorithm based optimum detailed design of reinforced concrete plane frames subject to ACI 318-05 provisions. Comput. Struct. **147**, 79–95 (2015). https://doi.org/10.1016/j.compstruc.2014.10.003

6. Angelo, J.S., Bernardino, H.S., Barbosa, H.J.C.: Ant colony approaches for multiobjective structural optimization problems with a cardinality constraint. Adv. Eng. Softw. **80**, 101–115 (2015). https://doi.org/10.1016/j.advengsoft.2014.09.015

7. Aydoğdu, İ., Akın, A., Saka, M.P.: Design optimization of real world steel space frames using artificial bee colony algorithm with Lévy flight distribution. Adv. Eng. Softw. **92**, 1–14 (2016). https://doi.org/10.1016/j.advengsoft.2015.10.013

8. Basha, B.M., Babu, G.L.S.: Target reliability based design optimization of anchored cantilever sheet pile walls. Can. Geotech. J. **45**(4), 535–548 (2008). https://doi.org/10.1139/T08-004

9. Bekdaş, G., Nigdeli, S.M.: Estimating optimum parameters of tuned mass dampers using harmony search. Eng. Struct. **33**(9), 2716–2723 (2011). https://doi.org/10.1016/j.engstruct. 2011.05.024

10. Bekdaş, G., Nigdeli, S.M.: Cost optimization of T-shaped reinforced concrete beams under flexural effect according to ACI 318. In: Dmitriev, A., Camarinhas, C.L. (eds.) Proceedings of the 3rd European Conference of Civil Engineering (ECCIE'12), pp. 122–126. WSEAS Press, Paris (2012)

11. Bekdaş, G., Nigdeli, S.M.: Optimization of tuned mass damper with harmony search. In: Gandomi, A.H., Yang, X.S., Talatahari, S., Alavi, A.H. (eds.) Metaheuristic Applications in Structures and Infrastructures, pp. 345–371. Elsevier, Oxford (2013). https://doi.org/10.1016/ B978-0-12-398364-0.00014-0

12. Bekdaş, G., Nigdeli, S.M.: The optimization of slender reinforced concrete columns. In: Steinmann, P., Leugering, G. (eds.) 85th Annual Meeting of the International Association of Applied Mathematics and Mechanics, Erlangen, vol. 14, pp. 183–184 (2014). https://doi.org/ 10.1002/pamm.201410079

13. Bekdaş, G., Nigdeli, S.M., Yang, X.S.: Metaheuristic optimization for the design of reinforced concrete beams under flexure moments. In: Proceedings of the 5th European Conference of Civil Engineering (ECCIE'14), pp. 184–188 (2014)

14. Bekdaş, G., Nigdeli, S.M., Yang, X.S.: Sizing optimization of truss structures using flower pollination algorithm. Appl. Soft Comput. **37**, 322–331 (2015). https://doi.org/10.1016/j. asoc.2015.08.037

15. Ben-Tal, A., Nemirovski, A.: Robust truss topology design via semidefinite programming. SIAM J. Optim. **7**(4), 991–1016 (1997). https://doi.org/10.1137/S1052623495291951

16. Bookbinder, J.H., Désilets, A.: Transfer optimization in a transit network. Transport. Sci. **26**(2), 106–118 (1992). https://doi.org/10.1287/trsc.26.2.106

17. Camp, C.V.: Design of space trusses using big bang-big crunch optimization. J. Struct. Eng. **133**(7), 999–1008 (2007). https://doi.org/10.1061/(ASCE)0733-9445(2007)133:7(999)

18. Camp, C.V., Akin, A.: Design of retaining walls using big bang-big crunch optimization. J. Struct. Eng. **138**(3), 438–448 (2012). https://doi.org/10.1061/(ASCE)ST.1943-541X. 0000461

19. Camp, C.V., Bichon, B.J.: Design of space trusses using ant colony optimization. J. Struct. Eng. **130**(5), 741–751 (2004). https://doi.org/10.1061/(ASCE)0733-9445(2004)130:5(741)

20. Camp, C.V., Farshchin, M.: Design of space trusses using modified teaching-learning based optimization. Eng. Struct. **62–63**, 87–97 (2014). https://doi.org/10.1016/j.engstruct.2014.01. 020

21. Camp, C.V., Huq, F.: CO_2 and cost optimization of reinforced concrete frames using a big bang-big crunch algorithm. Eng. Struct. **48**, 363–372 (2013). https://doi.org/10.1016/j.engstruct.2012.09.004
22. Camp, C.V., Pezeshk, S., Hansson, H.: Flexural design of reinforced concrete frames using a genetic algorithm. J. Struct. Eng. **129**(1), 105–115 (2003). https://doi.org/10.1061/(ASCE)0733-9445(2003)129:1(105)
23. Camp, C.V., Bichon, B.J., Stovall, S.P.: Design of steel frames using ant colony optimization. J. Struct. Eng. **131**(3), 369–379 (2005). https://doi.org/10.1061/(ASCE)0733-9445(2005)131:3(369)
24. Castillo, E., Mínguez, R., Terán, A.R., Fernández-Canteli, A.: Design and sensitivity analysis using the probability-safety-factor method. An application to retaining walls. Struct. Saf. **26**(2), 159–179 (2004). https://doi.org/10.1016/S0167-4730(03)00039-0
25. Ceranic, B., Fryer, C., Baines, R.W.: An application of simulated annealing to the optimum design of reinforced concrete retaining structures. Comput. Struct. **79**(17), 1569–1581 (2001). https://doi.org/10.1016/S0045-7949(01)00037-2
26. Černý, V.: Thermodynamical approach to the traveling salesman problem: an efficient simulation algorithm. J. Optim. Theory Appl. **45**(1), 41–51 (1985). https://doi.org/10.1007/BF00940812
27. Chakroborty, P., Deb, K., Subrahmanyam, P.S.: Optimal scheduling of urban transit systems using genetic algorithms. J. Transport. Eng. **121**(6), 544–553 (1995). https://doi.org/10.1061/(ASCE)0733-947X(1995)121:6(544)
28. Chan, W.T., Chua, D.K.H., Kannan, G.: Construction resource scheduling with genetic algorithms. J. Constr. Eng. Manag. **122**(2), 125–132 (1996). https://doi.org/10.1061/(ASCE)0733-9364(1996)122:2(125)
29. Chang, F.S., Wu, J.S., Lee, C.N., Shen, H.C.: Greedy-search-based Multi-objective genetic algorithm for emergency logistics scheduling. Expert Syst. Appl. **41**(6), 2947–2956 (2014). https://doi.org/10.1016/j.eswa.2013.10.026
30. Cheng, Y.M.: Global optimization analysis of slope stability by simulated annealing with dynamic bounds and dirac function. Eng. Optim. **39**(1), 17–32 (2007). https://doi.org/10.1080/03052150600916294
31. Cheng, F.Y., Li, D.: Multiobjective optimization design with Pareto genetic algorithm. J. Struct. Eng. **123**(9), 1252–1261 (1997). https://doi.org/10.1061/(ASCE)0733-9445(1997)123:9(1252)
32. Cheng, Y.M., Li, L., Chi, S.C.: Performance studies on six heuristic global optimization methods in the location of critical slip surface. Comput. Geotech. **34**(6), 462–484 (2007). https://doi.org/10.1016/j.compgeo.2007.01.004
33. Cheng, Y.M., Li, L., Lansivaara, T., Chi, S.C., Sun, Y.J.: An improved harmony search minimization algorithm using different slip surface generation methods for slope stability analysis. Eng. Optim. **40**(2), 95–115 (2008). https://doi.org/10.1080/03052150701618153
34. Chien, S.I., Qin, Z.: Optimization of bus stop locations for improving transit accessibility. Transp. Plan. Technol. **27**(3), 211–227 (2004). https://doi.org/10.1080/0308106042000226899
35. Coello, C.A., Christiansen, A.D.: Multiobjective optimization of trusses using genetic algorithms. Comput. Struct. **75**(6), 647–660 (2000). https://doi.org/10.1016/S0045-7949(99)00110-8
36. Costa, A.L., da Conceição Cunha, M., Coelho, P.A.L.F., Einstein, H.H.: Solving high-speed rail planning with the simulated annealing algorithm. J. Transport. Eng. **139**(6), 635–642 (2013). https://doi.org/10.1061/(ASCE)TE.1943-5436.0000542
37. D'Ariano, A., Pacciarelli, D., Pranzo, M.: A branch and bound algorithm for scheduling trains in a railway network. Eur. J. Oper. Res. **183**(2), 643–657 (2007). https://doi.org/10.1016/j.ejor.2006.10.034
38. Dede, T., Ayvaz, Y.: Combined size and shape optimization of structures with a new metaheuristic algorithm. Appl. Soft Comput. **28**, 250–258 (2015). https://doi.org/10.1016/j.asoc.2014.12.007

39. Degertekin, S.O., Hayalioglu, M.S.: Sizing truss structures using teaching-learning-based optimization. Comput. Struct. **119**, 177–188 (2013). https://doi.org/10.1016/j.compstruc.2012.12.011
40. Desrochers, M., Desrosiers, J., Solomon, M.: A new optimization algorithm for the vehicle routing problem with time windows. Oper. Res. **40**(2), 342–354 (1992). https://doi.org/10.1287/opre.40.2.342
41. Desu, N.B., Deb, S.K., Dutta, A.: Coupled tuned mass dampers for control of coupled vibrations in asymmetric buildings. Struct. Control Health Monit. **13**(5), 897–916 (2006). https://doi.org/10.1002/stc.64
42. Dezani, H., Bassi, R.D.S., Marranghello, N., Gomes, L., Damiani, F., da Silva, I.N.: Optimizing urban traffic flow using genetic algorithm with Petri net analysis as fitness function. Neurocomputing **124**, 162–167 (2014). https://doi.org/10.1016/j.neucom.2013.07.015
43. Dias, J.C., Machado, P., Silva, D.C., Abreu, P.H.: An inverted ant colony optimization approach to traffic. Eng. Appl. Artif. Intell. **36**, 122–133 (2014). https://doi.org/10.1016/j.engappai.2014.07.005
44. Dong, Y., Frangopol, D.M., Saydam, D.: Pre-earthquake multi-objective probabilistic retrofit optimization of bridge networks based on sustainability. J. Bridge Eng. **19**(6), 04014, 018 (2014). https://doi.org/10.1061/(ASCE)BE.1943-5592.0000586
45. Dorigo, M.: Optimization, learning and natural algorithms. Ph.D. thesis, Politecnico di Milano, Milan (1992)
46. Dou, S., Jensen, J.S.: Optimization of hardening/softening behavior of plane frame structures using nonlinear normal modes. Comput. Struct. **164**, 63–74 (2016). https://doi.org/10.1016/j.compstruc.2015.11.001
47. El-Rayes, K., Moselhi, O.: Optimizing resource utilization for repetitive construction projects. J. Constr. Eng. Manag. **127**(1), 18–27 (2001). https://doi.org/10.1061/(ASCE)0733-9364(2001)127:1(18)
48. Erbatur, F., Hasançebi, O., Tütüncü, İ., Kılıç, H.: Optimal design of planar and space structures with genetic algorithms. Comput. Struct. **75**(2), 209–224 (2000). https://doi.org/10.1016/S0045-7949(99)00084-X
49. Erol, O.K., Eksin, I.: A new optimization method: big bang–big crunch. Adv. Eng. Softw. **37**(2), 106–111 (2006). https://doi.org/10.1016/j.advengsoft.2005.04.005
50. Esmaeili, M., Zakeri, J.A., Kaveh, A., Bakhtiary, A., Khayatazad, M.: Designing granular layers for railway tracks using ray optimization algorithm. Sci. Iran. Trans. A **22**(1), 47–58 (2015)
51. Farshidianfar, A., Soheili, S.: ABC optimization of TMD parameters for tall buildings with soil structure interaction. Interact. Multiscale Mech. **6**, 339–356 (2013). https://doi.org/10.12989/imm.2013.6.4.339
52. Farshidianfar, A., Soheili, S.: Ant colony optimization of tuned mass dampers for earthquake oscillations of high-rise structures including soil–structure interaction. Soil Dyn. Earthq. Eng. **51**, 14–22 (2013). https://doi.org/10.1016/j.soildyn.2013.04.002
53. Farshidianfar, A., Soheili, S.: Optimization of TMD parameters for earthquake vibrations of tall buildings including soil structure interaction. Int. J. Optim. Civil Eng. **3**, 409–429 (2013)
54. Fedghouche, F., Tiliouine, B.: Minimum cost design of reinforced concrete T-beams at ultimate loads using Eurocode2. Eng. Struct. **42**, 43–50 (2012). https://doi.org/10.1016/j.engstruct.2012.04.008
55. Fesanghary, M., Mahdavi, M., Minary-Jolandan, M., Alizadeh, Y.: Hybridizing harmony search algorithm with sequential quadratic programming for engineering optimization problems. Comput. Methods Appl. Mech. Eng. **197**(33), 3080–3091 (2008). https://doi.org/10.1016/j.cma.2008.02.006
56. Frangopol, D.M., Lin, K.Y., Estes, A.C.: Life-cycle cost design of deteriorating structures. J. Struct. Eng. **123**(10), 1390–1401 (1997). https://doi.org/10.1061/(ASCE)0733-9445(1997)123:10(1390)

57. Fwa, T.F., Chan, W.T., Hoque, K.Z.: Multiobjective optimization for pavement mainte-nance programming. J. Transport. Eng. **126**(5), 367–374 (2000). https://doi.org/10.1061/(ASCE)0733-947X(2000)126:5(367)
58. Gandomi, A.H., Alavi, A.H.: Krill Herd: a new bio-inspired optimization algorithm. Com-mun. Nonlinear Sci. Numer. Simul. **17**(12), 4831–4845 (2012). https://doi.org/10.1016/j.cnsns.2012.05.010
59. Gandomi, A.H., Yang, X.S., Alavi, A.H.: Mixed variable structural optimization using firefly algorithm. Comput. Struct. **89**(23), 2325–2336 (2011). https://doi.org/10.1016/j.compstruc.2011.08.002
60. Gandomi, A.H., Yang, X.S., Alavi, A.H., Talatahari, S.: Bat algorithm for constrained optimization tasks. Neural Comput. Appl. **22**(6), 1239–1255 (2013). https://doi.org/10.1007/s00521-012-1028-9
61. Gandomi, A.H., Yang, X.S., Alavi, A.H.: Cuckoo search algorithm: a metaheuristic approach to solve structural optimization problems. Eng. Comput. **29**(1), 17–35 (2013). https://doi.org/10.1007/s00366-011-0241-y
62. Gandomi, A.H., Talatahari, S., Yang, X., Deb, S.: Design optimization of truss structures using cuckoo search algorithm. Struct. Design Tall Spec. Build. **22**(17), 1330–1349 (2013). https://doi.org/10.1002/tal.1033
63. Gandomi, A.H., Yang, X.S., Talatahari, S., Alavi, A.H.: Metaheuristic Applications in Structures and Infrastructures. Elsevier, London (2013)
64. Gandomi, A.H., Kashani, A.R., Mousavi, M., Jalalvandi, M.: Slope stability analyzing using recent swarm intelligence techniques. Int. J. Numer. Anal. Methods Geomech. **39**(3), 295–309 (2015). https://doi.org/10.1002/nag.2308
65. Gao, W.: Determination of the noncircular critical slip surface in slope stability analysis by meeting ant colony optimization. J. Comput. Civil Eng. **30**(2), 06015001–1 (2016). https://doi.org/10.1061/(ASCE)CP.1943-5487.0000475
66. Geem, Z.W.: Multiobjective optimization of time-cost trade-off using harmony search. J. Constr. Eng. Manag. **136**(6), 711–716 (2010). https://doi.org/10.1061/(ASCE)CO.1943-7862.0000167
67. Geem, Z.W., Kim, J.H., Loganathan, G.V.: A new heuristic optimization algorithm: harmony search. Simulation **76**(2), 60–68 (2001). https://doi.org/10.1177/003754970107600201
68. Geem, Z.W., Kim, J.H., Loganathan, G.V.: Harmony search optimization: application to pipe network design. Int. J. Model. Simul. **22**(2), 125–133 (2002). https://doi.org/10.1080/02286203.2002.11442233
69. Ghazavi, M., Salavati, V.: Sensitivity analysis and design of reinforced concrete cantilever retaining walls using bacterial foraging optimization algorithm. In: Vogt, N., Schuppener, B., Straub, D., Bräu, G. (eds.) Proceedings of the 3rd International Symposium on Geotechnical Safety and Risk (ISGSR), München, pp. 307–314 (2011)
70. Gholizadeh, S., Shahrezaei, A.M.: Optimal placement of steel plate shear walls for steel frames by bat algorithm. Struct. Design Tall Spec. Build. **24**(1), 1–18 (2015). https://doi.org/10.1002/tal.1151
71. Goldberg, D.E., Samtani, M.P.: Engineering optimization via genetic algorithm. In: Proceed-ings of Ninth Conference on Electronic Computation, pp. 471–482. ASCE, New York (1986)
72. Govindaraj, V., Ramasamy, J.V.: Optimum detailed design of reinforced concrete continuous beams using genetic algorithms. Comput. Struct. **84**(1), 34–48 (2005). https://doi.org/10.1016/j.compstruc.2005.09.001
73. Govindaraj, V., Ramasamy, J.V.: Optimum detailed design of reinforced concrete frames using genetic algorithms. Eng. Optim. **39**(4), 471–494 (2007). https://doi.org/10.1080/03052150601180767
74. Hadi, M.N.S., Arfiadi, Y.: Optimum design of absorber for MDOF structures. J. Struct. Eng. **124**(11), 1272–1280 (1998). https://doi.org/10.1061/(ASCE)0733-9445(1998)124:11(1272)
75. Hajebi, S., Temate, S., Barrett, S., Clarke, A., Clarke, S.: Water distribution network sectorisation using structural graph partitioning and multi-objective optimization. Proc. Eng. **89**, 1144–1151 (2014). https://doi.org/10.1016/j.proeng.2014.11.238. 16th Water Distribution System Analysis Conference, WDSA2014

76. Hartog, J.P.D.: Mechanical Vibrations, 4th edn. McGraw-Hill, New York (1956)
77. Hasançebi, O.: Optimization of truss bridges within a specified design domain using evolution strategies. Eng. Optim. **39**(6), 737–756 (2007). https://doi.org/10.1080/03052150701335071
78. Hasançebi, O., Azad, S.K.: An exponential big bang-big crunch algorithm for discrete design optimization of steel frames. Comput. Struct. **110–111**, 167–179 (2012). https://doi.org/10.1016/j.compstruc.2012.07.014
79. Hasançebi, O., Azad, S.K.: Discrete size optimization of steel trusses using a refined big bang–big crunch algorithm. Eng. Optim. **46**(1), 61–83 (2014). https://doi.org/10.1080/0305215X.2012.748047
80. Ho-Huu, V., Nguyen-Thoi, T., Le-Anh, L., Nguyen-Trang, T.: An effective reliability-based improved constrained differential evolution for reliability-based design optimization of truss structures. Adv. Eng. Softw. **92**, 48–56 (2016). https://doi.org/10.1016/j.advengsoft.2015.11.001
81. Holland, J.H.: Adaptation in Natural and Artificial Systems: An Introductory Analysis with Applications to Biology, Control and Artificial Intelligence. MIT, Cambridge (1975)
82. Jin, Y.X., Cheng, H.Z., yong Yan, J., Zhang, L.: New discrete method for particle swarm optimization and its application in transmission network expansion planning. Electr. Power Syst. Res. **77**(3), 227–233 (2007). https://doi.org/10.1016/j.epsr.2006.02.016
83. Junghans, L., Darde, N.: Hybrid single objective genetic algorithm coupled with the simulated annealing optimization method for building optimization. Energy Build. **86**, 651–662 (2015). https://doi.org/10.1016/j.enbuild.2014.10.039
84. Kameshki, E.S., Saka, M.P.: Optimum design of nonlinear steel frames with semi-rigid connections using a genetic algorithm. Comput. Struct. **79**(17), 1593–1604 (2001). https://doi.org/10.1016/S0045-7949(01)00035-9
85. Kang, Z., Bai, S.: On robust design optimization of truss structures with bounded uncertainties. Struct. Multidiscip. Optim. **47**(5), 699–714 (2013). https://doi.org/10.1007/s00158-012-0868-3
86. Kanno, Y.: Mixed-integer second-order cone programming for global optimization of compliance of frame structure with discrete design variables. Struct. Multidiscip. Optim. **54**(2), 301–316 (2016). https://doi.org/10.1007/s00158-016-1406-5
87. Karovic, O., Mays, L.W.: Sewer system design using simulated annealing in excel. Water Resour. Manag. **28**(13), 4551–4565 (2014). https://doi.org/10.1007/s11269-014-0750-8
88. Kaveh, A.: Advances in Metaheuristic Algorithms for Optimal Design of Structures, 1st edn. Springer International Publishing, Cham (2014)
89. Kaveh, A., Abadi, A.S.M.: Harmony search based algorithms for the optimum cost design of reinforced concrete cantilever retaining walls. Int. J. Civil Eng. **9**(1), 1–8 (2011)
90. Kaveh, A., Bakhshpoori, T.: Optimum design of steel frames using cuckoo search algorithm with Lévy flights. Struct. Design Tall Spec. Build. **22**(13), 1023–1036 (2013). https://doi.org/10.1002/tal.754
91. Kaveh, A., Khayatazad, M.: A new meta-heuristic method: ray optimization. Comput. Struct. **112–113**, 283–294 (2012). https://doi.org/10.1016/j.compstruc.2012.09.003
92. Kaveh, A., Khayatazad, M.: Ray optimization for size and shape optimization of truss structures. Comput. Struct. **117**, 82–94 (2013). https://doi.org/10.1016/j.compstruc.2012.12.010
93. Kaveh, A., Maniat, M.: Damage detection in skeletal structures based on charged system search optimization using incomplete modal data. Int. J. Civil Eng. **12**(2), 291–298 (2014)
94. Kaveh, A., Massoudi, M.S.: Multi-objective optimization of structures using charged system search. Sci. Iran. **21**(6), 1845–1860 (2014)
95. Kaveh, A., Sabzi, O.: A comparative study of two meta-heuristic algorithms for optimum design of reinforced concrete frame. Int. J. Civil Eng. **9**(3), 193–206 (2011)
96. Kaveh, A., Shokohi, F.: Cost optimization of castellated beams using charged system search algorithm. Iran. J. Sci. Technol. Trans. Civil Eng. **38**(C1+), 235–249 (2014). https://doi.org/10.22099/ijstc.2014.1866

97. Kaveh, A., Soleimani, N.: CBO and DPSO for optimum design of reinforced concrete cantilever retaining walls. Asian J. Civil Eng. **16**(6), 751–774 (2015)
98. Kaveh, A., Talatahari, S.: Particle swarm optimizer, ant colony strategy and harmony search scheme hybridized for optimization of truss structures. Comput. Struct. **87**(5), 267–283 (2009). https://doi.org/10.1016/j.compstruc.2009.01.003
99. Kaveh, A., Talatahari, S.: Size optimization of space trusses using big bang–big crunch algorithm. Comput. Struct. **87**(17), 1129–1140 (2009). https://doi.org/10.1016/j.compstruc.2009.04.011
100. Kaveh, A., Talatahari, S.: A discrete big bang-big crunch algorithm for optimal design of skeletal structures. Asian J. Civil Eng. **11**(1), 103–122 (2010)
101. Kaveh, A., Talatahari, S.: A novel heuristic optimization method: charged system search. Acta Mech. **213**(3), 267–289 (2010). https://doi.org/10.1007/s00707-009-0270-4
102. Kaveh, A., Zakian, P.: Enhanced bat algorithm for optimal design of skeletal structures. Asian J. Civil Eng. **15**(2), 179–212 (2014)
103. Kaveh, A., Kalateh-Ahani, M., Fahimi-Farzam, M.: Constructability optimal design of reinforced concrete retaining walls using a multi-objective genetic algorithm. Struct. Eng. Mech. **47**(2), 227–245 (2013). https://doi.org/10.12989/sem.2013.47.2.227
104. Kaveh, A., Sheikholeslami, R., Talatahari, S., Keshvari-Ilkhichi, M.: Chaotic swarming of particles: a new method for size optimization of truss structures. Adv. Eng. Softw. **67**, 136–147 (2014). https://doi.org/10.1016/j.advengsoft.2013.09.006
105. Kaveh, A., Kaveh, A., Nasrollahi, A.: Charged system search and particle swarm optimization hybridized for optimal design of engineering structures. Sci. Iran. **21**(2), 295–305 (2014)
106. Kaveh, A., Mohammadi, S., Khademhosseini, O., Keyhani, A., Kalatjari, V.: Optimum parameters of tuned mass dampers for seismic applications using charged system search. Iran. J. Sci. Technol. Trans. Civil Eng. **39**, 21–40 (2015)
107. Kelesoglu, O.: Fuzzy multiobjective optimization of truss-structures using genetic algorithm. Adv. Eng. Softw. **38**(10), 717–721 (2007). https://doi.org/10.1016/j.advengsoft.2007.03.003
108. Kennedy, J., Eberhart, R.: Particle swarm optimization. In: Proceedings of IEEE International Conference on Neural Networks, vol. 4, pp. 1942–1948 (1995). https://doi.org/10.1109/ICNN.1995.488968
109. Khadem-Hosseini, O., Pirgholizadeh, S., Kaveh, A.: Semi-active tuned mass damper performance with optimized fuzzy controller using CSS algorithm. Asian J. Civil Eng. (BHRC) **16**(5), 587–606 (2015)
110. Kirkpatrick, S., Gelatt, C.D., Vecchi, M.P.: Optimization by simulated annealing. Science **220**(4598), 671–680 (1983). https://doi.org/10.1126/science.220.4598.671
111. Kociecki, M., Adeli, H.: Two-phase genetic algorithm for topology optimization of free-form steel space-frame roof structures with complex curvatures. Eng. Appl. Artif. Intell. **32**, 218–227 (2014). https://doi.org/10.1016/j.engappai.2014.01.010
112. Kong, J.S., Frangopol, D.M.: Life-cycle reliability-based maintenance cost optimization of deteriorating structures with emphasis on bridges. J. Struct. Eng. **129**(6), 818–828 (2003). https://doi.org/10.1061/(ASCE)0733-9445(2003)129:6(818)
113. Koumousis, V.K., Georgiou, P.G.: Genetic algorithms in discrete optimization of steel truss roofs. J. Comput. Civil Eng. **8**(3), 309–325 (1994). https://doi.org/10.1061/(ASCE)0887-3801(1994)8:3(309)
114. Krishnamoorthy, C.S., Venkatesh, P.P., Sudarshan, R.: Object-oriented framework for genetic algorithms with application to space truss optimization. J. Comput. Civil Eng. **16**(1), 66–75 (2002). https://doi.org/10.1061/(ASCE)0887-3801(2002)16:1(66)
115. Kroon, L.G., Dekker, R., Vromans, M.J.C.M.: Cyclic railway timetabling: a stochastic optimization approach. In: Geraets, F., Kroon, L., Schoebel, A., Wagner, D., Zaroliagis, C.D. (eds.) Algorithmic Methods for Railway Optimization, pp. 41–66. Springer, Berlin (2007)
116. Kuan, S.N., Ong, H.L., Ng, K.M.: Solving the feeder bus network design problem by genetic algorithms and ant colony optimization. Adv. Eng. Softw. **37**(6), 351–359 (2006). https://doi.org/10.1016/j.advengsoft.2005.10.003

117. Kusano, I., Baldomir, A., Jurado, J.Á., Hernández, S.: Probabilistic Optimization of the main cable and bridge deck of long-span suspension bridges under flutter constraint. J. Wind Eng. Ind. Aerodyn. **146**, 59–70 (2015). https://doi.org/10.1016/j.jweia.2015.08.001

118. Lee, C., Ahn, J.: Flexural design of reinforced concrete frames by genetic algorithm. J. Struct. Eng. **129**(6), 762–774 (2003). https://doi.org/10.1061/(ASCE)0733-9445(2003)129:6(762)

119. Lee, T.Y., Kim, Y.H., Kang, S.W.: Optimization of tensioning strategy for asymmetric cable-stayed bridge and its effect on construction process. Struct. Multidiscip. Optim. **35**(6), 623–629 (2008). https://doi.org/10.1007/s00158-007-0172-9

120. Lepš, M., Šejnoha, M.: New approach to optimization of reinforced concrete beams. Comput. Struct. **81**(18), 1957–1966 (2003). https://doi.org/10.1016/S0045-7949(03)00215-3

121. Leung, A.Y.T., Zhang, H.: Particle swarm optimization of tuned mass dampers. Eng. Struct. **31**(3), 715–728 (2009). https://doi.org/10.1016/j.engstruct.2008.11.017

122. Leung, A.Y.T., Zhang, H., Cheng, C.C., Lee, Y.Y.: Particle swarm optimization of TMD by non-stationary base excitation during earthquake. Earthq. Eng. Struct. Dyn. **37**(9), 1223–1246 (2008). https://doi.org/10.1002/eqe.811

123. Leung, Y.F., Klar, A., Soga, K.: Theoretical study on pile length optimization of pile groups and piled rafts. J. Geotech. Geoenviron. Eng. **136**(2), 319–330 (2010). https://doi.org/10.1061/(ASCE)GT.1943-5606.0000206

124. Li, J.P.: Truss topology optimization using an improved species-conserving genetic algorithm. Eng. Optim. **47**(1), 107–128 (2015). https://doi.org/10.1080/0305215X.2013.875165

125. Li, W., Li, Q., Steven, G.P., Xie, Y.M.: An evolutionary approach to elastic contact optimization of frame structures. Finite Elem. Anal. Design **40**(1), 61–81 (2003). https://doi.org/10.1016/S0168-874X(02)00179-8

126. Li, L.J., Huang, Z.B., Liu, F., Wu, Q.H.: A heuristic particle swarm optimizer for optimization of pin connected structures. Comput. Struct. **85**(7), 340–349 (2007). https://doi.org/10.1016/j.compstruc.2006.11.020

127. Lievens, K., Lombaert, G., Roeck, G.D., den Broeck, P.V.: Robust design of a tmd for the vibration serviceability of a footbridge. Eng. Struct. **123**, 408–418 (2016). https://doi.org/10.1016/j.engstruct.2016.05.028

128. Liu, W., Ye, J.: Collapse optimization for domes under earthquake using a genetic simulated annealing algorithm. J. Constr. Steel Res. **97**, 59–68 (2014). https://doi.org/10.1016/j.jcsr.2014.01.015

129. Liu, L., Burns, S.A., Feng, C.W.: Construction time-cost trade-off analysis using LP/IP hybrid method. J. Constr. Eng. Manag. **121**(4), 446–454 (1995). https://doi.org/10.1061/(ASCE)0733-9364(1995)121:4(446)

130. Liu, C., Gao, Z., Zhao, W.: A new path planning method based on firefly algorithm. In: 2012 Fifth International Joint Conference on Computational Sciences and Optimization, pp. 775–778 (2012). https://doi.org/10.1109/CSO.2012.174

131. Lucas-Simarro, J.L., Moreno-Vozmediano, R., Montero, R.S., Llorente, I.M.: Cost optimization of virtual infrastructures in dynamic multi-cloud scenarios. Concurrency Comput. Pract. Exp. **27**(9), 2260–2277 (2015). https://doi.org/10.1002/cpe.2972

132. Lund, J.R., Ferreira, I.: Operating rule optimization for Missouri river reservoir system. J. Water Resour. Plan. Manag. **122**(4), 287–295 (1996). https://doi.org/10.1061/(ASCE)0733-9496(1996)122:4(287)

133. Maier, H.R., Simpson, A.R., Zecchin, A.C., Foong, W.K., Phang, K.Y., Seah, H.Y., Tan, C.L.: Ant colony optimization for design of water distribution systems. J. Water Resour. Plan. Manag. **129**(3), 200–209 (2003). https://doi.org/10.1061/(ASCE)0733-9496(2003)129:3(200)

134. Marano, G.C., Greco, R., Chiaia, B.: A comparison between different optimization criteria for tuned mass dampers design. J. Sound Vib. **329**(23), 4880–4890 (2010). https://doi.org/10.1016/j.jsv.2010.05.015

135. Martínez, F.J., González-Vidosa, F., Hospitaler, A., Yepes, V.: Heuristic optimization of RC bridge piers with rectangular hollow sections. Comput. Struct. **88**(5), 375–386 (2010). https://doi.org/10.1016/j.compstruc.2009.11.009

136. Martínez, F.J., González-Vidosa, F., Hospitaler, A., Alcalá, J.: Design of tall bridge piers by ant colony optimization. Eng. Struct. **33**(8), 2320–2329 (2011). https://doi.org/10.1016/j.engstruct.2011.04.005

137. Martinez-Martin, F.J., Gonzalez-Vidosa, F., Hospitaler, A., Yepes, V.: Multi-objective optimization design of bridge piers with hybrid heuristic algorithms. J. Zhejiang Univ. Sci. A **13**(6), 420–432 (2012). https://doi.org/10.1631/jzus.A1100304

138. Mauder, T., Sandera, C., Stetina, J., Seda, M.: Optimization of the quality of continuously cast steel slabs using the firefly algorithm. Mater. Technol. **45**(4), 4551–4565 (2011)

139. Middleton, R.S., Brandt, A.R.: Using infrastructure optimization to reduce greenhouse gas emissions from oil sands extraction and processing. Environ. Sci. Technol. **47**(3), 1735–1744 (2013). https://doi.org/10.1021/es3035895

140. Miguel, L.F.F., Lopez, R.H., Miguel, L.F.F.: Multimodal size, shape, and topology optimisation of truss structures using the firefly algorithm. Adv. Eng. Softw. **56**, 23–37 (2013). https://doi.org/10.1016/j.advengsoft.2012.11.006

141. Mijar, A.R., Swan, C.C., Arora, J.S., Kosaka, I.: Continuum topology optimization for concept design of frame bracing systems. J. Struct. Eng. **124**(5), 541–550 (1998). https://doi.org/10.1061/(ASCE)0733-9445(1998)124:5(541)

142. Mohammadi, K., Eslami, H.R., Kahawita, R.: Parameter estimation of an ARMA model for river flow forecasting using goal programming. J. Hydrol. **331**(1), 293–299 (2006). https://doi.org/10.1016/j.jhydrol.2006.05.017. Water Resources in Regional Development: The Okavango River

143. Montalvo, I., Izquierdo, J., Pérez-García, R., Herrera, M.: Water distribution system computer-aided design by agent swarm optimization. J. Comput. Aided Civ. Infrastruct. Eng. **29**(6), 433–448 (2014). https://doi.org/10.1111/mice.12062

144. Mugisha, S.: Infrastructure optimization and performance monitoring: empirical findings from the water sector in Uganda. Afr. J. Bus. Manag. **2**(1), 13–25 (2008)

145. Nicholson, J.C.: Design of wind turbine tower and foundation systems: optimization approach. Master's thesis, University of Iowa, Iowa (2011)

146. Nigdeli, S.M., Bekdaş, G.: Optimum tuned mass damper design for preventing brittle fracture of RC buildings. Smart Struct. Syst. **12**, 137–155 (2013). https://doi.org/10.12989/sss.2013.12.2.137

147. Nigdeli, S.M., Bekdaş, G.: Optimum tuned mass damper approaches for adjacent structures. Earthq. Struct. **7**, 1071–1091 (2014). https://doi.org/10.12989/eas.2014.7.6.1071

148. Nigdeli, S.M., Bekdaş, G.: Optimum design of RC continuous beams considering unfavourable live-load distributions. KSCE J. Civil Eng. **21**(4), 1410–1416 (2017). https://doi.org/10.1007/s12205-016-2045-5

149. Nigdeli, S.M., Bekdaş, G., Kim, S., Geem, Z.W.: A novel harmony search based optimization of reinforced concrete biaxially loaded columns. Struct. Eng. Mech. **54**(6), 1097–110 (2015). https://doi.org/10.12989/sem.2015.54.6.1097

150. Nigdeli, S.M., Bekdaş, G., Yang, X.S.: Application of the Flower Pollination Algorithm in Structural Engineering, pp. 25–42. Springer International Publishing, Cham (2016)

151. Nikolić, M., Teodorović, D., Vukadinović, K.: Disruption management in public transit: the bee colony optimization approach. Transport. Plann. Technol. **38**(2), 162–180 (2015). https://doi.org/10.1080/03081060.2014.997447

152. Paya, I., Yepes, V., González-Vidosa, F., Hospitaler, A.: Multiobjective optimization of concrete frames by simulated annealing. Comput. Aided Civ. Inf. Eng. **23**(8), 596–610 (2008). https://doi.org/10.1111/j.1467-8667.2008.00561.x

153. Paya-Zaforteza, I., Yepes, V., Hospitaler, A., González-Vidosa, F.: CO_2-optimization of reinforced concrete frames by simulated annealing. Eng. Struct. **31**(7), 1501–1508 (2009). https://doi.org/10.1016/j.engstruct.2009.02.034

154. Perea, C., Alcala, J., Yepes, V., Gonzalez-Vidosa, F., Hospitaler, A.: Design of reinforced concrete bridge frames by heuristic optimization. Adv. Eng. Softw. **39**(8), 676–688 (2008). https://doi.org/10.1016/j.advengsoft.2007.07.007

155. Perez, R.E., Behdinan, K.: Particle swarm approach for structural design optimization. Comput. Struct. **85**(19), 1579–1588 (2007). https://doi.org/10.1016/j.compstruc.2006.10.013

156. Pezeshk, S., Camp, C.V., Chen, D.: Design of nonlinear framed structures using genetic optimization. J. Struct. Eng. **126**(3), 382–388 (2000). https://doi.org/10.1061/(ASCE)0733-9445(2000)126:3(382)

157. Pourzeynali, S., Lavasani, H.H., Modarayi, A.H.: Active control of high rise building structures using fuzzy logic and genetic algorithms. Eng. Struct. **29**(3), 346–357 (2007). https://doi.org/10.1016/j.engstruct.2006.04.015

158. Putha, R., Quadrifoglio, L., Zechman, E.: Comparing ant colony optimization and genetic algorithm approaches for solving traffic signal coordination under oversaturation conditions. Comput. Aided Civ. Inf. Eng. **27**(1), 14–28 (2012). https://doi.org/10.1111/j.1467-8667.2010.00715.x

159. Rajan, S.D.: Sizing, shape, and topology design optimization of trusses using genetic algorithm. J. Struct. Eng. **121**(10), 1480–1487 (1995). https://doi.org/10.1061/(ASCE)0733-9445(1995)121:10(1480)

160. Rajeev, S., Krishnamoorthy, C.S.: Discrete optimization of structures using genetic algorithms. J. Struct. Eng. **118**(5), 1233–1250 (1992). https://doi.org/10.1061/(ASCE)0733-9445(1992)118:5(1233)

161. Rajeev, S., Krishnamoorthy, C.S.: Genetic algorithm-based methodology for design optimization of reinforced concrete frames. Comput. Aided Civ. Inf. Eng. **13**(1), 63–74 (1998). https://doi.org/10.1111/0885-9507.00086

162. Reed, M., Yiannakou, A., Evering, R.: An ant colony algorithm for the multi-compartment vehicle routing problem. Appl. Soft Comput. **15**, 169–176 (2014). https://doi.org/10.1016/j.asoc.2013.10.017

163. Richardson, J.N., Adriaenssens, S., Bouillard, P., Filomeno Coelho, R.: Multiobjective topology optimization of truss structures with kinematic stability repair. Struct. Multidiscip. Optim. **46**(4), 513–532 (2012). https://doi.org/10.1007/s00158-012-0777-5

164. Robelin, C.A., Madanat, S.M.: History-dependent bridge deck maintenance and replacement optimization with Markov decision processes. J. Inf. Syst. **13**(3), 195–201 (2007). https://doi.org/10.1061/(ASCE)1076-0342(2007)13:3(195)

165. Saad, L., Aissani, A., Chateauneuf, A., Raphael, W.: Reliability-based optimization of direct and indirect LCC of RC bridge elements under coupled fatigue-corrosion deterioration processes. Eng. Fail. Anal. **59**, 570–587 (2016). https://doi.org/10.1016/j.engfailanal.2015.11.006

166. Sabzkouhi, A.M., Haghighi, A.: Uncertainty analysis of pipe-network hydraulics using a many-objective particle swarm optimization. J. Hydraul. Eng. **142**(9), 04016030 (2016). https://doi.org/10.1061/(ASCE)HY.1943-7900.0001148

167. Sadek, F., Mohraz, B., Taylor, A.W., Chung, R.M.: A method of estimating the parameters of tuned mass dampers for seismic applications. Earthq. Eng. Struct. Dyn. **26**(6), 617–635 (1997). https://doi.org/10.1002/(SICI)1096-9845(199706)26:6<617::AID-EQE664>3.0.CO;2-Z

168. Sahab, M.G., Ashour, A.F., Toropov, V.V.: Cost optimisation of reinforced concrete flat slab buildings. Eng. Struct. **27**(3), 313–322 (2005). https://doi.org/10.1016/j.engstruct.2004.10.002

169. Sahab, M.G., Toropov, V.V., Gandomi, A.H.: A review on traditional and modern structural optimization: problems and techniques. In: Gandomi, A.H., Yang, X.S., Talatahari, S., Alavi, A.H. (eds.) Metaheuristic Applications in Structures and Infrastructures, pp. 25–47. Elsevier, Oxford (2013). https://doi.org/10.1016/B978-0-12-398364-0.00002-4

170. Saka, M.P.: Optimum design of steel frames using stochastic search techniques based on natural phenomena: a review. In: Topping, B.H.V. (ed.) Civil Engineering Computations: Tools and Techniques, Stirlingshire, pp. 105–147 (2007). https://doi.org/10.4203/csets.16.6

171. Schutte, J.F., Groenwold, A.A.: Sizing design of truss structures using particle swarms. Struct. Multidiscip. Optim. **25**(4), 261–269 (2003). https://doi.org/10.1007/s00158-003-0316-5

172. Senouci, A.B., Eldin, N.N.: Use of genetic algorithms in resource scheduling of construction projects. J. Constr. Eng. Manag. **130**(6), 869–877 (2004). https://doi.org/10.1061/(ASCE)0733-9364(2004)130:6(869)
173. Šešok, D., Belevičius, R.: Global optimization of trusses with a modified genetic algorithm. J. Civil Eng. Manag. **14**(3), 147–154 (2008). https://doi.org/10.3846/1392-3730.2008.14.10
174. Sgambi, L., Gkoumas, K., Bontempi, F.: Genetic algorithms for the dependability assurance in the design of a long-span suspension bridge. Comput. Aided Civ. Inf. Eng. **27**(9), 655–675 (2012). https://doi.org/10.1111/j.1467-8667.2012.00780.x
175. Singh, M.P., Singh, S., Moreschi, L.M.: Tuned mass dampers for response control of torsional buildings. Earthq. Eng. Struct. Dyn. **31**(4), 749–769 (2002). https://doi.org/10.1002/eqe.119
176. Smith, C.C., Cubrinovski, M.: Pseudo-static limit analysis by discontinuity layout optimization: application to seismic analysis of retaining walls. Soil Dyn. Earthq. Eng. **31**(10), 1311–1323 (2011). https://doi.org/10.1016/j.soildyn.2011.03.014
177. Sonmez, M.: Artificial bee colony algorithm for optimization of truss structures. Appl. Soft Comput. **11**(2), 2406–2418 (2011). https://doi.org/10.1016/j.asoc.2010.09.003
178. Steinbuch, R.: Bionic optimisation of the earthquake resistance of high buildings by tuned mass dampers. J. Bionic Eng. **8**(3), 335–344 (2011). https://doi.org/10.1016/S1672-6529(11)60036-X
179. Talatahari, S., Sheikholeslami, R., Shadfaran, M., Pourbaba, M.: Optimum design of gravity retaining walls using charged system search algorithm. Math. Prob. Eng. **2012**, 1–10 (2012). https://doi.org/10.1155/2012/301628
180. Talatahari, S., Gandomi, A.H., Yun, G.J.: Optimum design of tower structures using firefly algorithm. Struct. Design Tall Spec. Build. **23**(5), 350–361 (2014). https://doi.org/10.1002/tal.1043
181. Talatahari, S., Gandomi, A.H., Yang, X.S., Deb, S.: Optimum design of frame structures using the eagle strategy with differential evolution. Eng. Struct. **91**, 16–25 (2015). https://doi.org/10.1016/j.engstruct.2015.02.026
182. Talatahariand, S., Kaveh, A.: Improved bat algorithm for optimum design of large-scale truss structures. Int. J. Optim. Civil Eng. **5**(2), 241–254 (2015)
183. Tang, H., Zhou, J., Xue, S., Xie, L.: Big bang-big crunch optimization for parameter estimation in structural systems. Mech. Syst. Signal Process. **24**(8), 2888–2897 (2010). https://doi.org/10.1016/j.ymssp.2010.03.012
184. Temur, R., Bekdaş, G.: Constructability optimal design of reinforced concrete retaining walls using a multi-objective genetic algorithm. Struct. Eng. Mech. **57**(4), 763–783 (2016). https://doi.org/10.12989/sem.2016.57.4.763
185. Temür, R., Türkan, Y.S., Toklu, Y.C.: Geometrically Nonlinear Analysis of Trusses Using Particle Swarm Optimization, pp. 283–300. Springer International Publishing, Cham (2015). https://doi.org/10.1007/978-3-319-13826-8_15
186. Toğan, V.: Design of planar steel frames using teaching-learning based optimization. Eng. Struct. **34**, 225–232 (2012). https://doi.org/10.1016/j.engstruct.2011.08.035
187. Toğan, V., Daloğlu, A.T.: An improved genetic algorithm with initial population strategy and self-adaptive member grouping. Comput. Struct. **86**(11), 1204–1218 (2008). https://doi.org/10.1016/j.compstruc.2007.11.006
188. Toklu, Y.C., Bekdas, G., Temur, R.: Analysis of trusses by total potential optimization method coupled with harmony search. Struct. Eng. Mech. **45**(2), 183–199 (2013). https://doi.org/10.12989/sem.2013.45.2.183
189. Tong, K.H., Bakhary, N., Kueh, A.B.H., Yassin, A.Y.M.: Optimal sensor placement for mode shapes using improved simulated annealing. Smart Struct. Syst. **13**, 389–406 (2014). https://doi.org/10.12989/sss.2014.13.3.389
190. Torii, A.J., Lopez, R.H., Miguel, L.F.F.: Design complexity control in truss optimization. Struct. Multidiscip. Optim. **54**(2), 289–299 (2016). https://doi.org/10.1007/s00158-016-1403-8
191. Tran, D.H., Cheng, M.Y., Cao, M.T.: Hybrid multiple objective artificial bee colony with differential evolution for the time-cost-quality tradeoff problem. Knowl. Based Syst. **74**, 176–186 (2015). https://doi.org/10.1016/j.knosys.2014.11.018

192. Tran, D.H., Cheng, M.Y., Prayogo, D.: A novel multiple objective symbiotic organisms search (MOSOS) for time-cost-labor utilization tradeoff problem. Knowl. Based Syst. **94**, 132–145 (2016). https://doi.org/10.1016/j.knosys.2015.11.016

193. Walraven, E., Spaan, M.T.J., Bakker, B.: Traffic flow optimization: a reinforcement learning approach. Eng. Appl. Artif. Intell. **52**, 203–212 (2016). https://doi.org/10.1016/j.engappai.2016.01.001

194. Warburton, G.B.: Optimum absorber parameters for various combinations of response and excitation parameters. Earthq. Eng. Struct. Dyn. **10**(3), 381–401 (1982). https://doi.org/10.1002/eqe.4290100304

195. Wiering, M., Vreeken, J., van Veenen, J., Koopman, A.: Simulation and optimization of traffic in a city. In: IEEE Intelligent Vehicles Symposium, pp. 453–458 (2004). https://doi.org/10.1109/IVS.2004.1336426

196. Woo, S., Yeo, H.: Optimization of pavement inspection schedule with traffic demand prediction. Procedia Soc. Behav. Sci. **218**, 95–103 (2016). https://doi.org/10.1016/j.sbspro.2016.04.013. International Institute for Infrastructure Renewal and Reconstruction (I3R2)

197. Wurbs, R.A.: Reservoir-system simulation and optimization models. J. Water Resour. Plan. Manag. **119**(4), 455–472 (1993). https://doi.org/10.1061/(ASCE)0733-9496(1993)119:4(455)

198. Xiong, Y., Kuang, Y.: Applying an ant colony optimization algorithm-based multiobjective approach for time-Cost Trade-Off. J. Constr. Eng. Manag. **134**(2), 153–156 (2008). https://doi.org/10.1061/(ASCE)0733-9364(2008)134:2(153)

199. Yang, I.T.: Using elitist particle swarm optimization to facilitate bicriterion time-cost trade-off analysis. J. Constr. Eng. Manag. **133**(7), 498–505 (2007). https://doi.org/10.1061/(ASCE)0733-9364(2007)133:7(498)

200. Yang, X.S.: Nature-Inspired Metaheuristic Algorithms. Luniver Press, Frome (2008)

201. Yang, X.S.: Firefly algorithms for multimodal optimization. In: Watanabe, O., Zeugmann, T. (eds.) Stochastic Algorithms: Foundations and Applications, pp. 169–178. Springer, Berlin (2009)

202. Yang, X.S.: A New Metaheuristic Bat-Inspired Algorithm, pp. 65–74. Springer, Berlin (2010). https://doi.org/10.1007/978-3-642-12538-6_6

203. Yang, X.L.: Seismic passive pressures of earth structures by nonlinear optimization. Arch. Appl. Mech. **81**(9), 1195–1202 (2011). https://doi.org/10.1007/s00419-010-0478-8

204. Yang, X.S.: Flower pollination algorithm for global optimization. In: Durand-Lose, J., Jonoska, N. (eds.) Unconventional Computation and Natural Computation, pp. 240–249. Springer, Berlin (2012)

205. Yang, X.S., Deb, S.: Cuckoo search via Lévy flights. In: 2009 World Congress on Nature Biologically Inspired Computing (NaBIC), pp. 210–214 (2009). https://doi.org/10.1109/NABIC.2009.5393690

206. Yang, X., Gandomi, A.H.: Bat algorithm: a novel approach for global engineering optimization. Eng. Comput. **29**(5), 464–483 (2012). https://doi.org/10.1108/02644401211235834

207. Yang, L., Li, K., Gao, Z., Li, X.: Optimizing trains movement on a railway network. Omega **40**(5), 619–633 (2012). https://doi.org/10.1016/j.omega.2011.12.001

208. Yang, L., Zhou, X., Gao, Z.: Credibility-based rescheduling model in a double-track railway network: a fuzzy reliable optimization approach. Omega **48**, 75–93 (2014). https://doi.org/10.1016/j.omega.2013.11.004

209. Yang, X.S., Bekdaş, G., Nigdeli, S.M.: Metaheuristics and Optimization in Civil Engineering, 1st edn. Modeling and Optimization in Science and Technologies. Springer International Publishing, Cham (2016)

210. Yepes, V., Alcala, J., Perea, C., González-Vidosa, F.: A parametric study of optimum earth-retaining walls by simulated annealing. Eng. Struct. **30**(3), 821–830 (2008). https://doi.org/10.1016/j.engstruct.2007.05.023

211. Yepes, V., Gonzalez-Vidosa, F., Alcala, J., Villalba, P.: CO_2-optimization design of reinforced concrete retaining walls based on a VNS-threshold acceptance strategy. J. Comput. Civil Eng. **26**(3), 378–386 (2012). https://doi.org/10.1061/(ASCE)CP.1943-5487.0000140

212. Yoo, D.G., Kim, J.H., Geem, Z.W.: Overview of harmony search algorithm and its applications in civil engineering. Evol. Intel. **7**(1), 3–16 (2014). https://doi.org/10.1007/s12065-013-0100-4
213. Yu, D., Tian, X., Xing, X., Gao, S.: Signal timing optimization based on fuzzy compromise programming for isolated signalized intersection. Math. Probl. Eng. **2016**, 1–12 (2016). https://doi.org/10.1155/2016/1682394
214. Zhang, Y., Ng, S.T.: An ant colony system based decision support system for construction time-cost optimization. J. Civil Eng. Manag. **18**(4), 580–589 (2012). https://doi.org/10.3846/13923730.2012.704164
215. Zhao, F., Zeng, X.: Simulated annealing-genetic algorithm for transit network optimization. J. Comput. Civil Eng. **20**(1), 57–68 (2006). https://doi.org/10.1061/(ASCE)0887-3801(2006)20:1(57)
216. Zheng, D.X.M., Ng, S.T., Kumaraswamy, M.M.: Applying Pareto ranking and niche formation to genetic algorithm-based multiobjective time-cost optimization. J. Constr. Eng. Manag. **131**(1), 81–91 (2005). https://doi.org/10.1061/(ASCE)0733-9364(2005)131:1(81)
217. Zhu, W., Hu, H., Huang, Z.: Calibrating rail transit assignment models with genetic algorithm and automated fare collection data. Comput. Aided Civ. Inf. Eng. **29**(7), 518–530 (2014). https://doi.org/10.1111/mice.12075

Chapter 7
A Bioreactor Fault Diagnosis Based on Metaheuristics

Lídice Camps Echevarría, Orestes Llanes-Santiago, and Antônio José Silva Neto

7.1 Introduction

Fault Diagnosis, also known as Fault Detection and Isolation (FDI), is the area of knowledge related to methods for detecting, isolating, and identifying faults in control systems or processes [21, 30, 31]. It is a very important issue in the industry. Important topics, such as reliability, safety, efficiency, and maintenance, depend on the correct diagnosis of the industrial processes.

In order to prevent the propagation of faults, which may cause serious damages in industrial systems, the faults need to be diagnosed as fast as possible. This issue is very important for online processes. The FDI methods also need to be able to avoid false alarms attributable to external causes such as uncertainties affecting the measurements, external disturbances or spurious signals, which commonly affect systems or processes. This characteristic is called robustness. The sensitivity to incipient faults is also a desirable feature of the FDI methods. As a consequence, robustness to external disturbances affecting the system, sensitivity to incipient faults, and a proper diagnosis time, in order to prevent propagation of faults, are desired characteristics of the diagnosis systems. No individual characteristic is more

L. Camps Echevarría (✉)
Mathematics Department, Universidad Tecnológica de La Habana (Cujae), La Habana, Cuba

O. Llanes-Santiago
Automatic and Computing Department, Universidad Tecnológica de La Habana (Cujae), La Habana, Cuba
e-mail: orestes@tesla.cujae.edu.cu

A. J. Silva Neto
Department of Mechanical Engineering and Energy, Polytechnic Institute, IPRJ-UERJ, Nova Friburgo, Brazil
e-mail: ajsneto@iprj.uerj.br

© Springer Nature Switzerland AG 2019
G. Mendes Platt et al. (eds.), *Computational Intelligence, Optimization and Inverse Problems with Applications in Engineering*,
https://doi.org/10.1007/978-3-319-96433-1_7

important than the others. An adequate balance among them is the key for practical applications of FDI methods [21, 30, 31].

Within FDI methods, the model based methods are a very important group [38]. An analytical model of the system can incorporate the dynamics of the faults that can eventually affect the system [13, 17]. Such dynamics can be modeled by means of a fault vector.

Recent results presented in Refs. [1, 5–7, 10, 11] have shown that a formulation of Fault Diagnosis as an optimization problem, i.e. the estimation of the fault vector being obtained by solving an optimization problem, allows to achieve an appropriate balance between robustness and sensitivity. In the works just listed, the optimization problems are solved with metaheuristics.

Many of the initial works in FDI were linked with chemical processes [22]. For the chemical industry, the development of FDI methods is a very current topic. In the particular case of the chemical and biochemical industries, the use of nonlinear bioreactors is very common. Moreover, the propagation of faults can occur very fast, with drastic consequences. Therefore, the fault diagnosis of the systems for both industries is of upmost importance. Some works related to fault diagnosis in nonlinear bioreactors are described in Refs. [1, 18]. This chapter presents the application of three metaheuristics with different search and evolution strategies: Ant Colony Optimization with Dispersion (ACO-d), Differential Evolution with Particle Collisions (DEwPC), and the Covariance Matrix Adaptation Evolution Strategy (CMA-ES), to the fault diagnosis in a nonlinear bioreactor benchmark problem. For that purpose, the fault diagnosis is formulated as an optimization problem, as described in Refs. [6, 10, 11]. This formulation allows the use of metaheuristics for solving the optimization problem. The analysis of the diagnosis quality is based on robustness and diagnosis time. Furthermore, the results are compared with other reported in the literature.

Based on the comparison among the three metaheuristics when diagnosing the bioreactor, it is presented a proposal for collecting information regarding the quality of the diagnosis obtained with metaheuristics. This information can be organized in tables which can be used by experts.

This chapter also shows how to improve the metaheuristics stopping criteria, when they are applied to FDI.

The chapter is organized as follows. The fault diagnosis approach applied to diagnosing the chosen benchmark problem appears in Sect. 7.2. Section 7.3 describes the nonlinear bioreactor benchmark problem under study. Section 7.4 presents the three metaheuristics used for diagnosing the bioreactor. Section 7.5 shows the test cases considered and the parameters used in the metaheuristics. Section 7.6 presents the numerical experiments performed and their results. Finally, Sect. 7.7 presents the conclusions and a proposal for future works.

7.2   Fault Diagnosis Formulated as an Optimization Problem

Models for describing control systems vary depending on the dynamics of the process, and objectives to be reached with the simulation [13, 17, 21, 31]. In the formulation of the fault diagnosis as an inverse problem, it is considered a model of the system that directly includes fault dynamics. For that purpose, the faults are classified as [13, 17]:

- Actuator faults: $f_u \in \mathbb{R}^p$.
- Process faults: $f_p \in \mathbb{R}^q$.
- Sensor faults: $f_y \in \mathbb{R}^m$.

With this classification, the fault dynamics is represented in the models by means of a fault vector $f = (f_u \ f_p \ f_y)^t$, with $f \in \mathbb{R}^{p+q+m}$ [13]. The model with the fault dynamics is described by means of:

$$\dot{x}(t) = g(x(t), u(t), \theta, f) \tag{7.1}$$

$$y(t) = h(x(t), f) \tag{7.2}$$

$$x(t_0) = x_0 \tag{7.3}$$

where $x(t) \in \mathbb{R}^n$ is the state variable vector; $u(t) \in \mathbb{R}^p$ is the input vector; $\theta \in \mathbb{R}^l$ is the parameter vector of the model and $t \in [t_0, t_1]$. The output vector $y(t) \in \mathbb{R}^m$ is measured with sensors which can be affected by faults represented by f_y.

Considering that the faults do not change with time, the following optimization problem is formulated:

$$\min F(\hat{f}) = \sum_{t=1}^{N_s} \left[y_t(u, f) - \hat{y}_t(u, \hat{f}) \right]^2 \tag{7.4}$$

$$\text{s.t.} \qquad f_{\min} \le \hat{f} \le f_{\max}$$

where N_s is the number of sampling instants; $\hat{y}_t(u, \hat{f})$ is the vector of estimated values of the output obtained with the model given by Eqs. (7.1)–(7.3) at each time instant t; $y_t(u, f)$ is the output vector measured by the sensors at the same time instant t.

Solving the optimization problem given by Eq. (7.4), estimates for the fault vector f can be obtained. With the estimated values, the system can be directly diagnosed. An estimated value different from zero for a component f_i of the fault vector f indicates that the fault is affecting the system and its magnitude is the estimated value.

7.3 Description of the Bioreactor Benchmark Problem

This section describes the main characteristics of the bioreactor benchmark problem that will be diagnosed with the three metaheuristics described in Sect. 7.4. Its simulation is based on the description given in Ref. [12].

The bioreactor contains microorganisms and substrate with concentration values represented by ξ_1 and ξ_2, respectively. The state vector is $x \in \mathbb{R}^2$, $x = [x_1 \ x_2]^t$ being

$$x_1 = \xi_1 \tag{7.5}$$

$$x_2 = \frac{a_1 \xi_1 \xi_2}{a_2 \xi_1 + \xi_2} \tag{7.6}$$

The processes is described by means of the system of equations:

$$\dot{x}_1 = x_2 - \eta x_1 \tag{7.7}$$

$$\dot{x}_2 = -\frac{a_2 x_2 \left(x_2^2 - a_1 \eta x_1^2 \right) + (a_1 x_1 - x_2)^2 (a_4 \eta - a_3 x_2)}{a_1 a_2 x_1^2} \tag{7.8}$$

$$x(t_0) = x_0 \tag{7.9}$$

being $a_1, a_2, a_3, a_4 \in \mathbb{R}$ model parameters and $\eta \in \mathbb{R}$ is the dilution rate.

The concentration of microorganism is measured with the help of a sensor, being represented by the equation $y = x_1$.

The process can be affected by two process faults f_{p1} and f_{p2} that represent substances disturbing the concentration of the microorganisms. These faults can be modelled as:

$$f_{p1} = \theta_1 \Psi_1 \tag{7.10}$$

$$f_{p2} = \theta_2 \Psi_2 \tag{7.11}$$

being $\theta_1, \theta_2 \in \mathbb{R}$ unknown concentration coefficients; Ψ_1 and Ψ_2 are functions used for simulating the time dependent effect of the faults f_{p1} and f_{p2}, respectively.

The system with faults is modelled by means of the following nonlinear equations:

$$\dot{x}_1 = x_2 - \eta x_1 + \theta_1 \Psi_1 - \theta_2 \Psi_2 \tag{7.12}$$

$$\dot{x}_2 = -\frac{a_2 x_2 \left(x_2^2 - a_1 \eta x_1^2 \right) + (a_1 x_1 - x_2)^2 (a_4 \eta - a_3 x_2)}{a_1 a_2 x_1^2} \tag{7.13}$$

$$x(t_0) = x_0 \tag{7.14}$$

$$y = x_1 \tag{7.15}$$

Table 7.1 Parameter values
for the bioreactor benchmark
problem

a_1	a_2	a_3	a_4
1.0	1.0	1.0	0.1

The values for the parameters model and input signal are shown in Table 7.1 and
they coincide with the ones used in Refs. [18, 39].

The initial condition is $x(0) = [0.05\ 0.025]^t$ and the input $u(t)$ follows the
following law:

$$u(t) = \begin{cases} 0.08 \text{ if } & 0\,h \le t < 10\,h \\ 0.02 \text{ if } & 10\,h \le t < 20\,h \\ 0.08 \text{ if } & t \ge 20\,h \end{cases} \tag{7.16}$$

In order to make comparisons with reported results, the experiments consider
the faulty situation described in Ref. [39]. Therefore, it is used $\theta_1 = 0.01$, $\theta_2 = 0.015$ and:

$$\Psi_1(t) = \begin{cases} 0 \text{ if } 0\,h \le t < 20\,h \\ 1 \text{ if } \quad t \ge 20\,h \end{cases} \tag{7.17}$$

and

$$\Psi_2(t) = \begin{cases} 0 \text{ if } 0\,h \le t < 30\,h \\ 1 \text{ if } \quad t \ge 30\,h \end{cases} \tag{7.18}$$

In all test cases performed, the output y is affected by Gaussian noise whose
empirical standard deviation is a percent of x_1. The addition of noise is aimed to
simulate more realistic conditions. The noise affecting the systems is one of the
recognized causes of a wrong diagnosis, and leads to the necessity of robust FDI
methods. All implementations were made using Matlab®.

7.3.1 FDI Formulation for the Benchmark Problem

Due to the fact that the functions Ψ_1, Ψ_2 are known, the diagnosis of the two
described faults, which can eventually affect the system, is completed when the
two parameters θ_1, θ_2 are estimated. It is reached solving the optimization problem:

$$\min F(\hat{\theta}_1, \hat{\theta}_2) = \sum_{t=1}^{N_s} \left[y_t(\theta_1, \theta_2) - \hat{y}_t(\hat{\theta}_1, \hat{\theta}_2, u) \right]^2$$
$$\text{s.t} \quad \theta_{1(\min)} \le \hat{\theta}_1 \le \theta_{1(\max)} \tag{7.19}$$
$$\theta_{2(\min)} \le \hat{\theta}_2 \le \theta_{2(\max)}$$

where $\hat{y}_t(\hat{\theta}_1, \hat{\theta}_2, u)$ is the estimated output vector at each instant of time t, and it is obtained from the model described by Eqs. (7.12)–(7.15) considering the values of $u(t)$ given by Eq. (7.16); $y_t(\theta_1, \theta_2)$ is the output, which is measured by the sensor at the same time instant t.

7.4 Description of the Metaheuristics for the Benchmark Problem Fault Diagnosis

This section describes the three metaheuristics applied to fault diagnosis in the benchmark problem.

7.4.1 Ant Colony Optimization with Dispersion (ACO-d)

Ant Colony Optimization with dispersion, ACO-d,, [2, 3], is a variant of the algorithm Ant Colony Optimization, ACO [14–16]. ACO was initially proposed for integer programming problems [15, 16], but it has been successfully extended to continuous optimization problems [4, 29, 32–34].

ACO is inspired on the behavior of ants, specifically when they are seeking a path between their colony and a food source. This behavior is explained with the deposition and evaporation of a substance: the pheromone [4, 15, 16, 29, 33]. The paths with greater concentration of this substance attract the other ants; and they will deposit their pheromone in such paths.

The idea of the deposit and evaporation of the pheromone is simulated by means of a pheromone matrix $\mathbf{F}$ which is updated at each iteration and is accessible to all the ants in the colony [4, 15, 16, 29, 33].

ACO-d attempts to simulate the deposit of pheromone, as close to the real phenomena as possible: the pheromone affects the path where it is deposited, and the paths next to it [3]. This is called dispersion. As a consequence, the difference between ACO and ACO-d is how they update the pheromone matrix.

7.4.1.1 Description of the Algorithm for Continuous Problems

The algorithm of ACO-d for the continuous optimization problems is based on the adaptation of ACO for continuous problems reported in Refs. [29, 34].

At each iteration $Iter$, ACO-d generates a new population (colony) with Z solutions (ants): $P_{Iter} = \{X_{Iter}^1, X_{Iter}^2, ...X_{Iter}^Z\}$. The generation of P_{Iter} is based on a probability matrix PC which depends on $\mathbf{F}$.

For the continuous problem, the first step is to divide the feasible interval of each variable x_n of the problem into k possible discrete values x_n^k. The generation of the colony at each iteration uses the information that was obtained from the previous colonies. This is saved in the pheromone accumulative probability matrix $PC \in M_{n \times k}(\mathbb{R})$, whose elements are updated at each iteration $Iter$ as a function of $\mathbf{F} \in M_{n \times k}(\mathbb{R})$:

$$pc_{ij}(Iter) = \frac{\sum_{l=1}^{j} f_{il}(Iter)}{\sum_{l=1}^{k} f_{il}(Iter)} \qquad (7.20)$$

being f_{ij} elements of $\mathbf{F}$ and they express the pheromone level of the discrete jth value of the ith variable.

The elements of $\mathbf{F}$ are also updated at each iteration based on the evaporation factor C_{evap} and the incremental factor C_{inc}:

$$f_{ij}(Iter) = (1 - C_{evap})f_{ij}(Iter - 1) + \delta_{ij,best} C_{inc} f_{ij}(Iter - 1) \qquad (7.21)$$

being

$$\delta_{ij,best} = \begin{cases} 1 \text{ if } x_i^j = x_i^{(best)} \\ 0 \text{ otherwise} \end{cases} \qquad (7.22)$$

and $x_i^{(best)}$ is the ith component of the best solution X^{best}.

The updating of the elements in matrix $\mathbf{F}$ in ACO-d incorporates a coefficient of dispersion, C_{dis}. This extra element is used for a certain number of solutions which are *near* to the best solution. Therefore, the first step is to define the maximum number of neighbors of X^{best}, which receive extra deposition of pheromone. For that, it is adopted a scheme for which each component x_n^{best} has a maximum number of neighbors for receiving pheromone. Let's denote such set of neighbors $V(x_n^{best})$ and let's introduce its definition:

Definition 7.1 Let's denote by X^{best} the vector that represents the best solution provided by ACO-d until a given iteration and x_n^{best} its n components. Neighborhood for dispersion of x_n^{best} is denoted by $V(x_n^{best})$ and it is defined as:

$$V(x_n^{best}) = \left\{ x_n^m \; : \; d(x_n^{best}, x_n^m) < d_{\max}, \; 1 \le m \le k \right\} \tag{7.23}$$

The distance $d_{\max}$, that appears in Definition 7.1, is computed taking the average of the half of all the possible distances between values x_n^m and x_n^r with $m, r = 1, 2, \ldots k$, in ascending order [3]. As a consequence of the discretization, the value for $d_{\max}$ is computed by means of:

$$d_{\max} = \frac{h + 2h + 3h + \cdots + \left[\frac{k}{2}\right] h}{\left[\frac{k}{2}\right]} \tag{7.24}$$

where $h = \frac{b-a}{k}$ with $x_n \in (a, b)$, and $\left[\frac{k}{2}\right]$ represents the nearest integer to $\frac{k}{2}$.

Working with Eq. (7.24), considering the sum of the n first integers, observing that $d(x_n^m, x_n^{m+1}) = h$, and making $x_n^m = a + hm$ and $x_n^{best} = a + h\bar{m}$, the Definition 7.1 can be expressed as:

$$V\left(x_n^{best}\right) = \left\{ x_n^m \; : \; \bar{m} - \frac{\left[\frac{k}{2}\right] + 1}{2} < m < \bar{m} + \frac{\left[\frac{k}{2}\right] + 1}{2}, \; 1 \le m \le k \right\} \tag{7.25}$$

Therefore, the scheme for the pheromone deposit for $x_n^m \in V(x_n^{best})$ is expressed as:

$$f_{nm}(Iter) = f_{nm}(Iter) + \frac{C_{dis}}{|\bar{m} - m|} \tag{7.26}$$

where $f_{nm}(Iter)$ is the value from ACO, see Eq. (7.21).

Matrix $\mathbf{F}$ is randomly initialized. All its elements take random values within the interval $[0, 1]$.

The scheme for generating the new colony considers a parameter q_0. Each zth ant to be generated uses the following scheme:

- Generate n random numbers $q_1^{rand}, q_2^{rand}, \ldots, q_n^{rand}$.
- The value of the nth component of the zth ant is:

$$x_n^{(z)} = \begin{cases} x_n^{\bar{m}} & \text{if } q_n^{rand} < q_0 \\ x_n^{\hat{m}} & \text{if } q_n^{rand} \ge q_0 \end{cases} \tag{7.27}$$

where $\bar{m} \; : \quad f_{n\bar{m}} \ge f_{nm} \; \forall \; m = 1, 2, \ldots, k$ and $\hat{m} \; : \quad \left(pc_{n\hat{m}} > q_n^{rand}\right) \wedge \left(pc_{n\hat{m}} \le pc_{nm}\right) \forall m \ge \hat{m}$

The algorithm for ACO-d is given in Algorithm 7.1. The parameters for the ACO-d algorithm are: $C_{evap}, C_{inc}, C_{dis}, q_0, k$, and Z.

Algorithm 7.1 Algorithm for ACO-d
___
Data: $C_{evap}, C_{inc}, C_{dis}, q_0, k, Z, MaxIter$
Result: Best solution: X^{best}
Discretize the domain interval for each variable in k values;
    Generate a random initial pheromone matrix $\mathbb{F}$ with the same value for all the elements f_{ij};
    Compute matrix PC based on Eq. (7.20);
    Generate the initial colony with Eq. (7.27) and update X^{best};
for $l = 1$ *to* $l = MaxIter$ **do**
    Update matrix $\mathbb{F}$ based on Eq. (7.26);
    Update matrix PC based on Eq. (7.20);
    Generate a new colony with Eq. (7.27);
    Update X^{best};

    Verify stopping criteria;
end for
Solution: X^{best}

7.4.2 Differential Evolution with Particle Collision (DEwPC)

The algorithm Differential Evolution with Particle Collision, DEwPC, was proposed in 2012 [5, 8]. It is intended to improve the performance of the algorithm Differential Evolution, DE [24, 26, 35, 36], with the incorporation of some ideas from the Particle Collision Algorithm, PCA [23, 27, 28].

Differential Evolution is an evolutionary algorithm, and it is based on three operators: Mutation, Crossover, and Selection [35, 36]. The algorithm has three control parameters: size of the population, Z; Crossover Factor, C_{cross}; and Scaling Factor, C_{scal}.

Differential Evolution with Particle Collision keeps the same structure of the operators Mutation and Crossover from DE, while introduces a modification in the Selection operator [5, 8]. This modification adds a new parameter $MaxIter_c$ to the original version of DE.

Let's consider the optimization problem:

$$\min f$$

being the objective function $f : \mathbb{D} \subset \mathbb{R}^n \to \mathbb{R}$.

DEwPC generates at each iteration $Iter$ a new population of Z feasible candidate solutions $P_{Iter} = \left\{ X^1_{Iter}, X^2_{Iter}, \dots, X^Z_{Iter} \right\}$ for the minimization problem just described, based on three operators: Mutation, Crossover, and Selection [35, 36]. These operators have the same names as the operators from GA, but they are based on vector operations [26, 35, 36].

7.4.3   Description of the Algorithm

The initial population $P_0 = \{X_0^1, X_0^2, \ldots, X_0^Z\}$ is formed by Z feasible candidate solutions for the minimization problem. These solution candidates are vectors in $\mathbb{R}^n$.

At each iteration, the operators Mutation, Crossover, and Selection are applied to each member of the population from the previous iteration, following this order of application. As a result, a new population for the current iteration is generated, and the best solution is updated. This scheme is executed until the best solution provided by the algorithm is *good enough* according to a stopping criterion established a priori. The mechanism of generation of DEwPC, as in DE, is summarized by the notation:

$$DE/X_{Iter-1}^{\delta}/\gamma/\lambda$$

where X_{Iter-1}^{δ} and γ summarizes the characteristics of the Mutation and λ summarizes the characteristics of the Crossover. γ is the number of pair of solutions from P_{Iter} to be used for perturbing the last solution X_{Iter-1}^{δ}, and λ represents the distribution function to be used during the Crossover [26].

7.4.3.1   Mutation

The DEwPC Mutation operator coincides with the Mutation operator from DE. It is described as a perturbation at each member of the population from the previous iteration. For the perturbation, other members of the population are used. The scheme for the Mutation is summarized as $X_{Iter-1}^{\delta}/\gamma$ where X_{Iter-1}^{δ} indicates the vector to be perturbed and γ the amount of pairs of vectors (which are randomly selected) to be used for this aim. This process is applied Z times in order to generate a candidate population $\hat{P}_{Iter} = \{\hat{X}_{Iter}^1, \hat{X}_{Iter}^2, \ldots, \hat{X}_{Iter}^Z\}$ of Z solutions. The members of $\hat{P}_{Iter}$ are usually called donor vector. There are different Mutation schemes for DE. The algorithm DEwPC as proposed in [5, 8] uses the scheme:

- $X^{best}/1$: the best solution until the current iteration and one pair of vectors from P_{Iter-1} are used for the generation of a donor vector $\hat{X}_{Iter}^z$:

$$\hat{X}_{Iter}^z = X^{best} + C_{scal}\left(X_{Iter-1}^{\alpha} - X_{Iter-1}^{\beta}\right) \qquad (7.28)$$

where $X^{best}, X_{Iter-1}^{\alpha}, X_{Iter-1}^{\beta} \in \mathbb{R}^n$ are solutions from P_{Iter-1}.

The Mutation operator needs to have a value set for its parameter C_{scal}.

7.4.3.2   Crossover

The DEwPC Crossover operator also coincides with the Crossover of DE. It affects the vectors $\hat{X}^z_{Iter}$, which were obtained after applying the Mutation. The Crossover operator is applied to each component $\hat{x}^z_{(Iter)n}$ of the Z donor vectors $\hat{X}^z_{Iter}$ from population $\hat{P}_{Iter} = \left\{ \hat{X}^1_{Iter}, \hat{X}^2_{Iter}, \ldots, \hat{X}^Z_{Iter} \right\}$. It is described as:

$$\hat{x}^z_{(Iter)n} = \begin{cases} \hat{x}^z_{(Iter)n} & \text{if } q_{rand} \leq C_{cross} \\ x^\delta_{(Iter-1)n} & \text{otherwise} \end{cases} \tag{7.29}$$

where $0 \leq C_{cross} \leq 1$ is the Crossover Factor, another parameter of DE and as a consequence also a parameter of DEwPC; q_{rand} is a random number, which is generated by means of a distribution function.

Equation (7.29) represents how components of a donor vector $\hat{X}^z_{Iter}$ can be incorporated to the vector X^z_{Iter-1} with probability C_{cross}. The vectors obtained after the Crossover operator application are called Trial Vectors.

7.4.3.3   Selection

Finally, the Selection is applied in order to decide which of the Z trial vectors obtained after the Crossover will be part of the population of the current iteration.

The Selection operator of DEwPC is called Selection with Absorption-Scattering with a calculated probability, and takes the ideas of the Absorption and Scattering operators from the algorithm PCA. It was established as:

7.4.3.4   Selection with Absorption-Scattering with a Calculated Probability

- If $f(\hat{X}^z_{Iter}) \leq f(X^\delta_{Iter-1})$, then apply the Absorption operator to $\hat{X}^z_{Iter}$, see Algorithm 7.2.
- If $f(\hat{X}^z_{Iter}) > f(X^\delta_{Iter-1})$, then apply the Scattering operator with a calculated probability to $\hat{X}^z_{Iter}$, see Algorithm 7.3.

Algorithm 7.2 Algorithm for the Absorption operator

Data: $\hat{X}_{Iter}$
Result: X_{Iter}
$X_{Iter} = \hat{X}_{Iter}$;
SmallSearch(X_{Iter}, $MaxIter_c$);

Algorithm 7.3 Algorithm for the Scattering operator with a calculated probability

---

Data: $\hat{X}_{Iter}$, $f(X^{best})$
Result: X_{Iter}
Compute $f(\hat{X}_{Iter})$;
Compute $p_{r(Iter)} = 1 - \frac{f(X^{best})}{f(\hat{X}_{Iter})}$;
if $rand < p_{r(Iter)}$ **then**
> $X_{Iter} = \hat{X}_{Iter}$;
> Search(X_{Iter}, $MaxIter_c$);
> **else**;
> $X_{Iter} = X_{Iter-1}$;
end if

The Small Search function in the Absorption operator shown in Algorithm 7.2 indicates a small stochastic perturbation around a solution candidate. The Search function in the Scattering operator algorithm with a calculated probability in Algorithm 7.3 indicates a stochastic perturbation around a solution [5, 8, 9].

The Selection with Absorption-Scattering with a calculated probability is not applied to all the solution candidates at each iteration. It is combined with the Selection operator of DE which is described by:

The vector X^z_{Iter}, to become part of the population of the current iteration, is selected following the rule:

$$X^z_{Iter} = \begin{cases} \hat{X}^z_{Iter} & \text{if } f(\hat{X}^z_{Iter}) \le f(X^\delta_{Iter-1}) \\ X^\delta_{Iter-1} & \text{otherwise} \end{cases} \tag{7.30}$$

The algorithm for DEwPC is represented in Algorithm 7.4.

7.4.4 Covariance Matrix Adaptation Evolution Strategy (μ_w, λ) CMA-ES

The Covariance Matrix Adaptation Evolution Strategy, CMA-ES, is also an evolutionary algorithm [19, 20]. It was initially introduced to improve the local search performance of evolution strategies but it has revealed itself competitive in global search performances [19, 25].

The algorithm generates at each iteration $Iter$ a new population of λ feasible solutions $X^1_{Iter}, X^2_{Iter}, \ldots, X^\lambda_{Iter}$ with the application of a mutation operator. Here, the mutation uses the information of the μ better solutions generated in the former iteration $X^{\mu(1)}_{Iter-1}, X^{\mu(2)}_{Iter-1}, \ldots, X^{\mu(\mu)}_{Iter-1}$. This mechanism can be summarized with the notation:

$$(\mu_w, \lambda)$$

Algorithm 7.4 Algorithm for DEwPC

Data: C_{cross}, C_{scal}, Z, $MaxIter$, Generation Mechanism, $MaxIter_c$
Result: X^{best}

Generate an initial population of Z solutions;
Select best solution X^{best};
for $Iter = 1$ *to* $Iter = MaxIter$ **do**
  Apply Mutation, Eq. (7.28);
  Apply Crossover, Eq. (7.29);
  for $j = 1$ *to* $j = Z$ **do**
    if $rand < 0.7$ **then**
      Apply Selection with Absorption-Scattering with probability to $\hat{X}_{Iter}^{(j)}$, in
      Algorithms 7.2 and 7.3;
      else;
      Apply Selection to $\hat{X}_{Iter}^{(j)}$;
    end if
  end for
  Update X^{best};
  Verify stopping criteria;
end for
Solution: X^{best}

The values of μ and λ are dependent on the dimension of the search space n [19], see Table 7.2.

7.4.4.1 Mutation

The Mutation operator for CMA-ES is described by the equation:

$$X_{Iter}^z = X_{Iter-1}^{wmean} + \sigma_{Iter-1}\mathcal{N}\left(\mathbf{0}, C_{Iter-1}\right), \text{ with } z = 1, 2, \ldots, \lambda \qquad (7.31)$$

where $X_{Iter-1}^{wmean} \in \mathbb{R}^n$ is the weighted mean of the better μ solutions generated at the former iteration $Iter - 1$; σ_{Iter-1} is a real value named step-size and C_{Iter-1} is a covariance matrix. Both parameters have been updated at the end of the former iteration $Iter - 1$. It is possible to see that X_{Iter}^z follows a normal distribution with parameters $\mathcal{N}(X_{Iter-1}^{wmean}, \sigma_{Iter-1}^2 C_{Iter-1})$.

- The evolution of the step-size σ_{Iter} is determined by the conjugate evolution path $p_{\sigma(Iter-1)}$:

Table 7.2 Default parameter setting for (μ_w, λ) CMA-ES

λ	μ	w_i	c_c, c_σ	c_{cov}	d_σ	c_c^u
$4 + [3\ln(n)]$	$\lambda/2$	$\ln\left(\frac{\lambda+1}{2}\right) - \ln(i)$	$\frac{4}{n+4}$	$\frac{2}{(n+\sqrt{2})^2}$	$c_\sigma^{-1} + 1$	$\sqrt{c_c(2 - c_c)}$

$$\sigma_{Iter} = \sigma_{Iter-1} \exp\left(\frac{1 - \|p_{\sigma(Iter-1)}\| - \hat{\xi}_n}{d_\sigma} \frac{}{\hat{\xi}_n}\right) \tag{7.32}$$

being $d_\sigma \geq 1$ a damping parameter that affects the feasible change rate of σ_{Iter}; $\hat{\xi}_n = E[\|\mathcal{N}(0, I)\|]$ represents the expectation of the length of a $(0, I)$-normally distributed random vector, i.e. $\mathcal{N}(0, I)$ and $p_{\sigma(Iter-1)}$ is the conjugate evolution path.

- The conjugate evolution path that appears in the computation of the step-size σ_{iter} is calculated by means of:

$$p_\sigma(Iter) = (1 - c_\sigma)p_{\sigma(Iter-1)}$$
$$+ c_\sigma^u B_{Iter-1}(D_{Iter-1})^{-1}(B_{Iter-1})^{-1} \frac{c_w}{\sigma_{Iter-1}} \left[X_{Iter}^{wmean}\right.$$
$$\left. - X_{Iter-1}^{wmean}\right] \tag{7.33}$$

being $c_\sigma \in (0, 1]$ a parameter that determines the cumulation time for p_σ; $c_\sigma^u = \sqrt{c_\sigma(2 - c_\sigma)}$ is another parameter, which is derived from c_σ, and B and D are matrices that satisfy $C_{Iter} = B_{Iter}D_{Iter}(B_{Iter}D_{Iter})^t$, see Ref. [19] for their computation.

- The matrix C_{Iter} is updated at the end of the iteration $Iter - 1$ by means of the evolution path $p_{c(Iter)}$:

$$C_{Iter} = (1 - c_{cov})C_{Iter-1} + c_{cov} p_{c(Iter)}(p_{c(Iter)})^t \tag{7.34}$$

where $c_{cov} \in [0, 1]$ is the change rate of the covariance matrix C, and it is the same throughout the algorithm, see Table 7.2 for its value.

- The value of $p_{c(Iter)}$ is updated at each iteration before computing C_{Iter} by means of:

$$p_{c(Iter)} = (1 - c_c)p_{c(Iter-1)} + c_c^u \frac{c_w}{\sigma_{Iter}} \left[X_{Iter}^{wmean} - X_{Iter-1}^{wmean}\right] \tag{7.35}$$

being $c_c \in (0, 1]$ a parameter that determines the cumulation time for p_c; c_c^u is another parameter that normalizes the variance of p_c and $c_w = \frac{\sum_{i=1}^{\mu} w_i}{\sum_{i=1}^{\mu} w_i^2}$ where w_i is the weight of the solution $X_{Iter}^{\mu(i)}$.

7.4.4.2 Selection

In the operator Selection, the better μ vectors at iteration $Iter$ are taken, i.e. $X_{Iter}^{\mu(1)}, X_{Iter}^{\mu(2)}, \ldots X_{Iter}^{\mu(\mu)}$, such that:

$$f\left(X_{Iter}^{\mu(1)}\right) \leq f\left(X_{Iter}^{\mu(2)}\right) \leq \cdots \leq f\left(X_{Iter}^{\mu(\mu)}\right) \qquad (7.36)$$

Each solution at the operator Selection has weight w_i, with $i = 1, \ldots \mu$. The value of $X_{Iter}^{wmean} = \frac{\sum_{i=1}^{\mu} w_i X_{Iter}^{\mu(i)}}{\sum_{i=1}^{\mu} w_i}$ is computed. This value will be used in the operator Mutation at the next iteration.

Table 7.2 shows the default parameters setting for (μ_w, λ) CMA-ES proposed in Ref. [19].

The algorithm for CMA-ES is presented in Algorithm 7.5.

Algorithm 7.5 Algorithm for CMA-ES

Data: $n, \sigma_0, X_0^{wmean}, C_0, B_0, C_0, p_{c(0)}, p_{\sigma(0)}, MaxIter$
Result: X^{best}
Compute the parameters: $\lambda, \mu, c_c, c_\sigma, c_{cov}, c_c^u, d_\sigma, w_i$ (see Table 7.2);
Make best solution $X^{best} = X_0^{wmean}$;
for $Iter = 1$ to $Iter = MaxIter$ **do**
   Apply Mutation, Eq. (7.31);
   Apply Selection, Eq. (7.36);
   Update C_{Iter}, Eq. (7.34);
   Update σ_{Iter}, Eq. (7.32);
   Update X^{best};
   Verify stopping criteria;
end for
Solution: X^{best}

7.5 Test Cases Considered and Parameters Used in the Metaheuristics

The robustness and the computational cost were used as the criteria for obtaining information about the reliability and viability of the diagnosis, respectively. Table 7.3 shows the five faulty situations that are considered in the numerical experiments. They differ in the percentual noise level that affects the output. In order to compute some descriptive statistics, each algorithm was run 25 times for each of the five faulty situations that appear in Table 7.3.

Table 7.3 Test cases

	Case 1	Case 2	Case 3	Case 4	Case 5
θ_1 (see Eq. (7.10))	0.01	0.01	0.01	0.01	0.01
θ_2 (see Eq. (7.11))	0.015	0.015	0.015	0.015	0.015
Noise level	2%	5%	8%	10%	15%

With the aim of comparing the performance of each algorithm, when diagnosing the bioreactor, the indicators Success Rate (SR) and Success Performance (SP) were computed. The indicator SR gives the percent of successful runs, here named EE, of the algorithm, while SP $= \frac{\overline{Eval}_{EE}}{SR}$, being $\overline{Eval}_{EE}$ the average of number of evaluations of the objective function for successful runs [37]. The SP indicator gives an idea about the number of evaluations of the objective function that an algorithm needs for reaching a required accuracy in the estimations [37]. In this case, a successful run is obtained when the fault estimated values have a percentual relative error lower than the prescribed values $Err_{\theta 1}$ and $Err_{\theta 2}$, respectively.

The mean for the fault estimated values, θ_1 and θ_2, were computed. The mean value of the objective function, and the mean of the number of iterations required for the successful runs, whose notations are $\bar{\theta}_1$, $\bar{\theta}_2$, $\bar{F}_{EE}$ and $\overline{Iter}_{EE}$, respectively, were also computed.

7.5.1   Implementation of ACO-d

It is based on Algorithm 7.1. The parameter values used in the ACO-d algorithm were: $C_{evap} = 0.10$, $C_{inc} = 0.30$, $C_{dis} = 0.10$, $q_0 = 0.15$, $k = 63$, and $Z = 10$.

7.5.2   Implementation of DEwPC

It is based on Algorithm 7.4. The parameter values used in the DEwPC algorithm were: $Z = 10$; $C_{cross} = 0.9$; $C_{scal} = 0.6$, and $MaxIter_c = 5$, following the recommendations given in Ref. [5].

7.5.3   Implementation of (μ_w, λ) CMA-ES

It is based on Algorithm 7.5. The initial parameter values used in the (μ_w, λ) CMA-ES algorithm were $n = 2$; $\sigma_0 = 1$; $C_0 = I_{2 \times 2}$; $B_0 = I_{2 \times 2}$; $C_0 = I_{2 \times 2}$; $p_{c(0)} = [0 \ 0]^t$; $p_{\sigma(0)} = [0 \ 0]^t$; $X_0^{wmean} = [1 \ 1]^t$; and the other parameters are based on the formulas given in Table 7.2 with $n = 2$. The symbol I just used indicates the identity matrix.

7.6   Results

In Table 7.4 the values of the indicators SR and SP obtained by each algorithm, when diagnosing Cases 1–5, are presented (see Table 7.3). The algorithms were executed under two stopping criteria: maximum number of iterations ($MaxIter = 200$) and maximum allowed percent of relative error in the fault estimated values $Err_{\theta 1} = 5\%$ and $Err_{\theta 2} = 5\%$). It was considered that a successful run is that in which the algorithm reached both stopping criteria, i.e. $Err_{\theta 1} \leq 5\%$ and $Err_{\theta 2} \leq 5\%$. It can be observed that the SR for each algorithm decreases from Case 1 to Case 5, which is related to the increase of the noise level, from Case 1 to Case 5. The algorithms ACO-d and DEwPC achieved SR $= 100\%$ when diagnosing Cases 1 and 2, i.e. the diagnosis of the faults was 100% successful for a percent of relative error in the estimated values less than 5% (the diagnosis was at least 95% reliable). The best SR obtained with (μ_w, λ) CMA-ES was SR $= 96\%$, which corresponds to its values for Cases 1 and 2.

Table 7.4 shows also that for Cases 3 and 4, the SR indicators of ACO-d and DEwPC are similar, and lower than for Cases 1 and 2. The (μ_w, λ) CMA-ES algorithm showed similar behavior to ACO-d and DEwPC for Case 3, but for Case 4 its SR is very small.

For Case 5, it can be observed in Table 7.4 that ACO-d and DEwPC did not achieve successful runs, but (μ_w, λ) CMA-ES achieved an SR $= 12\%$.

From the analysis of the SR values shown in Table 7.4, it is concluded that for a noise level up to 5%, the diagnosis of these faults with a 95% of reliability (up to 5% of relative error in the estimated values) is 100% successful with ACO-d and DEwPC. Instead for a noise up to 10%, the success rate decreases until 84% for ACO-d and DEwPC. For a noise higher than 10%, it is not possible to diagnose the faults with a 95 % of reliability using DEwPC and ACO-d. The (μ_w, λ) CMA-ES algorithm was able to do that, but with a lower percent of success (12%).

The indicator SP can be used for analyzing the computational cost, which can be linked with the viability of the diagnosis. Table 7.4 shows that for Cases 1–3 the SP values obtained with the three algorithms are similar for each case. It means that the computational cost for diagnosing these faults with a percent of relative error less

Table 7.4 Values of the indicators SR and SP obtained with the three algorithms ACO-d, DEwPC, and (μ_w, λ) CMA-ES, when diagnosing Cases 1–5 from Table 7.3, with maximum percent of relative error in the fault estimated values $Err_{\theta 1} = 5\%$ and $Err_{\theta 2} = 5\%$

Case	ACO-d		DEwPC		(μ_w, λ) CMA-ES	
	SR	SP	SR	SP	SR	SP
1	100%	320	100%	311	96%	305
2	100%	280	100%	290	96%	321
3	84%	453	88%	340	72%	361
4	84%	381	84%	363	16%	2137
5	0%	–	0%	–	12%	2250

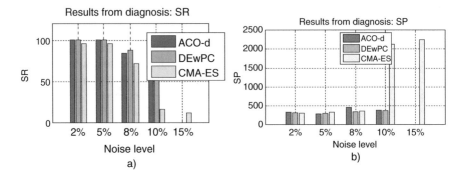

Fig. 7.1 Graphical representation of the results presented in Table 7.4. (**a**) SR. (**b**) SP

than 5% and with a noise level affecting the measurements up to 8% is similar for the three algorithms. But, the success rate obtained with (μ_w, λ) CMA-ES is always smaller than the values provided by the other two algorithms. For Case 4, the value of SR obtained with the algorithm (μ_w, λ) CMA-ES is almost six times higher than the values obtained with the other two algorithms. For Case 5, which corresponds to a noise level up to 15%, the algorithm (μ_w, λ) CMA-ES shows an SP similar to its results for Case 4. It is summarized that for a level noise up to 10%, the diagnosis obtained with the algorithms ACO-d and DEwPC is good, with an SR of 84% and an SP less or equal to 381. For a noise level up to 15%, the algorithms ACO-d and DEwPC are not a viable choice for diagnosing the faults. The algorithm (μ_w, λ) CMA-ES provides not so good results as the other two algorithms for the noise level up to 10%, but its performance for a noise level up to 15% is similar to its results for the noise level up to 10%. Therefore, for a higher noise level, the algorithm (μ_w, λ) CMA-ES seems to be a viable choice. Figure 7.1 shows graphically the results presented in Table 7.4.

Table 7.5 shows the results of the diagnosis when only the successful runs are considered. The value of the objective function increases from Case 1 up to Case 5, due to the increase of the noise level. It is also shown that the smallest average number of iterations for successful runs is reached by DEwPC in four out of the five cases (Cases 1–4) considered. The best values obtained for all parameters shown in Table 7.5 are indicated boldface.

In Table 7.6 are shown the mean values for the fault estimates obtained with each algorithm. For computing this average, it is only considered the values obtained in the non-successful runs. For Cases 1 and 2, (μ_w, λ) CMA-ES was the only algorithm that did not reach 100% of success, but it can be noticed that it achieved a maximum percent of relative error in the fault estimation equal to 11.3%. For Cases 3–5, the lowest sum of the relative errors of the fault estimates for the non-successful runs, i.e. $(ErrRel_{\theta 1} + ErrRel_{\theta 2})$, was reached by DEwPC. This table shows also that the diagnosis of the faults in Cases 3 and 4 with a percent of relative error in each estimation less than 8% is 100% successful with DEwPC. As shown in Table 7.4, for Case 5 only (μ_w, λ) CMA-ES achieved successful runs, but with a low SR = 12%.

Table 7.5 Results from the diagnosis process: considering only successful runs

Case	Algorithm	$\bar{\theta}_1(\theta_1 = 0.01)$	$\bar{\theta}_2(\theta_2 = 0.015)$	$\bar{F}_{EE}$	$\bar{Iter}_{EE}$
1	ACO-d	**0.0100**	0.0145	1.7799e−004	16
	DEwPC	0.0101	**0.0148**	**1.4438e−004**	**11**
	(μ_w, λ) CMA-ES	0.0099	0.0153	1.6452e−004	51
2	ACO-d	0.0103	**0.0148**	9.4406e−004	14
	DEwPC	**0.0098**	**0.0152**	7.1911e−004	**11**
	(μ_w, λ) CMA-ES	0.0105	**0.0152**	**6.9974e−004**	52
3	ACO-d	0.0102	**0.0150**	**0.0023**	19
	DEwPC	**0.0101**	0.0152	**0.0023**	12
	(μ_w, λ) CMA-ES	0.0095	0.0156	0.0031	38
4	ACO-d	0.0099	0.0144	0.0045	16
	DEwPC	0.0101	**0.0154**	**0.0039**	12
	(μ_w, λ) CMA-ES	**0.0100**	0.0144	0.0049	57
5	(μ_w, λ) CMA-ES	**0.0098**	**0.0145**	**0.0075**	45

Table 7.6 Results from the diagnosis considering only non-successful runs

Case	Algorithm	$\bar{\theta}_1(\theta_1 = 0.01)$	$\bar{\theta}_2(\theta_2 = 0.015)$	$Err_{\theta1}$	$Err_{\theta2}$
1	(μ_w, λ) CMA-ES	0.0107	0.0167	7%	11.3%
2	(μ_w, λ) CMA-ES	0.0108	0.0160	8%	6.7%
3	ACO-d	0.0111	0.0146	11%	2.6%
	DEwPC	0.0093	0.0152	7%	1.3%
	(μ_w, λ) CMA-ES	0.0083	0.0151	17%	0.6%
4	ACO-d	0.0111	0.0149	11%	0.6%
	DEwPC	0.0092	0.0148	8%	1.3%
	(μ_w, λ) CMA-ES	0.0088	0.0149	12%	0.6%
5	ACO-d	0.0117	0.0174	17%	16%
	DEwPC	0.0082	0.0158	18%	5.3%
	(μ_w, λ) CMA-ES	0.0082	0.0166	18%	10.6%

However, the results in Table 7.6 show that for Case 5, (μ_w, λ) CMA-ES achieved a maximum sum of the percent of relative errors in the estimations equal to 28.6% and for DEwPC this value is 23.3%.

Figure 7.2 shows a comparison between the real output of the system and the one obtained after the fault diagnosis with the metaheuristics. In particular, Fig. 7.3a shows the comparison for Case 3 and the values obtained with DEwPC. Figure 7.3b shows the comparison for Case 5 and the values obtained with (μ_w, λ) CMA-ES. These graphics are useful for observing the robustness of the diagnosis. For both graphics it is observed that even when the output values used for the metaheuristics for diagnosing the system were corrupted by noise, the estimated output obtained

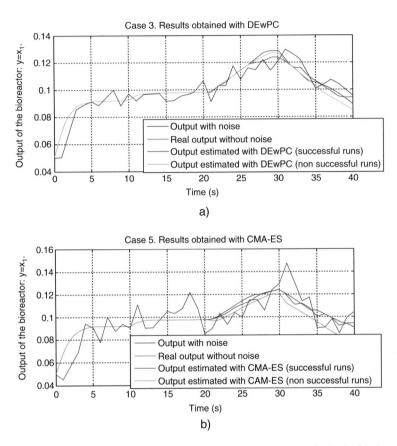

Fig. 7.2 Comparison between the real output and the estimated output obtained with the metaheuristics. (**a**) Case 3, Diagnosis with DEwPC. (**b**) Case 5, Diagnosis with CMA-ES

with the metaheuristics for the successful runs is very close to the real output of the system that is not affected by noise.

The experiments were repeated for those cases and algorithms that did not reach the SR $= 100\%$ in the previous experiments (see results in Table 7.4). This time, the values of the stopping criteria Err_{θ_1} and Err_{θ_2} used for the algorithms were increased to $Err_{\theta_1} = Err_{\theta_2} = 10\%$. The objective is to analyze the performances of the algorithms when a higher percent of relative error in the estimated values is allowed. Table 7.7 shows the results of the diagnosis performed. With the symbol $(-)$ it is indicated that the experiments were not repeated, because they reached an SR $= 100\%$ in the previous experiments (see results in Table 7.4). It can be observed that DEwPC and ACO-d reached SR $= 100\%$ for Cases 3 and 4 with values of SP similar to the values that were obtained for Cases 1 and 2 with smaller values of Err_{θ_1} and Err_{θ_2}, i.e. 5%. Moreover, the algorithm (μ_w, λ) CMA-ES reached SR $= 100\%$ for Cases 1–3 with similar values for SP to those obtained with ACO-d

Table 7.7 Values of SR and SP obtained with the three algorithms ACO-d, DEwPC, and (μ_w, λ) CMA-ES when the maximum allowed percentage of relative error in the fault estimated values Err_{θ_1} and Err_{θ_2} is increased up to $Err_{\theta_1} = Err_{\theta_2} = 10\%$

Case	ACO-d		DEwPC		(μ_w, λ) CMA-ES	
	SR	SP	SR	SP	SR	SP
1	–	–	–	–	100%	207
2	–	–	–	–	100%	212
3	100%	260	100%	253	100%	281
4	100%	280	100%	249	72%	408
5	30%	3125	32%	3051	72%	493

and DEwPC for Cases 3 and 4. When increasing the noise level up to 15%, the performance of ACO-d and DEwPC is degraded: the values of SR decreased to 30% and 32%, respectively, and the values of SP increased more than ten times the values obtained for Cases 3 and 4. It is then observed that for Case 5 the diagnosis is better with (μ_w, λ) CMA-ES. Moreover, the diagnosis performance with (μ_w, λ) CMA-ES is similar when the noise level increases from 10% up to 15%. For a noise level higher than 10%, (μ_w, λ) CMA-ES has a 72% of success of diagnosing the system with a maximum error in the estimated values equal to $Err_{\theta_1} = Err_{\theta_2} = 10\%$.

This robustness-computational cost study based on simulations of the system can be used for building decision tables which can be presented to the experts. These tables can be used with the aim of choosing an algorithm for diagnosing the system based on the desirable balance between reliability and computational cost of the diagnosis. For the bioreactor under study and the three algorithms applied, Tables 7.8 and 7.9 were obtained. Each table represents the results for a certain maximum percent of relative error in the fault estimated values, with different noise level in the output. For each noise level, it is presented the reliability (in this case understood as the maximum percent of success (SR) that can be obtained), the computational cost that corresponds to this case, and which algorithm provides the diagnosis with these characteristics. The following examples help to understand the results shown in Tables 7.8 and 7.9:

- Example 1: Suppose that it is necessary a diagnosis with a high reliability (100% of success) with a maximum percent of error in the fault estimated values of $Err_{\theta_1} = Err_{\theta_2} = 5\%$. Due to the desirable maximum percent of error in the fault estimated values, it is necessary to have a look in Table 7.8. This table shows that a diagnosis with this characteristics is only feasible for noise level up to 5% in the output, and the algorithms DEwPC and ACO-d are able to provide it.
- Example 2: If it is known that the output of system is affected by a noise level up to 15%, a reliable diagnosis cannot be obtained for a maximum percent of error in the fault estimation equal to $Err_{\theta_1} = Err_{\theta_2} = 5\%$. This is observed from Table 7.8: the SR for this situation is 12% and moreover, the computational cost for this not reliable diagnosis is high (this best result is provided by the algorithm

Table 7.8 Expected behavior of diagnosis when the maximum percent of relative error in fault estimated values is $Err_{\theta_1} = Err_{\theta_2} = 5\%$

Noise level	Success rate	Computational cost	Algorithm
2%	100%	Low	DEwPC, ACO-d
5%	100%	Low	DEwPC, ACO-d
8%	88%	Medium	DEwPC
10%	84%	Medium	DEwPC, ACO-d
15%	12%	High	(μ_w, λ) CMA-ES

Table 7.9 Expected behavior of diagnosis when the maximum percent of error in fault estimated values is $Err_{\theta_1} = Err_{\theta_2} = 10\%$

Noise level	Success rate	Computational cost	Algorithm
2%	100%	Low	DEwPC, ACO-d
5%	100%	Low	DEwPC, ACO-d
8%	100%	Low	DEwPC, ACO-d
10%	100%	Low	DEwPC, ACO-d
15%	72%	Medium	(μ_w, λ) CMA-ES

(μ_w, λ) CMA-ES). For this noise level in the output, Table 7.9 shows that when a maximum percent of relative error in the fault estimations is increased up to $Err_{\theta_1} = Err_{\theta_2} = 10\%$, then, the reliability of the diagnosis is increased up to SR $= 72\%$ and the computational cost for this diagnosis is medium (this diagnosis with such characteristics is provided by (μ_w, λ) CMA-ES).

Furthermore, the simulations also allow to establish better stopping criteria in order to improve the performance of the algorithms when diagnosing the system. For that purpose, the level of noise in the output (Err_y) and the percent of error accepted in fault estimated values (Err_{θ_i}) can be incorporated in the stopping criteria of the algorithms as follows:

- Value of the objective function: $f(\theta_1, \theta_2) \leq M(Err_y, Err_{\theta_i})$
- Number of objective function evaluations: $MaxEval \leq$ SP

7.6.1 Comparison with Other FDI Methods

The results of this study are compared with those obtained from the application of a new adaptive estimation algorithm for recursive estimation of the parameters related to faults, see results in Ref. [39]. Such algorithm was designed in a constructive manner through a nontrivial combination of a high gain observer and a linear adaptive observer [39]. The application of this method requires the design of an observer. The observer parameters for the bioreactor are $m = 2, \delta = 0.5, \rho = 8, \Gamma = diag([4; 2]), \Omega = diag([5; 5])$, see Ref. [39].

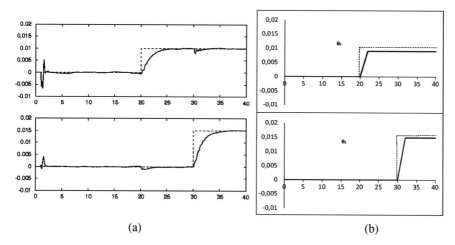

(a)   (b)

Fig. 7.3 Comparison between the results reported in Ref. [39] and those presented in this chapter, for Case 4, in Table 7.3. The dashed lines represent the exact fault values, and the solid lines are the estimated values. (**a**) Results reported in Ref. [39]. (**b**) Results based on metaheuristics

The results reported in Ref. [39] are for Case 4 described in Table 7.3. In Fig. 7.3 it is shown a comparison between the results from Ref. [39] and the results of this chapter. Both methods give good fault estimated values, but the diagnosis with metaheuristics does not require the design of further elements. Moreover, the use of metaheuristics can be easy generalized, and it has less parameters that depend on the system or process to be adjusted, while the observer most be designed for every system. This implies that the diagnosis based on metaheuristics is easier to implement than the observer based diagnosis.

7.7   Conclusions

In this chapter was presented the formulation of Fault Diagnosis for a bioreactor as an optimization problem, and its solution by means of metaheuristics. The study was developed using simulated data of a nonlinear bioreactor showing its suitability. Furthermore, the results were compared with other reported in the literature.

The influence of the metaheuristic stopping criteria allowed to analyze the robustness-computational cost balance. For that two indicators used for analyzing the performance of metaheuristics were computed (SR and SP). The results were used for building decision tables which can be presented to the experts. These tables can be used with the aim of choosing a metaheuristic for diagnosing the system based on the desirable balance between reliability (related to the robustness) and diagnosis time (related to computational cost).

The study also showed how to improve the stopping criteria of metaheuristics, when they are applied to FDI.

Acknowledgements The authors acknowledge the Brazilian Research supporting agencies CAPES—Fundação Coordenação de Aperfeiçoamento de Pessoal de Nível Superior, CNPq—Conselho Nacional de Desenvolvimento Científico e Tecnológico, and FAPERJ—Fundação Carlos Chagas Filho de Amparo à Pesquisa do Estado do Rio de Janeiro, as well as UERJ, Universidade do Estado do Rio de Janeiro and CUJAE, Universidad Tecnológica de La Habana José Antonio Echeverría.

References

1. Acosta Díaz, C., Camps Echevarría, L., Prieto Moreno, A., Silva Neto, A.J., Llanes Santiago, O.: A model-based fault diagnosis in a nonlinear bioreactor using an inverse problem approach. Chem. Eng. Res. Des. **114**, 18–29 (2016). https://doi.org/10.1016/j.cherd.2016.08.005
2. Becceneri, J.C., Zinober, A.: Extraction of energy in a nuclear reactor. In: XXXIII Brazilian Symposium on Operational Research, Campos do Jordão (2001)
3. Becceneri, J.C., Sandri, S., Luz, E.F.P.: Using ant colony systems with pheromone dispersion in the traveling salesman problem. In: Proceedings of the 11th International Conference of the Catalan Association for Artificial Intelligence, Sant Martí d'Empúries (2008)
4. Blum, C.: Ant colony optimization: introduction and recent trends. Phys. Life Rev. **2**(4), 353–373 (2005)
5. Camps Echevarría, L.: Fault diagnosis based on inverse problems. Ph.D. thesis, Instituto Superior Politécnico José Antonio Echeverría (2012)
6. Camps Echevarría, L., Llanes Santiago, O., Silva Neto, A.J.: A proposal to fault diagnosis in industrial systems using bio-inspired strategies. Ingeniare. Rev. Chil. Ing. **19**(2), 240–252 (2011)
7. Camps Echevarría, L., Llanes Santiago, O., Silva Neto, A.J.: Fault diagnosis based on inverse problem solution. In: 2011 International Conference on Inverse Problems in Engineering (ICIPE), Orlando (2011)
8. Camps Echevarría, L., Llanes Santiago, O., Silva Neto, A.J., Campos Velho, H.F.: An approach of fault diagnosis using meta-heuristics: a new variant of the differential evolution algorithm. Revista Computación y Sistemas (2012)
9. Camps Echevarría, L., Llanes Santiago, O., Silva Neto, A.J., Campos Velho, H.F.: Meta heuristics in the faults diagnosis: modification of the algorithm differential evolution. In: 2nd International Conference on Computational and Informatics Sciences, Havana (2013)
10. Camps Echevarría, L., Silva Neto, A.J., Llanes Santiago, O., Hernández Fajardo, J.A., Saánchez, D.J.: A variant of the particle swarm optimization for the improvement of fault diagnosis in industrial systems via faults estimation. Eng. Appl. Artif. Intell. **28**, 36–51 (2014)
11. Camps Echevarría, L., Campos Velho, H.F., Becceneri, J.C., Silva Neto, A.J., Llanes Santiago, O.: The fault diagnosis inverse problem with ant colony optimization and fuzzy ant colony optimization. Appl. Math. Comput. **227**(15), 687–700 (2014)
12. Contois, D.: Kinetics of bacteria growth relationship between population density and specific growth rate of continuous cultures. J. Genet. Macrobiol. **21**, 40–50 (1959)
13. Ding, S.X.: Model-Based Fault Diagnosis Techniques. Springer, Berlin (2008)
14. Dorigo, M.: Ottimizzazione, Apprendimento Automatico, Ed Algoritmi Basati su Metafora Naturale. Ph.D. thesis, Politécnico di Milano (1992)
15. Dorigo, M., Blum, C.: Ant colony optimization theory: a survey. Theor. Comput. Sci. **344**(2–3), 243–278 (2005)

16. Dorigo, M., Maniezzo, V., Colorni, A.: The ant system: optimization by a colony of cooperating agents. IEEE Trans. Syst. Man Cybern. B **26**(1), 29–41 (1996)
17. Frank, P.M.: Analytical and qualitative model-based fault diagnosis – a survey and some new results. Eur. J. Control **2**(1), 6–28 (1996)
18. Gauthier, J.P., Hammouri, H., Othman, S.: A simple observer for nonlinear systems, application to bioreactors. IEEE Trans. Autom. Control **37**(6), 875–880 (1992)
19. Hansen, N., Ostermeier, A.: Completely derandomized self-adaptation in evolution strategies. Evol. Comput. **9**(2), 159–195 (2001)
20. Hansen, N., Mueller, S.D., Koumoutsakos, P.: Reducing the time complexity of the derandomized evolution strategy with covariance matrix adaptation CMA-ES. Evol. Comput. **11**(1), 1–18 (2003)
21. Isermann, R.: Model based fault detection and diagnosis. status and applications. Annu. Rev. Control **29**(1), 71–85 (2005)
22. Isermann, R., Ballé, P.: Trends in the application of model-based fault detection and diagnosis of technical processes. Control Eng. Pract. **5**, 709–719 (1997)
23. Knupp, D.C., Silva Neto, A.J., Sacco, W.F.: Estimation of radiactive properties with the particle collision algorithm. In: Inverse Problems, Design and Optimization Symposium, Miami (2007)
24. Mezura-Montes, E., Velázquez-Reyes, J., Coello-Coello, C.A.: A comparative study of differential evolution variants for global optimization. In: GECCO 06, Seattle, Washington (2006)
25. Pavlidis, N.G., Parsopoulos, K.E., Vrahatis, M.N.: Computing Nash equilibria through computational intelligence methods. J. Comput. Appl. Math. **175**(1), 113–136 (2005)
26. Price, K.V., Storn, R.M., Lampinen, J.A.: Differential Evolution – A Practical Approach to Global Optimization. Springer, Berlin (2005)
27. Sacco, W.F., Oliveira, C.R.E.: A new stochastic optimization algorithm based on particle collisions. In: 2005 ANS Annual Meeting, Transactions of the American Nuclear Society (2005)
28. Sacco, W.F., Oliveira, C.R.E., Pereira, C.M.N.A.: Two stochastic optimization algorithms applied to nuclear reactor core design. Prog. Nucl. Energy **48**(6), 525–539 (2006)
29. Silva Neto, A.J., Llanes Santiago, O., Silva, G.N. (eds.): Mathematical Modelling and Computational Intelligence in Engineering Applications. Springer, Basel (2016)
30. Simani, S., Patton, R.J.: Fault diagnosis of an industrial gas turbine prototype using a system identification approach. Control. Eng. Pract. **16**(7), 769–786 (2008)
31. Simani, S., Fantuzzi, C., Patton, R.J.: Model-Based Fault Diagnosis in Dynamics Systems Using Identifications Techniques. Springer, London (2002)
32. Socha, K.: Ant colony optimization for continuous and mixed-variable domains. Ph.D. thesis, Universite Libre de Bruxelles (2008)
33. Socha, K., Dorigo, M.: Ant colony optimization for continuous domains. Eur. J. Oper. Res. **185**(3), 1155–1173 (2008)
34. Stephany, S., Becceneri, J.C., Souto, R.P., Campos Velho, H.F., Silva Neto, A.J.: A preregularization scheme for the reconstruction of a spatial dependent scattering albedo using a hybrid ant colony optimization implementation. Appl. Math. Model. **34**(3), 561–572 (2010)
35. Storn, R., Price, K.: Differential evolution – a simple and efficient adaptive scheme for global optimization over continuous spaces. Technical report, International Computer Science Institute (1995)
36. Storn, R., Price, K.: Differential evolution – a simple and efficient adaptive heuristic for global optimization over continuous spaces. J. Glob. Optim. **11**(4), 341–359 (1997)
37. Suganthan, P., Hansen, N., Liang, J., Deb, K., Chen, Y.P., Auger, A., Tiwari, S.: Problem definitions and evaluation criteria for the CEC 2005 special session on real parameter optimization. Technical Report, Nanyang Technological University (2005)
38. Venkatasubramanian, V., Rengaswamy, R., Yin, K., Kavuri, S.N.: A review of process fault detection and diagnosis. Part 1: quantitative model-based methods. Comput. Chem. Eng. **27**, 293–311 (2002)
39. Xu, A., Zhang, Q.: Nonlinear system fault diagnosis based on adaptive estimation. Automatica **40**, 1181–1193 (2004)

Chapter 8
Optimization of Nuclear Reactors Loading Patterns with Computational Intelligence Methods

Anderson Alvarenga de Moura Meneses, Lenilson Moreira Araujo, Fernando Nogueira Nast, Patrick Vasconcelos da Silva, and Roberto Schirru

8.1   Introduction

According to the International Energy Agency [15], worldwide nuclear energy production in 2014 was 2535 TWh, which is approximately 4.3 times Brazil's global energy consumption (590.5 TWh [37]). Thus, nuclear power corresponds to a relevant share in total electric generation in many countries such as France (78.4%), Ukraine (48.6%), and Sweden (42.3%) just to cite a few [15]. The optimization of resources and processes is therefore a primary and global scale goal in the nuclear power industry. Particularly the nuclear fuel cycle encompasses several processes such as conversion, enrichment, fabrication, recycling, and disposal.[1]

During the process of fabrication, the nuclear fuel is allocated in structures called Fuel Assemblies (FAs), according to the specificities of the Nuclear Power Plant (NPP), where it will be effectively used in the reactors for power generation. Nuclear reactors are from different types such as High Temperature Gas-Cooled

---

[1]The interested reader is referred to the website of the International Atomic Energy Agency (IAEA—https://www.iaea.org/). For more information on the nuclear fuel cycle, see https://infcis. iaea.org/NFCSS/NFCSSMain.asp?RightP=Modeling&EPage=1. For a more formal introduction to nuclear reactor core design, see Ref. [11, chapter 15].

A. A. d. M. Meneses (✉) · L. M. Araujo · F. N. Nast · P. V. da Silva
Institute of Engineering and Geosciences, Federal University of Western Pará, Santarém, Brazil
e-mail: anderson.meneses@pq.cnpq.br; lenilson.araujo@ufopa.edu.br

R. Schirru
Program of Nuclear Engineering, Federal University of Rio de Janeiro, Rio de Janeiro, Brazil
e-mail: schirru@lmp.ufrj.br

© Springer Nature Switzerland AG 2019
G. Mendes Platt et al. (eds.), *Computational Intelligence, Optimization and Inverse Problems with Applications in Engineering*,
https://doi.org/10.1007/978-3-319-96433-1_8

Fig. 8.1 Nuclear reactor core schematic representation (view from top): 121 Fuel Assemblies and symmetry lines

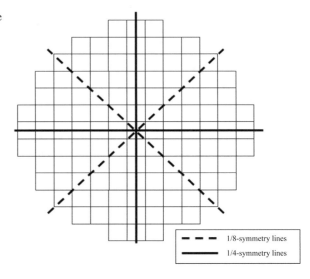

- - - 1/8-symmetry lines

——— 1/4-symmetry lines

Reactors, CANDU (heavy water reactors), and Light Water Reactors, which in turn are subdivided into Boiling Water Reactors and Pressurized Water Reactors (PWRs).[2]

One example of a PWR is in Angra 1 NPP, in the Southeast of Brazil. Angra 1 NPP is a 626 MW PWR designed by Westinghouse and operated by Eletronuclear, with a reactor core composed by 121 FAs, whose schematic representation with lines of symmetry is depicted in Fig. 8.1 (view from top). Thus the FAs are placed inside the reactor core and after a period of time called operation cycle,[3] part of the FAs are substituted in order to keep the NPP operating with nominal power (approximately 1/3 of them and generally the most burned FAs). Usually "fresh" FAs (without fission by-products such as Xenon and Samarium) obtained by the fabrication process join the remaining FAs (partially burned) in a new configuration called Loading Pattern (LP) inside the reactor core. The optimization of the nuclear LP then consists of obtaining an optimal (or near-optimal) LP according to specific criteria (also called objectives), subject to safety constraints. Levine [25] defines the goal of the LP optimization as determining the LP for producing full power within adequate safety margins. As already mentioned, the whole nuclear cycle is composed by several processes, and therefore the LP Optimization is also known as In-Core Fuel Management Optimization, since the main concern is finding an optimal (or near-optimal) arrangement of FAs inside the core.

Over the years many methods have been applied to the LP optimization including manual optimization and refueling patterns (such as zonal and scatter loading

---

[2]For more information on nuclear reactor types, the reader is referred to [11, 26, 59].

[3]The word "cycle" is used here in the context of the LP optimization (or in-core management) opposed to the context of the whole nuclear fuel cycle.

[11]), Mathematical Programming, Knowledge-Based Systems, and Optimization Metaheuristics. Early applications of Mathematical Programming include Dynamic Programming in 1965 [61], as well as Linear and Quadratic Programming in 1968 [56]. Knowledge-Based Systems incorporating expert knowledge have also been applied to the LP optimization including early uses of logical rules in 1972 [39]. The development and advance of optimization metaheuristics (or generic heuristic methods [57]) such as Genetic Algorithm in 1975 [14] and Simulated Annealing in 1983 [21], and their application to the LP optimization [7, 22, 43, 44, 55] inspired the application of other algorithms such as Tabu Search (see Refs. [13, 28]) and Ant Colony Optimization (see Refs. [27, 31]). Other developments include the metaheuristics Harmony Search [1, 51] and Cuckoo Search [62], among others [35].

The present chapter reviews the application of the metaheuristics Particle Swarm Optimization (PSO) [12, 19], Cross-Entropy (CE) [45, 46], Population-Based Incremental Learning (PBIL) [3], and Artificial Bee Colonies [16, 17] to the LP optimization. Results obtained for the optimization of Angra 1 NPP's 7th cycle of operation are described and discussed.

The remaining of the present chapter has the following structure. Related work is discussed in Sect. 8.2; Sect. 8.3 focuses on the LP optimization; the metaheuristics PSO, CE, PBIL, and ABC are described in Sect. 8.4; a review of the results in the literature and a discussion are presented in Sect. 8.5; conclusions and final remarks are in Sect. 8.6.

8.2 Related Work

In this section the principal works involving PSO, CE, PBIL, and ABC applied to the LP optimization are reviewed.

8.2.1 Particle Swarm Optimization

PSO was initially proposed in 1995 to optimize non-linear continuous functions [12], with a discrete binary version released in 1997 [18]. In 2002, Salman et al. [48] applied the continuous PSO algorithm to the Task Assignment Problem truncating real numbers in order to obtain feasible solutions, with the repetition of integers allowed in the candidate solutions for the problem. For problems in which the candidate solution cannot contain repetition of integers, Pang et al. [42] and Tasgetiren et al. [58] developed an adaptation of the PSO with the Random Keys (RK) model [4] proposed initially for the GA (see Appendix). Meneses et al. [33] adapted the PSO with RK to the LP optimization of a PWR. The algorithm PSO with RK was later implemented in a parallel version [60]. PSO with RK has also been used with heuristics, reducing the number of evaluations, and therefore the computational cost of the optimization [34]. Babazadeh et al. [2]

developed a discrete PSO for application to the LP optimization of a Water-Water Energetic Reactor (WWER, also transliterated as VVER) PWR. Khoshahval et al. [20] developed a continuous PSO also for application to a VVER PWR.

8.2.2 Cross-Entropy Algorithm

The CE algorithm was proposed for combinatorial optimization problems in 1999 [45]. Initially developed for rare events estimation in complex stochastic networks, the application to combinatorial problems was successful in several areas, such as telecommunications systems [10] and stochastic vehicle routing [8]. For a formal introduction and examples concerning the CE algorithm, the interested reader is referred to Ref. [46]. Schlünz et al. [49] applied a Multi-Objective (MO) version of the CE algorithm to the SAFARI-1 research reactor, comparing the performance with other MO algorithms in Ref. [50]. Meneses et al. [32] applied a CE algorithm to the optimization of a PWR.

8.2.3 Population-Based Incremental Learning

The algorithm PBIL was introduced by Baluja in 1994 [3]. Machado [30] applied a MO PBIL to the LP optimization of a PWR. Caldas and Schirru [5] implemented a Parameter-Free PBIL (FPBIL) also for the optimization of a PWR. Quantum versions of the PBIL algorithm were also applied to the optimization of a PWR [52, 53].

8.2.4 Artificial Bee Colonies

The bio-inspired ABC algorithm was first introduced by Karaboga [16, 17]. Oliveira and Schirru [41] applied the ABC with RK to the LP optimization of a PWR. Safarzadeh et al. [47] applied the ABC to a VVER PWR.

8.3 Loading Pattern Optimization

According to Levine [25], the goal of the LP optimization problem is to determine the LPs for producing full power within adequate safety margins. In addition, Hill and Parks [13] state that the diversity of objective functions is an interesting point found in LP optimization works, including: (1) Maximization of end-of-cycle (EOC) reactivity; (2) Maximization of discharge burn-up; (3) Minimization of feed

Fig. 8.2 Representation of
the 1/8-symmetry model for
Angra 1 NPP. Except for the
central FA in grey, the 20 FAs
are permuted. The continuous
line represents the 1/4 line of
symmetry and the dashed line
represents the 1/8 line of
symmetry in Fig. 8.1

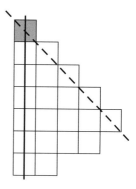

enrichment; (4) Minimization of power peaking; (5) and Minimization of the fresh
fuel inventory. Chapot et al. [7] and other authors (e.g., [5, 27, 30, 33]) maximize the
cycle length, based on the Boron concentration yielded at the Xenon equilibrium.
Thus, there exist multiple criteria that can be represented as objective functions for
the LP optimization problem. Therefore, stating the LP optimization problem as MO
is very usual. Those objective functions can also be used combined or aggregated,
even with constraints, forming an aggregated objective function (or an aggregated
fitness function, which is the term used in the context of several metaheuristics).
Schlünz et al. [51] proposed a unified methodology for handling single or multi-
objective LP optimization problems.

The LP optimization has been tackled in many different ways over the years,
including for a set of plants [39]. Single or multi-cycle approach (whether one or
more time intervals of operation are considered) may also be used in the statement
of the problem [55]. Concerning the symmetry of the core, quarter core or octave
symmetries (see Figs. 8.1 and 8.2) are used, for obtaining both a reduced time of
evaluation of a candidate solution by the Reactor Physics code and a symmetric
power distribution inside the core. Another advantage of such approaches is that the
search space of the combinatorial problem becomes reduced.

Hill and Parks [13] used three conditions specified by a candidate LP: the FA to
be loaded in each core location, the Burnable Poison (BP) loading with each fresh
FA, and the orientation of each burnt FA. Other authors [5, 7, 27, 30, 33] encode
their candidate solutions considering only the allocation of FAs to their position in
the core.

LP Optimization of Angra 1 NPP, in Brazil Angra 1 NPP is a 626 MW 2-loop
PWR located at Rio de Janeiro State, Brazil, operated by Eletronuclear, whose core
is composed by 121 FAs. The Reactor Physics code RECNOD is a simulator for
Angra 1 NPP [6, 7]. RECNOD is a nodal code based on the works described in
Refs. [24, 29, 38], applied to optimization surveys in several works (e.g. [5–7, 30–
33, 36]).

The octant-symmetry for the RECNOD simulation is depicted in Fig. 8.2. In our
simulations, FAs of the symmetry lines (quartets) are not exchanged with elements
out of the symmetry lines (octets). In addition, the central element is not permuted,

Table 8.1 Burnup and k_{inf} values for the FAs of RECNOD code

FAs	Burnup	k_{inf}
1	9603	1.069
2	13,045	0.906
3	7882	1.087
4	13,006	0.906
5	0	1.187
6	13,012	0.906
7	14,650	1.037
8	8622	1.079
9	13,181	0.903
10	0	1.193
11	14,068	1.026
12	13,115	0.906
13	13,135	0.904
14	0	1.188
15	0	1.194
16	11,404	1.050
17	7873	1.099
18	0	1.191
19	0	1.188
20	13,285	0.907

which yields $10! \times 10! \cong 1.3 \times 10^{13}$ possible permutations. Table 8.1 exhibits the burnup and k_{inf} values for the basic LP of RECNOD code depicted in Fig. 8.2 [6, 41].

The nuclear parameters yielded by the code are, among others, the Maximum Assembly Relative Power (P_{rm}) and the Boron Concentration (C_B). The value of P_{rm} is used as a constraint related to safety. For a maximum required radial power peak factor $F_{XYmax} = 1.435$ for Angra 1 NPP, the calculations yield a correspondent $P_{rm} = 1.395$. Any LP with $P_{rm} > 1.395$ is infeasible in the sense of the safety requirements.

The value for C_B yielded by the RECNOD code is given at the equilibrium of Xenon, which reduces the computational cost of the processing, without impairing its validity for optimization purposes. Chapot [6] demonstrated that it is possible to extrapolate and predict the cycle length based on the C_B at the equilibrium of Xenon, in such a way that 4 ppm are approximately equivalent to one Effective Full Power Day (EFPD). In addition, one more EFPD is equivalent to a profit of approximately hundreds of thousand dollars. The results in Sect. 8.5 refer to the C_B at the equilibrium of Xenon.

The LP optimization problem related to Angra 1, according to the parameters given by RECNOD, can therefore be stated as

$$\text{minimize} \quad S = \frac{1}{C_B} \tag{8.1}$$

$$\text{subject to} \quad P_{rm} \leq 1.395 \tag{8.2}$$

The aggregated fitness function used in [32–36] (considering that the values of P_{rm} are always greater than the reciprocal of the Boron Concentration) is

$$Fitness = \begin{cases} \dfrac{1}{C_B}, & \text{if } P_{rm} \leq 1.395 \\ P_{rm}, & \text{otherwise} \end{cases} \tag{8.3}$$

8.4   Some Optimization Metaheuristics Applied to the LP Optimization

In this section we present the main theoretical aspects of the algorithms PSO, CE, PBIL, and ABC for application to the LP optimization problem.

8.4.1   Particle Swarm Optimization (PSO)

A swarm with P particles optimizes in an n-dimensional search space. Each particle i has a position $\mathbf{x_i^t} = [x_{i1}, x_{i2}, \ldots, x_{in}]$ and a velocity $\mathbf{v_i^t} = [v_{i1}, v_{i2}, \ldots, v_{in}]$ at an iteration t, through the dimension j updated according to the equations

$$v_{ij}^{t+1} = w^t v_{ij}^t + c_1 r_1^t \left(pbest_{ij} - x_{ij}^t \right) + c_2 r_2^t \left(gbest_j - x_{ij}^t \right) \tag{8.4}$$

and

$$x_{ij}^{t+1} = x_{ij}^t + v_{ij}^{t+1} \tag{8.5}$$

The inertia weight w^t decreases linearly according to the equation

$$w^t = w - \frac{w - w_{min}}{t_{max}} t \tag{8.6}$$

where w is the maximum inertia constant, t_{max} is the maximum number of iterations, and t is the current iteration. High values of w^t lead to global search, whereas low values of w^t lead to local search.

At the right side of Eq. (8.4), the first term represents the influence of the own particle motion, acting as a memory of the particle's previous behavior; the second term represents the individual cognition, where the particle compares its position with its previous best position **pbest$_i$**; and the third term represents the social aspect of intelligence, based on a comparison between the particle's position and the best result obtained by the swarm **gbest** (global best position). Equation (8.5) describes how the positions are updated. Both c_1 and c_2 are acceleration constants: c_1 is related to the individual cognition, whereas c_2 is related to social learning; r_1 and r_2 are uniformly distributed random numbers. The initialization of the algorithm is at random, that is, the positions and velocities of the particles are initialized randomly at implementation. The algorithm is described in Ref. [33].

The positions $\mathbf{x}_i^t$ updated by Eqs. (8.4) and (8.5) are evaluated by an objective function $f(\mathbf{x}_i)$, which is the fitness of the problem. The positions vectors **gbest** $= [gbest_1, gbest_2, \ldots, gbest_n]$ and **pbest$_i$** $= [pbest_{i1}, pbest_{i2}, \ldots, pbest_{in}]$ are updated depending on the information acquired by the swarm, constructing its knowledge on the search space along the iterations.

8.4.2 Cross-Entropy Algorithm (CE)

The CE method [45, 46] is a stochastic method adapted to combinatorial optimization. The method's name is derived from the relative entropy or cross-entropy or Kullback-Leibler (KL) divergence D_{KL} [9, 23]. The KL divergence between two probability density functions (pdfs) $g(x)$ and $h(x)$ compares the entropy of two distributions over the same random variable and is given by

$$D_{KL}(g \parallel h) = \int g(x) \ln \left[\frac{g(x)}{h(x)} \right] dx = \int g(x) \ln[g(x)] dx - \int g(x) \ln[h(x)] dx \tag{8.7}$$

Generally speaking, CE encompasses two steps: (1) generation of candidate solutions according to a random mechanism; and (2) adjustment of the algorithm's parameters in order to generate a better sample in the next iteration. Thus the KL divergence plays a key role during the execution of the method which is the guidance of the search to more promising pdfs for the step (1) of the algorithm so that optimal or near-optimal candidate solutions may be found.

Consider a perform function $S(\mathbf{x})$, which corresponds to the objective function or fitness function. For the LP optimization, $\mathbf{x}$ corresponds to a candidate solution vector representing a candidate LP. For a minimization problem the minimum $\gamma^* \in \mathbb{R}$ is the evaluation of the best solution $\mathbf{x}^*$ and is given by

$$S(\mathbf{x}^*) = \gamma^* = min_{\mathbf{x}} \, S(\mathbf{x}) \tag{8.8}$$

which is known as a deterministic version of the optimization problem. The deterministic problem is associated with a stochastic (or expected) version in which given a random vector $\mathbf{X}$ with pdf $f(\cdot, \mathbf{u})$ for some $\mathbf{u} \in U$, the principal interest in the stochastic version to describe an estimator ℓ for the probability of obtaining $S(\mathbf{X}) \leq \gamma$

$$\ell(\gamma) = P_u(S(\mathbf{X}) \leq \gamma) \tag{8.9}$$

where γ is a threshold value that separates the candidate solutions of an iteration into two groups, those with lower evaluations which will contribute for the estimation of a new pdf, and those that will not. At each iteration for a minimization problem, the candidate solutions with lower evaluation values form an elite group, with the objective of obtaining the probability matrix $\hat{P}^{(t)}$, whose elements are given by

$$\hat{p}_{ij} = \frac{\sum_{k=1}^{N} I_{\{S(\mathbf{X}_k) \leq \gamma\}} \times I_{\{\mathbf{X}_k \in x_{ij}\}}}{\sum_{k=1}^{N} I_{\{S(\mathbf{X}_k) \leq \gamma\}}} \tag{8.10}$$

as in the following example.

Consider the hypothetical example with ten candidate LPs for the LP optimization of Angra 1 NPP in Table 8.2 (for more details see Ref. [32]). Consider that the evaluation of the LPs in Table 8.2 by a Reactor Physics code yields the values in Table 8.3. Sorting the LPs according to their values of $S(\mathbf{X}_k)$ gives as result the data in Table 8.4. The elite group based on Table 8.4 is given in Table 8.5.

Once the elite group is defined, the probability matrix $\hat{P}^{(t)}$ can be found. $\hat{P}^{(t)}$ will act as a discrete surrogate for the pdf in order to find the estimator $\ell(\gamma)$ in Eq. (8.9), and in practice it will be used for the random selection of new LPs in the next iteration. For example, consider Table 8.5. In the position **1**, the FA 9 occurs two times, then for the first column of Table 8.6, the 9th line will be 2/4. The FAs 5 and 10 occur one time each in the first column in Table 8.5, therefore the 5th and 10th lines of the 1st column in Table 8.6 will be 1/4 each. The values obtained for the hypothetical example are exhibited in Table 8.6. Once $\hat{P}^{(t)}$ is defined, the CE algorithm iterates and the process starts again.

Table 8.2 Hypothetical data for an example of calculation of $\hat{p}_{ij}$ for the LPO of Angra 1 NPP, with FAs in italic and positions in bold

LP	**1**	**2**	**3**	**4**	**5**	**6**	**7**	**8**	**9**	**10**	**11**	**12**	**13**	**14**	**15**	**16**	**17**	**18**	**19**	**20**
X_1	*9*	*5*	*6*	*8*	*1*	*7*	*2*	*4*	*10*	*3*	*14*	*20*	*18*	*15*	*12*	*13*	*16*	*17*	*19*	*11*
X_2	*9*	*3*	*10*	*8*	*7*	*4*	*6*	*1*	*2*	*5*	*13*	*14*	*18*	*19*	*17*	*12*	*15*	*11*	*16*	*20*
X_3	*8*	*1*	*6*	*2*	*7*	*3*	*10*	*5*	*9*	*4*	*12*	*18*	*20*	*16*	*15*	*14*	*17*	*13*	*19*	*11*
X_4	*9*	*7*	*4*	*2*	*5*	*10*	*3*	*1*	*8*	*6*	*12*	*18*	*20*	*17*	*16*	*14*	*19*	*15*	*13*	*11*
X_5	*9*	*5*	*10*	*8*	*1*	*7*	*4*	*6*	*3*	*2*	*14*	*18*	*10*	*19*	*13*	*12*	*16*	*15*	*17*	*11*
X_6	*8*	*7*	*9*	*10*	*1*	*3*	*6*	*5*	*2*	*4*	*13*	*20*	*18*	*19*	*12*	*14*	*16*	*15*	*17*	*11*
X_7	*5*	*2*	*8*	*6*	*1*	*9*	*7*	*4*	*3*	*10*	*13*	*20*	*18*	*17*	*12*	*14*	*19*	*11*	*15*	*16*
X_8	*10*	*1*	*8*	*9*	*5*	*4*	*2*	*6*	*7*	*3*	*12*	*20*	*18*	*15*	*16*	*14*	*19*	*17*	*13*	*11*
X_9	*9*	*1*	*8*	*6*	*7*	*3*	*5*	*2*	*10*	*4*	*11*	*20*	*18*	*17*	*12*	*14*	*19*	*15*	*16*	*13*
X_{10}	*10*	*5*	*6*	*8*	*9*	*1*	*2*	*7*	*4*	*3*	*14*	*13*	*20*	*16*	*19*	*12*	*15*	*18*	*17*	*11*

Table 8.3 Hypothetical values obtained for the LPs of Table 8.2

LP	C_B (ppm)	P_{rm}	$S(\mathbf{X}_k)$ (objective function)
$\mathbf{X}_1$	1250	1.410	1.410
$\mathbf{X}_2$	1300	1.389	0.000777
$\mathbf{X}_3$	1280	1.396	1.396
$\mathbf{X}_4$	1200	1.390	0.00083
$\mathbf{X}_5$	1270	1.415	1.415
$\mathbf{X}_6$	1310	1.405	1.405
$\mathbf{X}_7$	1150	1.392	0.00087
$\mathbf{X}_8$	1350	1.393	0.00074
$\mathbf{X}_9$	1260	1.425	1.425
$\mathbf{X}_{10}$	1100	1.380	0.00091

Table 8.4 Sort of the LPs of Table 8.3 according to their fitness evaluation $S(\mathbf{X}_k)$

#	LP	$S(\mathbf{X}_k)$
1	$\mathbf{X}_9$	1.425
2	$\mathbf{X}_5$	1.415
3	$\mathbf{X}_1$	1.410
4	$\mathbf{X}_6$	1.405
5	$\mathbf{X}_3$	1.396
6	$\mathbf{X}_{10}$	0.00091
7	$\mathbf{X}_7$	0.00087
8	$\mathbf{X}_4$	0.00083
9	$\mathbf{X}_2$	0.00077
10	$\mathbf{X}_8$	0.00074

Table 8.5 Elite group for the hypothetical database on Table 8.4, with FAs in italic and positions in bold

LP	**1**	**2**	**3**	**4**	**5**	**6**	**7**	**8**	**9**	**10**	**11**	**12**	**13**	**14**	**15**	**16**	**17**	**18**	**19**	**20**
$\mathbf{X}_2$	*9*	*3*	*10*	*8*	*7*	*4*	*6*	*1*	*2*	*5*	*13*	*14*	*18*	*19*	*17*	*12*	*15*	*11*	*16*	*20*
$\mathbf{X}_4$	*9*	*7*	*4*	*2*	*5*	*10*	*3*	*1*	*8*	*6*	*12*	*18*	*20*	*17*	*16*	*14*	*19*	*15*	*13*	*11*
$\mathbf{X}_7$	*5*	*2*	*8*	*6*	*1*	*9*	*7*	*4*	*3*	*10*	*13*	*20*	*18*	*17*	*12*	*14*	*19*	*11*	*15*	*16*
$\mathbf{X}_8$	*10*	*1*	*8*	*9*	*5*	*4*	*2*	*6*	*7*	*3*	*12*	*20*	*18*	*15*	*16*	*14*	*19*	*17*	*13*	*11*

8.4.3 Population-Based Incremental Learning (PBIL)

The PBIL algorithm associates a probability vector with real values between 0 and 1, at each position, that is used for generating binary individuals that represent candidate solutions for the problem to be optimized. The probability vector specifies the probability of obtaining at each position the value 1. An example of probability vector almost converged for an 8-bit problem is [0.10, 0.95, 0.05, 0.98, 0.99, 0.25, 0.03, 0.95].

Such vector of probability may generate binary individuals such as [0, 1, 0, 1, 1, 0, 0, 1] or even [1, 1, 0, 1, 0, 1, 0, 1], which will be decoded for representing candidate solutions for the optimization problem (for more details, see Refs. [3, 30]).

Table 8.6 The probability matrix $\hat{P}^{(t)} = (\hat{p}_{ij}^{(t)})$ for the hypothetical example (FAs in italic and positions in bold)

FAs	1	2	3	4	5	6	7	8	9	10	11	12	13	14	15	16	17	18	19	20
1	0	1/4	0	0	1/4	0	0	2/4	0	0	0	0	0	0	0	0	0	0	0	0
2	0	1/4	0	1/4	0	1/4	0	1/4	0	0	0	0	0	0	0	0	0	0	0	0
3	0	1/4	0	0	0	0	1/4	0	1/4	1/4	0	0	0	0	0	0	0	0	0	0
4	0	0	1/4	0	0	2/4	0	1/4	0	0	0	0	0	0	0	0	0	0	0	0
5	1/4	0	0	0	2/4	0	0	0	0	1/4	0	0	0	0	0	0	0	0	0	0
6	0	0	0	1/4	0	0	1/4	1/4	0	1/4	0	0	0	0	0	0	0	0	0	0
7	0	1/4	0	0	1/4	0	1/4	0	1/4	0	0	0	0	0	0	0	0	0	0	0
8	0	0	2/4	1/4	0	0	0	0	1/4	0	0	0	0	0	0	0	0	0	0	0
9	2/4	0	0	1/4	0	1/4	0	0	0	0	0	0	0	0	0	0	0	0	0	0
10	1/4	0	1/4	0	0	1/4	0	0	0	1/4	0	0	0	0	0	0	0	0	0	0
11	0	0	0	0	0	0	0	0	0	0	0	0	0	0	0	0	0	2/4	0	2/4
12	0	0	0	0	0	0	0	0	0	0	2/4	0	0	0	1/4	1/4	0	0	0	0
13	0	0	0	0	0	0	0	0	0	0	2/4	0	0	0	0	0	0	0	2/4	0
14	0	0	0	0	0	0	0	0	0	0	0	1/4	0	0	3/4	0	0	0	0	0
15	0	0	0	0	0	0	0	0	0	0	0	0	0	1/4	0	0	1/4	1/4	1/4	0
16	0	0	0	0	0	0	0	0	0	0	0	0	0	0	2/4	0	0	0	1/4	1/4
17	0	0	0	0	0	0	0	0	0	0	0	0	0	2/4	1/4	0	0	1/4	0	0
18	0	0	0	0	0	0	0	0	0	0	0	1/4	3/4	0	0	0	0	0	0	0
19	0	0	0	0	0	0	0	0	0	0	0	0	1/4	0	0	3/4	0	0	0	0
20	0	0	0	0	0	0	0	0	0	0	0	2/4	1/4	0	0	0	0	0	0	1/4

In the PBIL algorithm, the entire population of individuals is defined from the vector probability, and the operators employed for the evolution of this population do not act upon the population as in GA, but upon the PBIL probability vector.

The probability vector needs to be updated at each generation of a population. This update is made using two vectors, and possibly a mutation. The vectors are respectively the vector that has the best fitness evaluation and the vector that has the worst fitness evaluation. The best vector is the one that presents the highest fitness value for a maximization problem (or the lowest for a minimization problem) and changes the probability vector so that this one approaches its representation. The worst vector is the one with the lowest fitness value, for a maximization problem (or highest for a minimization problem) and changes the probability vector so that it distances itself from its representation.

The equations for updating the probability vector are described below.

With information of the best individual:

$$P(i) = P(i) \times (1 - Lr_{pos}) + v(i) \times Lr_{pos} \qquad (8.11)$$

with information of the worst individual:

$$P(i) = P(i) \times (1 - Lr_{neg}) + v(i) \times Lr_{neg} \tag{8.12}$$

with mutation:

$$P(i) = P(i) \times (1 - Ir_{mut}) + v_m(i) \times Ir_{mut} \tag{8.13}$$

where:

Lr_{pos} is the positive learning rate;
Lr_{neg} is the negative learning rate;
Ir_{mut} is the rate at which the mutation will affect the probability vector;
$P(i)$ is the value of the probability vector at position i;
$v(i)$ is the value of the best vector at position i; and
$v_m(i)$ is the value drawn for position i.

In this case, for the worst vector, according to [3], there is a condition. The positions of the worst vector are compared to the best vector, and if the value of the position (0 or 1) is different, then it is changed as in Eq. (8.11). If the values are equal, they remain the same, and the probability vector is distanced from this individual.

For the case of the mutation, there is a lottery of a random value between 0 and 1, where it is compared to the value of the ProbMut parameter. ProbMut is the parameter that governs the chance that mutation occurs in the population. If the value drawn is smaller than this parameter, then the mutation happens.

Since the previous condition is true, which makes the mutation possible, there is a new lottery but this one is to define a value that determines whether or not there will be an increment, that is, this new draw will imitate the creation process of an individual, with the possibility of incrementing or not the so-called rate of increase Ir_{mut}. The latter plays a role similar to that of Lr_{pos} and Lr_{neg}. Whereas Lr_{pos} and Lr_{neg} are related to the best and worst vector, respectively, Ir_{mut} is related to how much the mutation will change the value of the corresponding position, as shown in Eq. (8.13) [3].

8.4.4 Artificial Bee Colonies (ABC)

The ABC algorithm was initially proposed by Karaboga [16], as a metaheuristic of swarm intelligence applied to optimization of multimodal functions in continuous search spaces.

The ABC algorithm tries to simulate the behavior of real honey bees in search of food sources. In the algorithm, the food sources represent the solutions, where each source has an associated nectar and the amount of nectar defines the fitness of the food source, that is, the greater the amount of nectar the more suitable the source. The colony of bees is subdivided into three groups: worker bees, observer bees, and explorer bees.

One half of the colony consists of worker bees and the other half is made up of observer bees. Worker bees are bees that already explore a particular food source and share information about the source with the observer bees. Based on the shared information, the observer bees decide the best source to be explored, and become workers of that source. Explorer bees are worker bees who are positioned in stagnant sources, do not show improvement, and are re-sent in searches of new food sources at random. Thus, worker bees represent the aspects of local search, the observer bees represent the sharing of information, and exploratory bees represent aspects of global search. The following are the main steps of the algorithm:

1. Initialization of the population of food sources.
2. Each worker bee explores a new source in the vicinity of its initial source and starts to explore the best source between them.
3. Based on the amount of nectar the bees are placed in the sources. Then they explore a new source in the vicinity of their current position and explore the best source among them.
4. The decision criterion is tested to determine whether the worker bee will become an explorer bee. If so, then the explorer bee starts searching new sources of food.
5. The best source is memorized.

Steps (2–5) are repeated until the stopping criterion is reached.

In the first step we place each worker bee in a food source (x_{ij}) using Eq. (8.14):

$$x_{ij} = x_j^{min} + \left(x_j^{max} - x_i^{max}\right) \times rand \tag{8.14}$$

where $i = 1, \ldots, SN$ and $j = 1, \ldots, D$ being that SN represents the swarm size, D indicates the dimension of the problem, $rand$ is a random real number in the interval $[0, 1]$, x_j^{min} and x_j^{max} indicate the lower limit and the upper limit of the search space for each dimension j, respectively. At the second step is performed a local search, where each worker bee explores a source in the neighborhood v_{ij} using Eq. (8.15):

$$v_{ij} = x_{ij} + \phi_{ij}(x_{ij} - x_{kj}) \tag{8.15}$$

where ϕ_{ij} is a random number in the interval $[-1, 1]$, $k = 1, \ldots, SN$ with $k \neq i$. After its production, the new source $\mathbf{v_i}$ is compared to $\mathbf{x_i}$ and the worker bee starts to explore the best source between them.

At the third step, sources $\mathbf{x_i}$ are selected based on the probability given by Eq. (8.16), and a copy is created with the best sources, which will be the sources $\mathbf{x_i}$ of the observer bees. Then, Eq. (8.15) is used again to produce new neighboring sources $\mathbf{v_i}$ from these better sources, which will be compared to the $\mathbf{x_i}$ sources of the observer bees, and the observing bee will explore the best between them. These sources $\mathbf{x_i}$ of the observer bees are then compared with the sources $\mathbf{x_i}$ of the worker bees, and the worker bees start to explore the best solution between them.

$$P_i = fit_i / \sum_{j=1}^{SN} fit_j \qquad (8.16)$$

where fit_i is the fitness of the source i.

At the fourth step the stopping criterion is tested. If the source x_i does not show improvement after a predefined limit of cycles (Eq. (8.17)), it is abandoned, and the worker bee from that source becomes an explorer bee and is sent looking for a new source using Eq. (8.14).

The ABC algorithm presents, in addition to the parameters common to the algorithms of swarm intelligence, swarm size and number of iterations, only one more control parameter, the limit L, which is suggested in the literature as the product between swarm size (SN) and the dimension of the problem (D):

$$L = SN \times D \qquad (8.17)$$

The Modified ABC Algorithm The modification of ABC proposed by Oliveira and Schirru [41] (and also tested in Ref. [54]) modifies the original process of comparison, resulting in an immediate improvement in average swarm fitness. In the standard algorithm the neighboring sources v_i corresponding to the observer bees are compared to the sources x_i of the observer bees. If an observer v_i gets a larger amount of nectar, then it takes the place of the x_i observer. Then, observer bees are compared to the x_i worker bees which gave origin to them, and if the observers have a better fitness then they occupy the workers' place, resulting in a set of worker bees with the best solutions of the iteration. On the other hand in the modified algorithm, neighboring sources v_i corresponding to the observer bees are directly compared to the x_i worker bees, replacing them if it is obtained a better fitness, also resulting in a set of worker bees with the best solutions of the iteration, but with a reduced population mean (for a minimization problem).

8.5 Results and Discussion

In this section we review some results obtained for the algorithms PSO, CE, PBIL, and ABC considering the optimization of the 7th cycle of Angra 1 NPP with the Reactor Physics code RECNOD. We have divided the results into two sections, since during our experiments we have noticed that CE and PSO tend to obtain better results for approximately 10,000 iterations of the algorithms, which we have called "short run." PBIL and ABC have not attained good results on the short run so we tried a greater number of evaluations (respectively 100,000 and 50,000 evaluations), which we have called "long run."

8.5.1 Short-Run Results (PSO and CE)

Table 8.7 exhibits some results for the PSO and CE algorithms for the 7th cycle of Angra 1 NPP [32–34, 36]. RNAH stands for Reactive Neighborhood Acceptance Heuristic and verifies if a candidate LPs is worth to evaluate according to a degree of acceptance depending on the neighborhood between FAs, that is, RNAH discards candidate solutions that are possibly very reactive according to a set of rules, reducing the computational cost of evaluation during the execution of the algorithm [34].

Between PSO and CE we consider that the best algorithm for a short run for the optimization of the 7th cycle of Angra 1 NPP is the CE algorithm. CE's average, best, and even the worst results compare favorably regarding the PSO algorithm. Although PSO algorithm performs a good exploration of the search space, CE provided a robust and superior performance for the problem tested. However, for the BIBLIS-2D benchmark's data, which is another instance of the LPO problem, both CE and PSO yielded poor performances, as discussed by Meneses et al. [36].

Table 8.7 Some results in the literature for PSO with RK and CE algorithms for the cycle 7 of Angra 1 NPP obtained with the Reactor Physics code RECNOD (C_B in ppm; PSO-RNAH's numbers of evaluations are averaged because several supposedly reactive LPs are discarded during the execution of the algorithm)

Method	Average C_B	Max. C_B	Min. C_B	Number of evaluations	Number of tests
PSO[a] [33]	1254	1396	1068	10,000	15
PSO-RNAH[b] [34]	1131	1221	977	4197[c]	20
PSO-RNAH[d] [34]	1183	1325	1048	5087[c]	20
PSO-RNAH[e] [34]	1216	1342	1089	7583[c]	20
PSO[a] [36]	1240	1402	1024	10,000	50
CE[f] [32]	1322	1432	1245	10,000	50
CE[g] [32]	1340	1439	1250	10,000	50
CE[h] [32]	1339	1407	1184	10,000	50
CE[g] [36]	1318	1409	1155	10,000	50

[a] $c_1 = c_2 = 1.8$; $w = 0.8$–0.2
[b] Maximum degree 1
[c] On average
[d] Maximum degree 2
[e] Maximum degree 3
[f] $\rho = 0.05$; $\mu = 1000$
[g] $\rho = 0.10$; $\mu = 1000$
[h] $\rho = 0.15$; $\mu = 1000$

Table 8.8 Some results for PBIL and ABC algorithms for the cycle 7 of Angra 1 NPP obtained with the Reactor Physics code RECNOD (C_B in ppm; ABC's numbers of evaluations are approximated because the scout bees' flights depend on the evolution of the algorithm)

Method	Average C_B	Max. C_B	Min. C_B	Number of evaluations	Number of tests
PBIL[a] [40]	1311	1402	1237	100,000	10
PBIL[b] [40]	1308	1393	1237	100,000	10
PBIL[c] [40]	1282	1401	1225	100,000	10
PBIL[d] [40]	1311	1330	1288	100,000	10
PBIL[d] [36]	1303	1400	1209	100,000	50
ABC[e] [54]	1287	1398	1197	50,000[f]	10
ABC[g] [54]	1320	1439	1246	50,000[f]	10
ABC[h] [54]	1288	1404	1232	50,000[f]	10
ABC[g] [36]	1307	1435	1224	100,000[f]	50

[a]With the best individual—Eq. (8.11)
[b]With the best and worst individuals—Eqs. (8.11) and (8.12)
[c]With the best individual and mutation—Eqs. (8.11) and (8.13)
[d]With the best individual, the worst individual, and mutation—Eqs. (8.11)–(8.13)
[e]$Limit = 20$
[f]Approximately
[g]$Limit = 50$
[h]$Limit = 4000$

8.5.2   Long-Run Results (PBIL and ABC)

Table 8.8 exhibits some results for the PBIL and ABC algorithms for the 7th cycle of Angra 1 NPP [36, 40, 54].

Both ABC and PBIL have very competitive performance for average and maximum results in the problem tested. For such results, ABC is less expensive computationally since it obtains similar results when compared to PBIL in approximately half of the evaluations. Notwithstanding, when compared to CE, ABC and PBIL need, respectively, 5 times and 10 times more evaluations to obtain their results.

8.6   Conclusion

In the present chapter we reviewed the application of Computational Intelligence methods to the problem of the LP optimization, particularly the optimization metaheuristics PSO, CE, PBIL, and ABC. The LP optimization problem is a prominent and complex real-world problem in Nuclear Engineering. Meneses et al. [36] analyzed the results for the 7th cycle of operation of Angra 1 NPP, evaluated with the Reactor Physics code RECNOD, with evidence that CE obtains better results than PSO, with a more robust performance for short runs with 10,000 evaluations of candidate solutions. For long runs (50,000 and 100,000 evaluations of candidate solutions), the algorithms PBIL and ABC obtain very competitive results, although similar to those of CE obtained in approximately one fifth and one tenth of evaluations, respectively.

Acknowledgements A.A.M.M. acknowledges CNPq (Brazilian National Research Council— Project no. 472912/2013-5) and Federal University of Western Pará for financial support. P.V.S. acknowledges Federal University of Western Pará for supporting the research. F.N.N. acknowledges CNPq for supporting the research. R.S. acknowledges CNPq and FAPERJ for supporting the research. The authors acknowledge CNEN for the agreement of mutual cooperation with the Federal University of Western Pará (part of CAMP agreement). Portions of the present chapter were published in the journals Annals of Nuclear Energy and Progress in Nuclear Energy, as well as in the conference National Meeting in Computational Modeling in Brazil. The authors would like to thank the reviewers for their valuable comments and suggestions.

Appendix

Random Keys (RK)

The real numbers obtained in the algorithms PSO and ABC are transformed using the RK methodology. The Random Keys model was proposed by Bean [4] and in the context of the present research consists of transforming a real number vector into a candidate solution with no repetition of elements. In other words, it transforms a vector of real numbers that serve as the keys to ordering other integer numbers in order to form a possible solution to the optimization problem. To exemplify the operation of the RK, consider the real numbers vector [0.35, 0.61, 0.11, 0.86, 0,47] generated during the search. The vector is then associated with an integer vector [1, 2, 3, 4, 5]. The solution is transformed, and the integer vector [3, 1, 5, 2, 4] is obtained, since the real keys are sorted in ascending order, also sorting the integers associated with it. Adaptation made to the ABC is done taking in consideration that each food source obtained from the code is used as a key for the production of possible solutions, which are evaluated using the objective function of the problem. For more information see Refs. [7, 33].

References

1. Aghaie, M., Nazari, T., Zolfaghari, A., Minuchehr, A., Shirani, A.: Investigation of PWR core optimization using harmony search algorithms. Ann. Nucl. Energy **57**, 1–15 (2013)
2. Babazadeh, D., Boroushaki, M., Lucas, C.: Optimization of fuel core loading pattern design in a VVER nuclear power reactors using particle swarm optimization (PSO). Ann. Nucl. Energy **36**, 923–930 (2009)
3. Baluja, S.: Population-based incremental learning: a method for integrating genetic search based function optimization and competitive learning. Technical report, Carnegie Mellon University (1994)

4. Bean, J.C.: Genetic algorithms and random keys for sequencing and optimization. ORSA J. Comput. **6**, 154–160 (1994)
5. Caldas, G.H.F., Schirru, R.: Parameterless evolutionary algorithm applied to the nuclear reload problem. Ann. Nucl. Energy **35**, 583–590 (2008)
6. Chapot, J.L.C.: Pressurized water reactor's fuel management automatic optimization by using genetic algorithms. D.Sc. Thesis (in Portuguese), COPPE/UFRJ (2000)
7. Chapot, J.L.C., Silva, F.C., Schirru, R.: A new approach to the use of genetic algorithms to solve the pressurized water reactor's fuel management optimization problem. Ann. Nucl. Energy **26**, 641–655 (1999)
8. Chepuri, K., Homem de Mello, T.: Solving the vehicle routing problem with stochastic demands using the cross-entropy method. Ann. Oper. Res. **134**, 153–181 (2005)
9. Cover, T.M., Thomas, J.A.: Elements of Information Theory. Wiley, Hoboken (2006)
10. de Boer, P.T., Kroese, D.P., Rubinstein, R.Y.: A fast cross-entropy method for estimating buffer overflows in queuing networks. Manag. Sci. **50**, 883–895 (2004)
11. Duderstadt, J.J., Hamilton, L.J.: Nuclear Reactor Analysis. Wiley, Hoboken (1976)
12. Eberhart, R.C., Kennedy, J.: A new optimizer using particles swarm theory. In: Proceedings of 6th International Symposium on Micro Machine and Human Science, pp. 39–43 (1995)
13. Hill, N.J., Parks, G.T.: Pressurized water reactor in-core nuclear fuel management by tabu search. Ann. Nucl. Energy **75**, 64–71 (2015)
14. Holland, J.H.: Adaptation in Natural and Artificial Systems. University of Michigan Press, Ann Arbor (1975)
15. International Energy Agency: Key World Energy Statistics. https://www.iea.org/publications/freepublications/publication/KeyWorld2016.pdf (2016). Retrieved 17 Apr 2018
16. Karaboga, D.: An idea based on honey bee swarm for numerical optimization. Technical report - TR 06, Erciyes University (2005)
17. Karaboga, D., Basturk, B.: A powerful and efficient algorithm for numerical function optimization: artificial bee colony (ABC) algorithm. J. Glob. Optim. **39**, 459–471 (2007)
18. Kennedy, J., Eberhart R.C.: A discrete binary version of the particle swarm algorithm. In: IEEE International Conference on Systems, Man and Cybernetics, pp. 4104–4109 (1997)
19. Kennedy, J., Eberhart, R.C.: Swarm Intelligence. Morgan Kaufmann Publishers, San Francisco (2001)
20. Khoshahval, F., Zolfaghari, A., Minuchehr, H., Sadighi, M., Norouzi, A.: PWR fuel management optimization using continuous particle swarm intelligence. Ann. Nucl. Energy **37**, 1263–1271 (2010)
21. Kirkpatrick, S., Gelatt, C.D. Jr., Vecchi, M.P.: Optimization by simulated annealing. Science **220**, 671–680 (1983)
22. Kropaczek, D.J., Turinsky, P.J.: In-core nuclear fuel management optimization for pressurized reactors utilizing simulated annealing. Nucl. Technol. **39**, 9–32 (1991)
23. Kullback, S., Leibler, R.A.: On information and sufficiency. Ann. Math. Stat. **22**, 79–86 (1951)
24. Langenbuch, S., Maurer, W., Werner, W.: Coarse mesh flux expansion method for analysis of space-time effects in large water reactor cores. Nucl. Sci. Eng. **63**, 437–456 (1977)
25. Levine, S.: In-core fuel management of four reactor types. In: Ronen, Y. (ed.) Handbook of Nuclear Reactors Calculations, vol. II. CRC Press, Boca Raton (1987)
26. Lewis, E.E.: Fundamentals of Nuclear Reactor Physics. Elsevier, New York (2008)
27. Lima, A.M.M., Schirru, R., Silva, F.C., Medeiros, J.A.C.C.: A nuclear reactor core fuel reload optimization using ant colony connective networks. Ann. Nucl. Energy **35**, 1606–1612 (2008)
28. Lin, C., Yang, J., Lin, K., Wang, Z.: Pressurized water reactor loading pattern design using the simple tabu search. Nucl. Sci. Eng. **129**, 61–71 (1998)
29. Liu, Y.S., Meliksetian, A., Rathkopf, J.A., Little, D.C., Nakano, F., Poplaski, M.J.: ANC: a Westinghouse advanced nodal computer code. Technical report WCAP-10965, Westinghouse (1985)
30. Machado, M.D.: A multi-objective PBIL evolutionary algorithm applied to a nuclear reactor core reload optimization problem. D.Sc. Thesis (in Portuguese), COPPE/UFRJ (2005)

31. Machado, L., Schirru, R.: The Ant-Q algorithm applied to the nuclear reload problem. Ann. Nucl. Energy **29**, 1455–1470 (2002)
32. Meneses, A.A.M., Schirru, R.: A cross-entropy method applied to the in-core fuel management optimization of a pressurized water reactor. Prog. Nucl. Energy **83**, 326–335 (2015)
33. Meneses, A.A.M., Machado, M.D., Schirru, R.: Particle swarm optimization applied to the nuclear reload problem of a pressurized water reactor. Prog. Nucl. Energy **51**, 319–326 (2009)
34. Meneses, A.A.M., Gambardella, L.M., Schirru, R.: A new approach for heuristics-guided search in the in-core fuel management optimization. Prog. Nucl. Energy **52**, 339–351 (2010)
35. Meneses, A.A.M., Rancoita, P., Schirru, R., Gambardella, L.M.: A class-based search for the in-core fuel management optimization of a pressurized water reactor. Ann. Nucl. Energy **37**, 1554–1560 (2010)
36. Meneses, A.A.M., Araujo, L.M., Nast, F.N., Silva, P.V., Schirru, R.: Application of metaheuristics to loading pattern optimization problems based on the IAEA-3D and BIBLIS-2D data. Ann. Nucl. Energy **111**, 329–339 (2018)
37. Ministry of Mines and Energy: Brazilian Energy Balance 2015 Year 2014. https://ben.epe.gov.br/downloads/Relatorio_Final_BEN_2015.pdf (2015). Retrieved 17 Apr 2018
38. Montagnini, B., Soraperra, P., Trentavizi, C., Sumini, M., Zardini, D.M.: A well-balanced coarse mesh flux expansion method. Ann. Nucl. Energy **21**, 45–53 (1994)
39. Naft, B.N., Sesonske, A.: Pressurized water optimal fuel management. Nucl. Technol. **14**, 123–132 (1972)
40. Nast, F.N., Silva, P.V., Schirru, R., Meneses, A.A.M.: Population-based incremental learning applied to the problem of nuclear fuels recharge (in Portuguese). In: XIX National Meeting in Computational Modeling, João Pessoa (2016)
41. Oliveira, I.M.S., Schirru, R.: Swarm intelligence of artificial bees applied to in-core fuel management optimization. Ann. Nucl. Energy **38**, 1039–1045 (2011)
42. Pang, W., Wang, K.-P., Zhou, C.-G., Dong, L.-J., Liu, M., Zhang, H.-Y., Wang, J.-Y.: Modified particle swarm optimization based on space transformation for solving traveling salesman problem. In: Proceedings of the 3rd International Conference on Machine Learning and Cybernetics (2004)
43. Parks, G.T.: An intelligent stochastic optimization routine for nuclear fuel cycle design. Nucl. Technol. **89**, 233–246 (1990)
44. Poon, P.W., Parks, G.T.: Optimising PWR reload core designs. In: Parallel Problems Solving from Nature II, pp. 371–380. Elsevier, Amsterdam(1992)
45. Rubinstein, R.: The cross-entropy method for combinatorial and continuous optimization. Methodol. Comput. Appl. Probab. **1**, 127–190 (1999)
46. Rubinstein, R., Kroese, D.P.: The Cross-Entropy Method: A Unified Approach to Combinatorial Optimization, Monte-Carlo Simulation, and Machine Learning. Springer, Berlin (2004)
47. Safarzadeh, O., Zolfaghari, A., Norouzi, A., Minuchehr, H.: Loading pattern optimization of PWR reactors using artificial bee colony. Ann. Nucl. Energy **38**, 2218–2226 (2011)
48. Salman, A., Ahmad, I., Al-Madani, S.: Particle swarm optimization for task assignment problem. Microprocess. Microsyst. **26**, 363–371 (2002)
49. Schlünz, E.B., Bokov, P.M., van Vuuren, J.H.: Research reactor in-core fuel management optimisation using the multiobjective cross-entropy method. In: International Conference on Reactor Physics – PHYSOR 2014, Kyoto (2014)
50. Schlünz, E.B., Bokov, P.M., van Vuuren, J.H.: A comparative study on multiobjective metaheuristics for solving constrained in-core fuel management optimisation problems. Comput. Oper. Res. **75**, 174–190 (2016)
51. Schlünz, E.B., Bokov, P.M., Prinsloo, R.H., van Vuuren, J.H.: A unified methodology for single- and multiobjective in-core fuel management optimisation based on augmented Chebyshev scalarisation and a harmony search algorithm. Ann. Nucl. Energy **87**, 659–670 (2016)
52. Silva, M.H., Schirru, R.: Optimization of nuclear reactor core fuel reload using the new quantum PBIL. Ann. Nucl. Energy **38**, 610–614 (2011)

53. Silva, M.H., Schirru, R.: A self-adaptive quantum PBIL method for the nuclear reload optimization. Prog. Nucl. Energy **74**, 103–109 (2014)
54. Silva, P.V., Nast, F.N., Schirru, R., Meneses, A.A.M.: Artificial bee colony algorithm modified applied to fuel reload in nuclear reactors (in Portuguese). In: XIX National Meeting in Computational Modeling, João Pessoa (2016)
55. Stevens, J.G., Smith, K.S., Rempe, K.R., Downar, T.J.: Optimization of pressurized water reactor shuffling by simulated annealing with heuristics. Nucl. Sci. Eng. **121**, 68–77 (1995)
56. Tabak, D.: Optimization of nuclear reactor fuel recycle via linear and quadratic programming. IEEE Trans. Nucl. Sci. **15**, 60–64 (1968)
57. Taillard, E.D., Gambardella, L.M., Gendreau, M., Potvin, J.-Y.: Adaptive memory programming: a unified view of metaheuristics. Eur. J. Oper. Res. **135**, 1–6 (2001)
58. Tasgetiren, M.F., Sevkli, M., Liang, Y.C., Gencyilmaz, G.: Particle swarm optimization algorithm for single machine total weighted tardiness problem. In: Proceedings of the IEEE Congress on Evolutionary Computation, pp. 1412–1419 (2004)
59. Todreas, N.E., Kazimi, M.S.: Nuclear Systems I – Thermal Hydraulic Fundamentals. Taylor and Francis, Didcot (2004)
60. Waintraub, M., Schirru, R., Pereira, C.M.N.A.: Multiprocessor modeling of parallel particle swarm optimization applied to nuclear engineering problems. Prog. Nucl. Energy **51**, 680–688 (2009)
61. Wall, I., Fenech, H.: Application of dynamic programming to fuel management optimization. Nucl. Sci. Eng. **22**, 285–297 (1965)
62. Yarizadeh-Beneh, M., Mazaheri-Beni, H., Poursalehi, N.: Improving the refueling cycle of a WWER-1000 using cuckoo search method and thermal-neutronic coupling of PARCS v2.7, COBRA-EN and WIMSD-5B codes. Nucl. Eng. Des. **310**, 247–257 (2016)

Chapter 9
Inverse Problem of an Anomalous Diffusion Model Employing Lightning Optimization Algorithm

Luciano Gonçalves da Silva, Diego Campos Knupp, Luiz Bevilacqua, Augusto César Noronha Rodrigues Galeão, and Antônio José Silva Neto

9.1   Introduction

The spread of particles or microorganisms immersed in a given medium or deployed over a given substratum is frequently modeled as a diffusion process, using the well-known diffusion equation derived from the Fick's law. This model represents quite successfully the behavior of several physical phenomena related to dispersion processes; but for some cases, this approach fails to represent the real physical behavior. For instance, the spread of a population of particles may be partially and temporarily blocked when immersed in some particular media; an invading species may hold a fraction of the total population stationed on the conquered territory in order to guarantee territorial domain; and chemical reactions may induce adsorption processes for the diffusion of solutes in liquid solvents in the presence of adsorbent material [3, 5].

L. G. da Silva (✉)
Instituto Federal de Educação, Ciência e Tecnologia do Pará, IFPA, Paragominas, Brazil
e-mail: luciano.silva@ifpa.edu.br

D. C. Knupp · A. J. Silva Neto
Department of Mechanical Engineering and Energy, Polytechnic Institute, IPRJ-UERJ, Nova Friburgo, Brazil
e-mail: diegoknupp@iprj.uerj.br; ajsneto@iprj.uerj.br

L. Bevilacqua
Federal University of Rio de Janeiro, COPPE-UFRJ, Rio de Janeiro, Brazil
e-mail: bevilacqua@coc.ufrj.br

A. C. N. R. Galeão
Laboratório Nacional de Computação Científica, LNCC, Petrópolis, Brazil
e-mail: acng@lncc.br

© Springer Nature Switzerland AG 2019
G. Mendes Platt et al. (eds.), *Computational Intelligence, Optimization and Inverse Problems with Applications in Engineering*,
https://doi.org/10.1007/978-3-319-96433-1_9

185

Certain physicochemical phenomena need improvement on the analytical formulation due to other effects not taken into account in the classical diffusion theory. These include the flow through porous media [14]; diffusion processes for some dispersing substances immersed in particular supporting media [1, 6, 7, 9, 17], and the diffusion of hydrogen through metals, a case that is strongly influenced by the presence of hydrogen traps, such as grain boundaries, dislocations, carbides, and non-metallic particles [24]. In most published works on this matter, the well-known second order parabolic equation is assumed as the basic governing equation of the dispersion process, but the anomalous diffusion effect is modeled with the introduction of fractional derivatives [16], the imposition of a convenient variation of the diffusion coefficient with time or concentration [11, 25], or the introduction of complementary terms in the model to account for the retention phenomena [15].

Nevertheless, trying to overcome the anomalous diffusion issue by imposing an artificial dependence by the diffusion coefficient on the particle concentration, or by introducing extra differential terms, while the second order rank of the governing equation is kept unchanged, disguises the real physical phenomenon occurring in the process. In 2011, Bevilacqua et al. [3] derived a new analytical formulation to simulate the phenomena of anomalous diffusion. This formulation explicitly takes into account the retention effect in the dispersion process for the purpose of reducing all the diffusion processes with retention to a unifying phenomenon that may adequately simulate the retention effect. In addition to the diffusion coefficient, the newly introduced parameters characterize the blocking process. Specific experimental setups, together with an appropriate inverse analysis, need to be established to determine these complementary parameters.

This chapter investigates an inverse problem of anomalous diffusion that does not allow the simultaneous estimation of all parameters of the model. An estimation procedure is then presented in two steps: for example, an experimental procedure in which no anomalous diffusion takes place, and another one in which the anomalous effect is considered [8, 19]. Thus, the resulting inverse problem in Step 2 consists of a situation in which an anomalous diffusion occurs for the given problem, with the true diffusion coefficient already estimated in Step 1. The objective is to estimate the parameters related to the anomalous diffusion from this procedure. In this situation, it is critically important to consider how the uncertainty present in the supposedly known value of the true diffusion coefficient affects the estimation of the other parameters.

9.2 Direct Problem Formulation and Solution

Consider the process schematically represented in Fig. 9.1. The redistribution of the contents of each cell indicates that a fraction of the contents αp_n is retained in the

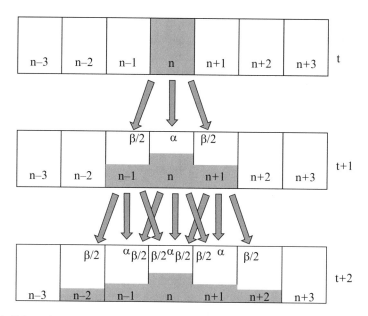

Fig. 9.1 Schematic representation of the symmetric distribution with retention, $\beta = 1 - \alpha$

nth cell and the remaining content is evenly transferred to the neighboring cells, that is, $0.5\beta p_n$ to the left, to the $(n-1)$th cell and $0.5\beta p_n$ to the right, to the $(n+1)$th cell, at each time step, where $\beta = 1 - \alpha$. This means that the dispersion runs slower than for the classical diffusion problem. Note that if $\beta = 1$, the problem is reduced to the classical distribution.

This process can be written as the following algebraic expressions:

$$p_n^t = (1 - \beta)p_n^{t-1} + \frac{1}{2}\beta p_{n-1}^{t-1} + \frac{1}{2}\beta p_{n+1}^{t-1} \tag{9.1a}$$

$$p_n^{t+1} = (1 - \beta)p_n^t + \frac{1}{2}\beta p_{n-1}^t + \frac{1}{2}\beta p_{n+1}^t \tag{9.1b}$$

Manipulating Eq. (9.1) in order to obtain finite difference terms yields:

$$\frac{\Delta p_n^{t+\Delta t}}{\Delta t} = \beta \left\{ \frac{1}{2}\frac{L_0^2}{T_0}\frac{\Delta^2 p_n}{\Delta x^2} + \frac{O\left(\Delta x^2\right)}{\Delta x^2} - (1 - \beta)\frac{1}{4}\frac{L_1^4}{T_0}\frac{\Delta^4 p_n}{\Delta x^4} \right\}^{t-\Delta t} \tag{9.2}$$

where T_0, L_0, and L_1 are integration parameters, described in more detail in [3]. Calling $K_2 = L_0^2/2T_0$ and $K_4 = L_1^4/4T_0$, both considered constant in this chapter, and taking the limit as $\Delta x \to 0$ and $\Delta t \to 0$, results [3]:

$$\frac{\partial p\left(x,t\right)}{\partial t} = \beta K_2 \frac{\partial^2 p\left(x,t\right)}{\partial x^2} - \beta\left(1 - \beta\right) K_4 \frac{\partial^4 p\left(x,t\right)}{\partial x^4} \tag{9.3a}$$

The fourth order term with negative sign introduces the anomalous diffusion effect, which shows up naturally, without any artificial assumption, as an immediate consequence of the temporary retention imposed by the redistribution law. Further discussion on the model derivation can be found in [3].

As the test case for the present work, consider the governing equation given by Eq. (9.3a) valid for $0 < x < 1$ and $t > 0$, with the following boundary and initial conditions:

$$p\,(0, t) = 1, \quad p\,(1, t) = 1, \quad \left.\frac{\partial p\,(x, t)}{\partial x}\right|_{x=0} = 0, \quad \left.\frac{\partial p\,(x, t)}{\partial x}\right|_{x=1} = 0, \quad t > 0$$

(9.3b)

$$p\,(x, 0) = f\,(x) = 2\sin^{100}\,(\pi x) + 1, \quad 0 \le x \le 1$$

(9.3c)

The problem given by Eq. (9.3) is solved in this work with an implicit finite difference scheme [20]. Concerning the inverse problem solution, observing the problem defined by Eq. (9.3), it is evident that the three parameters appearing in the model, β, K_2 and K_4 cannot be simultaneously estimated without any prior information, since there are three parameters defining two coefficients in Eq. (9.3a), i.e. there are infinite sets of values for the parameters $\mathbf{Z} = \{\beta, K_2, K_4\}$ that lead to the exactly same mathematical formulation, yielding non-uniqueness of the inverse problem solution, which was also illustrated by means of a sensitivity analysis in [20]. Since the most interesting aspect of this work is the identification of the three parameters appearing in the model, due to their direct physical interpretation [3, 4], we choose not to rewrite the problem in terms of two coefficients (which would multiply the second and fourth order differential terms). Instead, a two-step solution strategy is employed i.e., it is considered that a prior information can be obtained for the true diffusion coefficient, K_2. This prior information is obtained through an independent experiment. For example, an inverse problem in a physical situation in which the blocking process that characterizes the phenomenon of anomalous diffusion does not occur [8]. The inverse problem formulation and solution approaches used to solve this problem are discussed in the next section.

9.3   Inverse Problem

In order to investigate the inverse problem solution concerning the estimation of the model parameters $\mathbf{Z}$, a set of experimental data is considered available $\mathbf{p}^{exp}$, which are simulated in this chapter with the solution of Eq. (9.3) in which noise is added from a normal distribution with known variance σ_e,

$$p_i^{exp} = p_i\left(\mathbf{Z}_{exact}\right) + \epsilon_i, \quad \epsilon_i \sim N\left(0, \sigma_e^2\right) \tag{9.4}$$

Next, the inverse problem formulation and the solution by means of the maximum likelihood and Lightning Optimization Algorithm (LOA) are presented.

9.3.1   Maximum Likelihood

Assuming that the measurement errors related to the data $\mathbf{p}^{exp}$ are additive, uncorrelated, and have normal distribution, the probability density for the occurrence of the measurements $\mathbf{p}^{exp}$ with the given parameters values $\mathbf{Z}$ may be expressed as [12]

$$\pi\left(\mathbf{p}^{exp} \mid \mathbf{Z}\right) = (2\pi)^{-N_d/2} \left|\mathbf{W}\right|^{-1/2}$$
$$exp\left\{-\frac{1}{2}\left[\mathbf{p}^{exp} - \mathbf{p}^{calc}\left(\mathbf{Z}\right)\right]^T \mathbf{W}^{-1}\left[\mathbf{p}^{exp} - \mathbf{p}^{calc}\left(\mathbf{Z}\right)\right]\right\} \tag{9.5}$$

where N_d is the number of experimental data employed, and $\mathbf{p}^{calc}$ is the vector containing the quantities calculated through the direct problem solution employing the parameter values $\mathbf{Z}$. Hence, the likelihood estimates can be seen as the values of $\mathbf{Z}$ that maximizes the likelihood function given by Eq. (9.5), which may be achieved with the minimization of the argument of the exponential function in the referred equation. Assuming that the measurements errors are not correlated, and the variance is constant, then it is equivalent to minimize [2]:

$$Q_{ML}\left(\mathbf{Z}\right) = \left[\mathbf{p}^{exp} - \mathbf{p}^{calc}\left(\mathbf{Z}\right)\right]^T \left[\mathbf{p}^{exp} - \mathbf{p}^{calc}\left(\mathbf{Z}\right)\right] = \sum_{i=1}^{N_d}\left[p_i^{exp} - p_i^{calc}\left(\mathbf{Z}\right)\right]^2 \tag{9.6}$$

In this chapter, the Lightning Optimization Algorithm (LOA) is used to minimize Q_{ML}, in order to obtain $\mathbf{Z}$ from Eq. (9.6). There are cases in which the phenomenon of pure diffusion is temporary and subsequently the anomalous effect is verified [8, 13, 19]. As already mentioned, the simultaneous estimation of the parameters $\mathbf{Z} = \{\beta, K_2, K_4\}$ is not possible due to the correlation between these parameters in the model given by Eq. (9.3a). Therefore, a two-step solution strategy was used. In Step 1, the anomalous effect is not considered, and, thus, parameter K_2 is estimated. In Step 2, the anomalous effect is considered and, using the K_2 value estimated in Step 1, the values of β and K_4 are simultaneously estimated.

Table 9.1 A possible classification of nature-inspired algorithms

Evolutionary algorithms	Genetic algorithms
	Genetic programming
	Grammatical evolution
	Evolutionary strategy
Collective (swarm) intelligence	Ant colony
	Bee colony
	Algorithms of particle swarm
Neural networks	MLP—multi-layer perceptrons
	RBF—radial basis function network
	SOM—self-organizing maps
	ARTMap
Artificial immunology system	Negative selection algorithms
	Clonal expansion algorithms
	Network algorithms

9.3.2 Lightning Optimization Algorithm (LOA)

Many stochastic methods make analogies to phenomena observed in nature [22, 23], based on a population behavior or not. In general, the authors seek logical bases that allow the elaboration of an algorithm easy to be remembered and implemented. These algorithms can be divided into four classes, which are presented in Table 9.1.

There are in the literature methods based on the observation of nature, including with analogies to the phenomenon of electric discharges in the atmosphere, such as the Lightning Search Algorithm [18]. In this chapter we present a method of simple implementation, also based on the analogy with the lightning phenomenon, called Lightning Optimization Algorithm (LOA). This method was proposed and implemented by Silva in 2014 during the development of his doctoral thesis [19] and applied the solution of the inverse problem of the anomalous diffusion model proposed by Bevilacqua et al in 2011 [3]. At that time he was not aware of the Lightning Search Algorithm (LSA). LSA and LOA hold some similarity due to the natural phenomenon inspiration, but the latter may be considered of simpler implementation.

Following, the Lightning Optimization Algorithm—LOA [19] is presented, which is inspired by the observation of atmospheric discharges, whose algorithm is based on three main characteristics of such phenomenon.

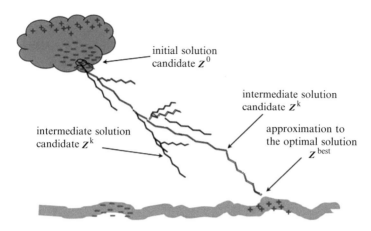

Fig. 9.2 Schematic representation of a cloud-soil atmospheric discharge

- A lightning bolt spreads along the "easy" path, describing a tortuous way in stages throughout its course. This path is called the stepped leader.
- Each step of the stepped leader ranges from 30 m up to 100 m.
- The atmospheric discharge source is at an unknown position (initial solution candidate) as well as the place of arrival (an approximation to the optimal solution).

In Fig. 9.2 is presented a schematic representation of an atmospheric discharge, indicating some of the elements considered in LOA.

9.3.2.1 Key Stages

Next are presented the LOA key stages implementation for a minimization problem. In Algorithm 9.1 LOA's implementation is described in a more detailed form, and in Fig. 9.3 its flowchart is shown.

Stage 1—Set a stopping criterion and make the iteration counter $k = 0$.
Stage 2—Set the maximum number of discharges divisions allowed Max_{div}, and generate an initial solution candidate $\mathbf{Z}^0$, make $\mathbf{Z}^{best} = \mathbf{Z}^0$, and calculate $Best = Q(\mathbf{Z}^0)$, with $Q(\mathbf{Z}) = Q_{ML}(\mathbf{Z})$ given by Eq. (9.6).
Stage 3—Generate a random number ray, between R_L and R_U (distance traveled in each step of the stepped leader). Make a disturbance vector $\mathbf{r}$:

$$\mathbf{r} = ray \times \mathbf{Z}^k \times (2r \text{ and } [0; 1] - 1) \qquad (9.7)$$

and determine a new candidate solution

$$\mathbf{Z}^{k+1} = \mathbf{Z}^k + \mathbf{r} \qquad (9.8)$$

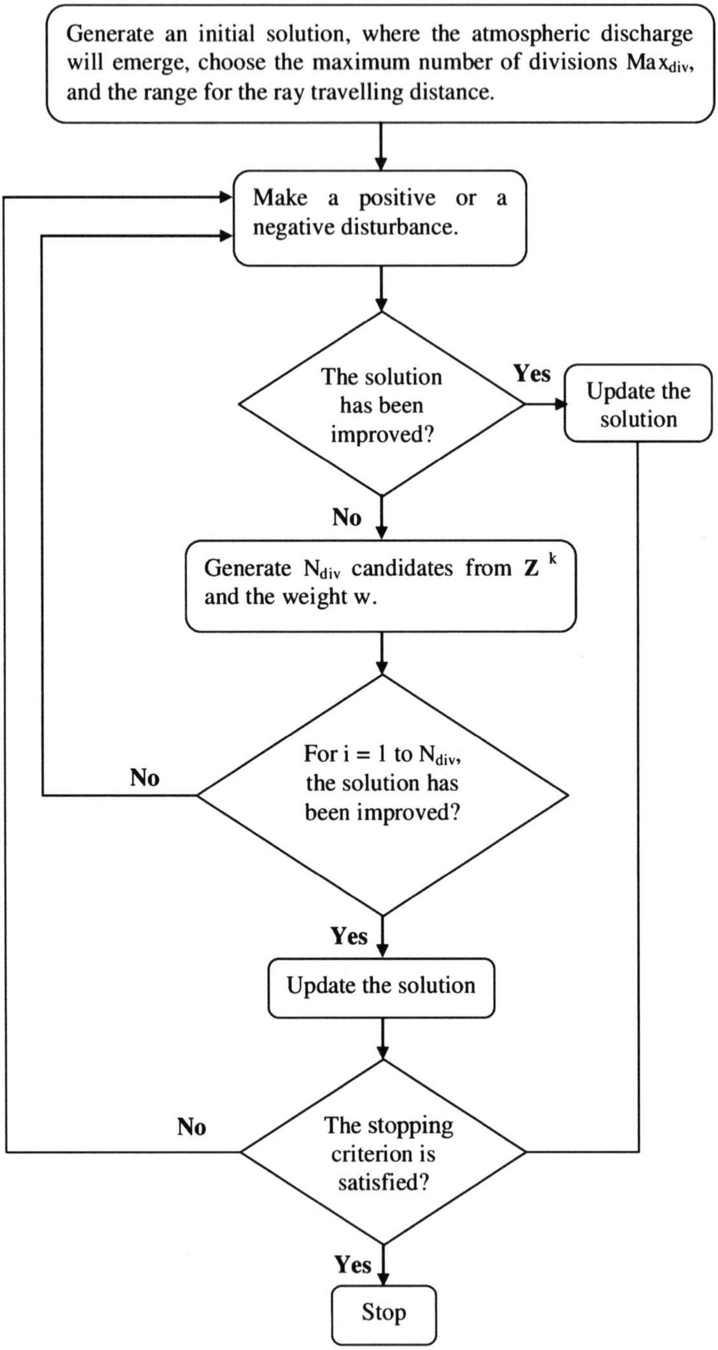

Fig. 9.3 Flowchart of LOA

If $Q(\mathbf{Z}^{k+1}) < Best$, make $\mathbf{Z}^{best} = \mathbf{Z}^{k+1}$, and $Best = Q(\mathbf{Z}^{k+1})$, otherwise make

$$\mathbf{Z}^{k+1} = \mathbf{Z}^k - \mathbf{r} \tag{9.9}$$

If $Q(\mathbf{Z}^{k+1}) < Best$, make $\mathbf{Z}^{best} = \mathbf{Z}^{k+1}$, and $Best = Q(\mathbf{Z}^{k+1})$, otherwise generate a random integer N_{div} between 2 and Max_{div}, and run the loop
  For $j = 1, \ldots, N_{div}$
    Generate a random number u in the range $[0; 1]$
    If $j = 1$, generate a random weight w in the range $[0; 1]$, or else

$$w = w \times (1 + u) \tag{9.10}$$

  make

$$\mathbf{Z}^{k+1} = \mathbf{Z}^k \times w \tag{9.11}$$

    If $Q(\mathbf{Z}^{k+1}) < Best$, make $\mathbf{Z}^{best} = \mathbf{Z}^{k+1}$ and $Best = Q(\mathbf{Z}^{k+1})$
  End For

Stage 4—If the prescribed stopping criterion is satisfied, then $\mathbf{Z}^{best}$ represents the solution to the problem, and Best the optimal value found, and stop the iterative procedure. Otherwise, make $k = k + 1$ and go back to Stage 3.

The method control parameters that must be adjusted for each type of problem to be solved are Max_{div}, R_L and R_U, representing the maximum number of divisions that can occur in the stepped leader, and the limits of the distance traveled in each step respectively. Keeping the analogy with the phenomenon of atmospheric discharges, R_L and R_U are chosen to be equal to 0.03 and 0.1, respectively. The method was tested using the nine Benchmark functions listed in Table 9.2 [10].

Table 9.2 Benchmark functions [10]

Number	Function
1	Ackley 1
2	Bent Cigar function
3	Discus function
4	Rosenbrock
5	Shifted and rotated Ackley's function
6	Weierstrass function
7	Griewank's function
8	High conditioned elliptic function
9	Rastrigin's function

Algorithm 9.1: Lightning Optimization Algorithm (LOA)

Data: Max_{div}, initial solution candidate $\mathbf{Z}^0$, range $[R_L; R_U]$
Result: $\mathbf{Z}^{best}$, $Best = Q(\mathbf{Z}^{best})$
initialization: $\mathbf{Z}^{best} = \mathbf{Z}^0$ and $Best = Q(\mathbf{Z}^0)$;
while *stopping criterion is not satisfied* **do**

    $ray = R_L + (R_U - R_L)rand[0; 1]$;
    $\mathbf{r} = ray \times \mathbf{Z}^{best} \times (2 \times rand[0; 1] - 1)$;
    $\mathbf{Z}^{k+1} = \mathbf{Z}^{best} + \mathbf{r}$;
    $Q_Z = Q(\mathbf{Z}^{k+1})$;
    if $Q_Z < Best$ **then**
        $\mathbf{Z}^{best} = \mathbf{Z}^{k+1}$;
        $Best = Q_Z$;
    else
        $\mathbf{Z}^{k+1} = \mathbf{Z}^{best} - \mathbf{r}$;
        $Q_Z = Q(\mathbf{Z}^{k+1})$;
    end if
    if $Q_Z < Best$ **then**
        $\mathbf{Z}^{best} = \mathbf{Z}^{k+1}$;
        $Best = Q_Z$;
    else
        generate a random integer N_{div} between 2 and Max_{div};
        for $j = 1$ *to* N_{div} **do**
            $u = rand[0; 1]$;
            if $j=1$ **then**
                $w = u$;
            else
                $w = w \times (1 + u)$;
            end if
            $\mathbf{Z}^{k+1} = \mathbf{Z}^k \times w$;
            $Q_Z = Q(\mathbf{Z}^{k+1})$;
            if $Q_Z < Best$ **then**
                $\mathbf{Z}^{best} = \mathbf{Z}^{k+1}$;
                $Best = Q_Z$;
            end if
        end for
    end if
end while

9.4   Results and Discussion

Observing Eq. (9.3a) it is clear that the three parameters appearing in the model, β, K_2 and K_4, cannot be simultaneously estimated without the availability of any prior information, since these parameters define only two coefficients. In the results presented next, a two-step solution is employed considering in the first experiment (Step 1) only the classical diffusion phenomenon, i.e., $\beta = 1$ in Eq. (9.3a), and then obtaining an estimate for K_2. Then, in the second independent experiment (Step 2), estimates for β and K_4 are obtained, assuming the solution obtained in Step 1 for K_2, and this time considering the anomalous effect in Eq. (9.3a). The synthetic experimental data used in Steps 1 and 2 were simulated at position $x = 0.5$, in transient measurements with $t = 0.1; 0.2; 0.3; \ldots; 10$ and $t = 10.1; 10.2; 10.3; \ldots; 20$, respectively, with the following exact values $\beta = 0.2$, $K_2 = 0.001$, $K_4 = 0.00001$ [13] and $\sigma_e = 0.02$ in Eq. (9.4), which results in errors of up to 4% in the experimental data. This case was investigated earlier in [13, 21], and its sensitivity analysis [2] indicates that $x = 0.5$ is the most appropriate choice for sensor location.

Figure 9.4 illustrates the behavior of the experimental data used in Steps 1 and 2 for the test case under study.

For the estimation of K_2 in Step 1 it was considered the search interval [0; 1]. We performed 100 LOA executions with 1000 objective function evaluations each. Table 9.3 shows the mean value μ, the standard deviation σ, the coefficient of variation $(\sigma/\mu) \times 100\%$ and the confidence interval, with a 95% confidence level, obtained from these 100 executions. Experimental data without noise, and with up to 4% noise level, were employed in the solution of the inverse problem. In Step 2, the search interval [0; 1] for β, and [0; 0.001] for K_4, were considered. In this step 100 LOA executions were also carried out with 10,000 objective function assessments each. Tables 9.4 and 9.5 show the mean value μ, the standard deviation

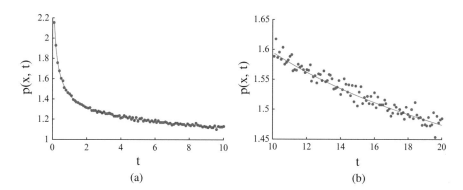

Fig. 9.4 Simulated experimental data (red dots) for the transient measurements obtained using a sensor located at $x = 0.5$. The solid line shows the numerical solution employed to simulate the experimental data. (**a**) Step 1; (**b**) Step 2

Table 9.3 Estimates for K_2 considering noiseless and noisy experimental data

K_2	Experimental error (0%)	Experimental error (4%)
μ	0.999997×10^{-3}	1.002456×10^{-3}
σ	1.148898×10^{-7}	2.438964×10^{-5}
$\frac{\sigma}{\mu} \times 100\%$	0.01149	2.432989
Confidence interval (95%)	$[0.999975; 1.00002] \times 10^{-3}$	$[0.997676; 1.007236] \times 10^{-3}$

(Step 1). Exact value $K_2 = 10^{-3}$

Table 9.4 Estimates for β considering noiseless and noisy experimental data

β	Experimental error (0%)	Experimental error (4%)
μ	0.199738	0.199451
σ	3.44407×10^{-3}	5.40807×10^{-3}
$\frac{\sigma}{\mu} \times 100\%$	1.72429	2.711475
Confidence interval (95%)	$[0.199063; 0.200413]$	$[0.198391; 0.200511]$

(Step 2). Exact value: $\beta = 0.2$

Table 9.5 Estimates for K_4 considering noiseless and noisy experimental data

K_4	Experimental error (0%)	Experimental error (4%)
μ	1.001800×10^{-5}	1.003513×10^{-5}
σ	1.952076×10^{-7}	2.68298×10^{-7}
$\frac{\sigma}{\mu} \times 100\%$	1.94857	2.67358
Confidence interval (95%)	$[0.99797; 1.00563] \times 10^{-5}$	$[0.998254; 1.008771] \times 10^{-5}$

(Step 2). Exact value: $K_4 = 10^{-5}$

σ, the coefficient of variation $(\sigma/\mu) \times 100\%$, and the confidence interval with a 95% confidence level, obtained from these 100 executions for the parameters β and K_4, respectively. Experimental data without noise, and with up to 4% noise level were employed in the solution of the inverse problem.

It is observed in Table 9.3 that the estimated value for K_2 in Step 1 is in good agreement with the exact value 0.001, and also the estimated confidence interval presented a small amplitude in the 95% confidence level. That holds for both noiseless and noisy (up to 4% level) experimental data. In Fig. 9.5 the results of Table 9.3 are graphically presented, and in Fig. 9.6 the evolution of the objective function value along LOA's iterative procedure is shown. Even though Steps 1 and 2 consider experimental data collected in independent experiments, we note that the estimate obtained for K_2 in Step 1 is essential for the estimation of β and K_4, since this value will be used as a priori information in Step 2. From the observation of the results in Table 9.3, we conclude that the LOA successfully fulfilled this task, i.e. a good estimate for K_2 was provided by LOA in Step 1.

Tables 9.4 and 9.5 show that with the use of the two-step procedure it was possible to obtain good estimates for β and K_4, as well as small amplitude confidence intervals considering a 95% confidence level. This indicates that even

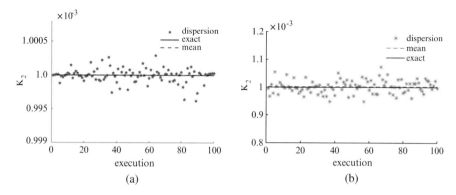

Fig. 9.5 Scatter plot for K_2 estimates (Step 1) obtained using LOA. (**a**) Noiseless experimental data; (**b**) noisy experimental data (up to 4% error)

Fig. 9.6 Evolution of the objective function value along LOA's iterative procedure for the estimation of K_2 in Step 1 (noiseless experimental data)

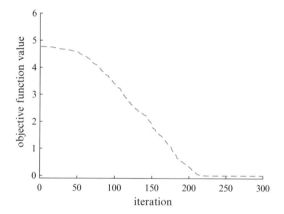

not being possible to estimate all the parameters in a single experiment, because of the correlation between them, the quality of the solution of the inverse problem is not compromised considering the two-step approach, using in Step 2 the information obtained a priori in Step 1. Therefore, LOA presented satisfactory results also in the solution of the inverse problem in Step 2. In Fig. 9.7, the results shown in Tables 9.4 and 9.5 are graphically presented.

Even though 1000 objective function evaluations have been used as a stopping criterion, it is observed in Fig. 9.6 that a much lower number of evaluations would be enough to achieve results as accurate as those obtained with 1000 evaluations. With just over 200 LOA iterations, which corresponds to approximately 500 evaluations of the objective function, the method would have already provided accurate solutions to the inverse problem.

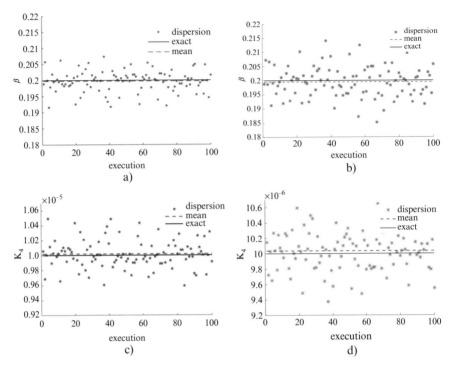

Fig. 9.7 Scatter plot for the estimates of β and K_2 (Step 2) obtained using LOA. (**a**) Estimation of β with noiseless experimental data; (**b**) estimation of β with noisy experimental data (up to 4% error); (**c**) estimation of K_4 with noiseless experimental data; (**d**) estimation of K_4 with noisy experimental data (up to 4% error)

9.5   Conclusions

In this chapter a new heuristic that makes an analogy to a natural phenomena is presented: the Lightning Optimization Algorithm (LOA), and it is used for the solution of an inverse problem involving a recently developed anomalous diffusion model. One of the peculiarities of this model for anomalous diffusion is the definition of two coefficients (diffusion and anomaly) from three parameters, which makes it impossible to estimate the three parameters simultaneously in a single experiment. Therefore, in this work a two-step solution was applied, that showed efficacy in the solution of the inverse problem. For this task, the LOA presented itself as an interesting tool that can even be improved in future works. The results were statistically analyzed and discussed. Both the good estimated mean values and the reduced amplitude confidence intervals indicate that the strategy adopted has been successfully applied.

Acknowledgements   The authors acknowledge the financial support provided by the Brazilian Agencies FAPERJ—Fundação Carlos Chagas Filho de Amparo à Pesquisa do Estado do Rio de Janeiro, CNPq—Conselho Nacional de Desenvolvimento Científico e Tecnológico, and CAPES—Coordenação de Aperfeiçoamento de Pessoal de Nível Superior.

References

1. Atsumi, H.: Hydrogen bulk retention in graphite and kinetics of diffusion. J. Nucl. Mater. **307/311**, 1466–1470 (2002)
2. Beck, J.V., Arnold, K.J.: Parameter Estimation in Engineering and Science. Wiley Interscience, New York (1977)
3. Bevilacqua, L., Galeão, A.C.N.R., Costa, F.P.: A new analytical formulation of retention effects on particle diffusion Process. Ann. Braz. Acad. Sci. **83**, 1443–1464 (1977)
4. Bevilacqua, L., Jiang, M., Silva Neto, A.J., Galeão, A.C.N.R.: An evolutionary model of bi-flux diffusion process. J. Braz. Soc. Mech. Sci. Eng. **38**, 1241–1432 (2016)
5. Brandani, S., Jama, M., Ruthven, D.: Diffusion, self-diffusion and counter-diffusion of benzene and p-xylene in Silicalite. Microporous Mesoporous Mater. **35/36**, 283–300 (2000)
6. D'Angelo, M.V., Fontana, E., Chertcoff, R., Rose, M.: Retention phenomena in non-Newtonian fluid flows. Physics A **327**, 44–48 (2003)
7. Deleersnijder, E., Beckers, J.M., Delhez, E.M.: The residence time of setting in the surface mixed layer. Environ. Fluid Mech. **6**, 25–42 (2006)
8. Derec, C., Smerlak, M., Servais, J., Bacri, J.-C.: Anomalous diffusion in microchannel under magnetic field. Phys. Procedia **9**, 109–112 (2010)
9. Green, P.F.: Translational dynamics of macromolecules in metals. In: Neogi, P. (ed.) Diffusion in Polymers. CRC Press, Boca Raton (1996)
10. Jamil, M., Yang, X.-S.: A literature survey of benchmark functions for global optimisation problems. Int. J. Math. Model. Numer. Optim. **4**, 150–194 (2013)
11. Joannès, S., Mazé, L., Bunsell, A.R.: A concentration-dependent diffusion coefficient model for water sorption in composite. Compos. Struct. **108**, 111–118 (2014)
12. Kaipio, J., Somersalo, E.: Statistical and Computational Inverse Problems. Applied Mathematical Sciences, vol. 160. Springer, Berlin (2004)
13. Knupp, D.C., Silva, L.G., Bevilacqua, L., Galeão, A.C.N.R., Silva Neto, A.J.: Inverse Analysis of a New Anomalous Diffusion Model Employing Maximum Likelihood and Bayesian Estimation. Mathematical Modeling and Computational Intelligence in Engineering Applications, vol. 1, pp. 89–104. Springer International Publishing, Berlin (2016)
14. Liu, H., Thompson, K.E.: Numerical modeling of reactive polymer flow in porous media. Comput. Chem. Eng. **26**, 1595–1610 (2002)
15. McNabb, A., Foster, P.K.: A new analysis of the diffusion of hydrogen in iron and ferritic steels. Trans. Metall. Soc. AIME **227**, 618–1963 (1963)
16. Metzler, R., Klafter, J.: The random walk's guide to anomalous diffusion: a fractional dynamics approach. Phys. Rep. **339**, 1–77 (2000)
17. Muhammad, N.: Hydraulic, diffusion, and retention characteristics of inorganic chemicals in bentonite. Ph.D. thesis. Department of Civil and Environmental Engineering, University of South Florida (2004)
18. Shareef, H., Ibrahim, A.A., Mutlag, A.H.: Lightning search algorithm. Appl. Soft Comput. J. **36**, 315–333 (2015)
19. Silva, L.G.: Direct and inverse problems in anomalous diffusion. Doctoral thesis, Rio de Janeiro State University (in Portuguese) (2016)
20. Silva, L.G., Knupp, D.C., Bevilacqua, L., Galeão, A.C.N.R., Simas, J.G., Vasconcellos, J.F., Silva Neto, A.J.: Investigation of a new model for anomalous diffusion phenomena by means

of an inverse analysis. In: Proceedings of 4th Inverse Problems, Design and Optimization Symposium, Albi (2013)

21. Silva, L.G., Knupp, D.C., Bevilacqua, L., Galeão, A.C.N.R., Silva Neto, A.J.: Inverse problem in anomalous diffusion with uncertainty propagation. Comput. Assist. Methods Mech. Sci. **21**, 245–255 (2014)

22. Silva Neto, A.J., Llanes Santiago, O., Silva, G.N.: Mathematical Modeling and Computational Intelligence in Engineering Applications. Springer, São Carlos (2016)

23. Silva Neto, A.J., Becceneri, J.C., Campos Velho, H.F. (eds.): Computational intelligence applied to inverse radiative transfer problems (in Portuguese). Ed UERJ, Rio de Janeiro (2016)

24. Turnbull, A., Carroll, M.W., Ferriss, D.H.: Analysis of hydrogen diffusion and trapping in a 13% chromium martensitic stainless steel. Acta Metall. **37**, 2039–2046 (1989)

25. Wu, J., Berland, K.M.: Propagators and time-dependent diffusion coefficients for anomalous diffusion. Biophys. J. **95**, 2049–2052 (2008)

Chapter 10
Study of the Impact of the Topology of Artificial Neural Networks for the Prediction of Meteorological Data

Roberto Luiz Souza Monteiro, Hernane Borges de Barros Pereira, and Davidson Martins Moreira

10.1 Introduction

Climate prediction via climatic variables is fundamental when considering the impacts of climate on human activities, especially those that add great economic value, such as agriculture or energy generation. Thus, selecting the method that produces the smallest possible errors and has the lowest computational cost becomes necessary. Research comparing the efficiency of artificial neural networks was performed in [9, 11]. Both used the multi-layered perceptron (MLP) [15] network to predict the next temperature step. In [11] are compared the results with traditional techniques using a linear combination, and concluded that MLP neural networks yield better results. The authors also used the same next-step prediction techniques to anticipate wind speed and humidity, reaching the same conclusion. In [10] it is obtained a small error in the next-step prediction, analyzing the mean absolute error (MAE), although they did not compare the results obtained with MLP networks with those of other prediction techniques.

In [5] are used MLP networks to predict precipitation in a city in Vietnam to predict the occurrence of flooding. In their study the authors verified that artificial neural networks yielded better results than the models used by them for comparison. In [1] it is noted that the nonlinear characteristics of climate variables make it difficult to build non-linear models using conventional statistics; the authors stated also that artificial neural networks constitute a more adequate technique. In [17] are used MLP networks, comparing the performance of different training methods, to predict climatic variables, concluding that these methods are applicable in this type of simulation.

R. L. S. Monteiro · H. B. d. B. Pereira · D. M. Moreira (✉)
Program of Computational Modeling, SENAI CIMATEC, Salvador, Brazil
e-mail: roberto@souzamonteiro.com

© Springer Nature Switzerland AG 2019
G. Mendes Platt et al. (eds.), *Computational Intelligence, Optimization and Inverse Problems with Applications in Engineering*,
https://doi.org/10.1007/978-3-319-96433-1_10

201

Studies on the next-step prediction of solar radiation were performed in [18], combining a self-regressive model of moving averages with artificial neural networks, and good results were obtained. More recently, in [4] are compared the results obtained with MLP neural networks, radial base function networks (RBF) and multiple linear regression (MLR) in the next-step prediction of temperature and humidity, and the authors obtained better results using MLP networks.

Although the previously cited authors demonstrated the applicability of artificial neural networks in the next-step prediction of the climatic variables, none of them compared the performance of different complex topologies of neural networks (complete, random, scale-free, small world, and hybrid), especially when applied to the prediction of these variables. In this chapter, we intend to fill this gap by analyzing the next-step predictive capacity of six neural network topologies— multilayer perceptron (MLP), complete (CP), random (RD), scale-free (SF), small world (SW), and hybrid (HY)—to predict temperature and solar radiation. In the next sections, we will present the methods used, and discuss the results obtained.

10.2 Materials and Methods

In this chapter, we use artificial neural networks for the next-step prediction of two climatic variables: solar radiation and temperature. We used data collected by the INMET (National Institute of Meteorology) meteorological station located in Salvador city, Bahia, Brazil. We used data of 1000 h for training and evaluation of the results that were collected from January to February of 2016.

10.2.1 Artificial Neural Networks

The most widely used artificial neuron model is the perceptron proposed in [15]. This model defines a neuron composed of inputs, a summation and an activation function. The value of each input is multiplied by a weight and the weighted values of the inputs are summed to yield the result of the sum which is used as the input of the activation function. To teach (train) the neuron, the weights are modified so that the output obtained corresponds to the desired value.

The original algorithm proposed in [15], to train a perceptron, allowed the training of only one neuron. This limitation discouraged interest in this computational model, since its application was very restricted. Consequently, a method called backpropagation, later generalized in [20], has been developed in [16]. This method has been used to train multilayered perceptron (MLP) networks since then, and is the most used method for predicting time series, such as those analyzed in this study. Figure 10.1 presents a multilayer perceptron, which will be used to explain the training method adopted in this study. To fix the weights, two formulas are used: $x_j = \sum_{i=1}^{n} x_i \cdot w_{i,j}$, where x_j is the neuron that calculates the output value, x_i is

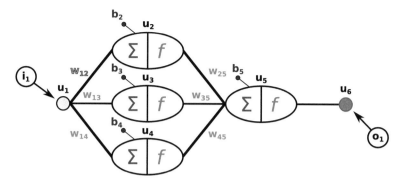

Fig. 10.1 A multilayer perceptron

an input synapse, $w_{i,j}$ is the weight of synapse i, and n is the number of synapses of the neuron j; $w_{i,j} = w_{i,j} + \eta \cdot \delta_j \cdot f'(x_j) \cdot x_i$, where $w_{i,j}$ is the weight of synapse i, η is the learning rate, δ_j is the error obtained at the output of neuron j, $f'(x_j)$ is the derivative of the activation function, and x_i is the value of synapse i. To calculate the error in the j neuron output, two other formulas are used: $\delta_j = z_j - y_j$ was used to calculate the error of the last layer, where z_j is the desired value in the output and y_j is the value obtained, and $\delta_j = \sum_{i=1}^{n} w_{i,j} \cdot \delta_i$ was used to calculate the error of each neuron of the intermediate layer, where n is the number of neurons in the next layer which receive the output of neuron j as input, and δ_i is the error obtained at the output of the next layer neuron after neuron j. The idea is to assign a neuron j an error proportional to its participation in the error of the next layer neuron. If the neuron connects to several other neurons in the next layer, we add up the weighted errors of each neuron of the next layer to get the error of the neuron in question.

Building programs to train complex topology neural networks by using the backpropagation method can be a difficult task as the code becomes more extensive and complex as the size of the network grows.

This problem of training complex topology neural networks motivated the authors of [13] to create a backpropagation-based algorithm for training and execution of complex topology neural networks, allowing the networks to be implemented as synaptic maps. This device, called a learning matrix, adapted for this study is presented in Fig. 10.2.

A learning matrix brings together in a single memory space all the data structures necessary to train the artificial neural network. In this way, access to the data occurs faster since all the elements are in a contiguous space. In addition, the training process becomes easy because the operation to update the network weights can be performed using a simple algorithm. However, the original algorithm presented in [13] allowed only networks designed to solve classification problems, since the network output was always binary and could not support neurons with biases (bias is a value that allows to move the neuron output curve). For the present work, we modified the original algorithm proposed by the authors, to allow the network to assume any real value in its output and to use biases to solve general problems. This

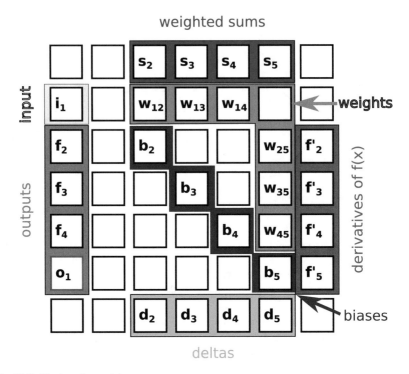

Fig. 10.2 The learning matrix

algorithm is presented in Fig. 10.3. The developed algorithm stores the bias values of the neurons in the network using the main diagonal of the matrix, thus allowing them to be saved after training the network.

To process data using the trained network, we adapted the processing algorithm developed in [13] to consider the biases determined during training. This algorithm is presented in Fig. 10.4.

10.2.1.1 Artificial Neural Networks of Complex Topology

Neural networks of MLP topology have a regular structure, since all the neurons of a layer connect to the neurons of the next layer, and only to them. In our study, we decided to compare results obtained with MLP networks with those coming from networks of complex topology.

The topologies of the most commonly cited complex networks in the literature are the complete network (CP), random network (RD) [7], scale-free networks (SF) [2], small world networks (SW) [19], and hybrid (HY) [8, 13, 14].

Neural networks of complete topology are those networks whose neurons are connected to all other neurons of the network. Neural networks with random

```
{
    Arguments: nn: learning matrix; in: input training data; out: output training data;
    ni: number of inputs; no: number of outputs; lrate: learning tax;
    af: activation function; af': derivative of activation function;
    oaf: output activation function; oaf': derivative of output activation function.
}
function Learn(nn, in, out, ni, no, lrate, af, af', oaf, oaf')
begin
    dim_nn = dim(nn)
    dim_i = dim_nn[0]
    dim_j = dim_nn[1]
    first_out = dim_j - 1 - no

    for i = 0 to dim_i - 2 step 1 do {Clear inputs and outputs.}
    begin
        nn[0, i] = 0.0
        nn[i, 0] = 0.0
        nn[i, dim_j - 1] = 0.0
        nn[dim_i - 1, i] = 0.0
    end

    for j = 0 to ni - 1 step 1 do {Assign inputs.}
        nn[j + 1, 0] = in[j]

    for j = ni + 1 to dim_j - 2 step 1 do {Calculate the neurons output.}
    begin
        nn[0, j] = 0.0
        for i = 1 to dim_i - 2 step 1 do {Weighted sums.}
            if i < j then
                if nn[i, j] != 0 then {x = x1 * w1 + x2 * w2 + ...}
                    nn[0, j] = nn[0, j] + nn[i, j] * nn[i, 0]
            else if i == j then
                if nn[i, j] != 0 then
                    nn[0, j] = nn[0, j] + nn[i, j]
            else
                break
        if j < first_out then {Activation function.}
        begin
            nn[j, 0] = @af(nn[0, j]) {Calculate y = f(x).}
            nn[j, dim_j - 1] = @af'(nn[0, j]) {Calculate df(x)/dx}
        end
        else {Activation function for the output layer.}
        begin
            nn[j, 0] = @oaf(nn[0, j]) {Calculate y = f(x).}
            nn[j, dim_j - 1] = @oaf'(nn[0, j]) {Calculate df(x)/dx}
        end
    end

    for i = 0 to no - 1 step 1 do {Calculate delta for the output neurons.}
        nn[dim_i - 1, first_out + i] = out[i] - nn[first_out + i, 0] {d = z - y}

    for j = dim_j - 2 to j ni + 1 step - 1 do {Calculate delta for hidden neurons.}
        for i = ni + 1 to i dim_i - 2 - no step 1 do
        begin
            if i == j then
                break
            if nn[i, j] != 0 then {d1 = w1 * d2 + w2 * d2 + ...}
                nn[dim_i - 1, i] = nn[dim_i - 1, i] + nn[i, j] * nn[dim_i - 1, j]
        end

    for j = no + 1 to dim_j - 2 step 1 do {Adjust weights.}
        for i = 1 to dim_i - 2 - no step 1 do
            if i < j then
                if nn[i, j] != 0 then {w1 = w1 + n * d * df(x)/dx * x1}
                    nn[i, j] = nn[i, j] + lrate * nn[dim_i - 1, j] * nn[j, dim_j - 1] * nn[i, 0]
            else if i == j then {Biases.}
                if nn[i, j] != 0 then
                    nn[i, j] = nn[i, j] + lrate * nn[dim_i - 1, j] * nn[j, dim_j - 1] * 1.0
            else
                break

    return nn
end
```

Fig. 10.3 The learning algorithm

```
{
    Arguments: nn: learning matrix; in: input data;
    ni: number of inputs; no: number of outputs;
    af: activation function; oaf: output activation function; of: output function.
}
function Process(nn, in, ni, no, af, oaf, of)
begin
    dim_nn = dim(nn)
    dim_i = dim_nn[0]
    dim_j = dim_nn[1]
    first_out = dim_j - 1 - no

    for i = 0 to dim_i - 2 step 1 do {Clear inputs and outputs.}
    begin
        nn[0, i] = 0.0
        nn[i, 0] = 0.0
        nn[i, dim_j - 1] = 0.0
        nn[dim_i - 1, i] = 0.0
    end

    for j = 0 to ni - 1 step 1 do {Assign inputs.}
        nn[j + 1, 0] = in[j]

    for j = ni + 1 to dim_j - 2 step 1 do {Calculate the neurons output.}
    begin
        nn[0, j] = 0.0
        for i = 1 to dim_i - 2 step 1 do {Weighted sums.}
            if i < j then
                if nn[i, j] != 0 then {x = x1 * w1 + x2 * w2 + …}
                    nn[0, j] = nn[0, j] + nn[i, j] * nn[i, 0]
            else if i == j then
                if nn[i, j] != 0 then
                    nn[0, j] = nn[0, j] + nn[i, j]
            else
                break
        if j < first_out then {Activation function.}
        begin
            nn[j, 0] = @af(nn[0, j]) {Calculate y = f(x).}
        end
        else {Activation function for the output layer.}
        begin
            nn[j, 0] = @oaf(nn[0, j]) {Calculate y = f(x).}
        end
    end

    for i = 0 to no - 1 step 1 do {Set the output matrix.}
        out[i] = @of(nn[first_out + i, 0])

    return out
end
```

Fig. 10.4 The algorithm for signal processing

topology have random connections between their neurons, which means that there is no known reason for these connections to be established. Scale-free neural networks have a preferential attachment between their neurons, which means that some neurons have more synaptic connections than others, and there is a criterion for this preference to occur. This criterion is a probability function presented in [2]. Neural networks with small-world topology have clusters, which are regions where an agglomeration of connected neurons occur, and this agglomeration follows the phenomenon described in [19]. A hybrid network is one that presents both scale-free and small-world network characteristics at the same time.

Our motivation for this work considers the results obtained in [13], who demonstrated that scale-free and hybrid networks yield better results when used to solve some classification problems. Studies in [3] showed a superior performance of small-world networks when used to implement associative memory. The research in [6] analyzed the influence of network topology on learning dynamics.

10.3   Methods

For this study, we created 100 networks of each topology, each with 26 neurons (five are inputs, 20 are processing, and one is the output). This distribution of neurons was obtained empirically by testing different values from the input neuron up to ten and varying the amount of processing neurons until obtaining the best results. For the creation of neural networks of complete topology, random, scale-free and small-world, we used the algorithms developed in [12]. For the creation of hybrid topology networks, we used the algorithm proposed in [13].

Each network was trained for 2000 epochs, and this number was also determined empirically, training an MLP network for 200,000 epochs and following the values of the mean square errors (MSE), and determining the lowest number of epochs that presented satisfactory results. The average values of the root mean square errors (RMSEs) were obtained and used to compute the means of learning times (ETL) and processing times (ETP). The solar radiation and the temperature were simulated, and the results were compared with those obtained by the meteorological station.

For each network topology, we performed one hundred simulations, each consisting of one training stage and one validation phase. For the training phase, we used 700 h of data, while for the validation step we used 300 h, using the same criteria as in [4], namely, 70% for training data and 30% for validation.

10.4   Results

After training and validation, we constructed the root mean square error (RMSE), mean learning time (ETL), and mean processing time (ETP) graphs, which were necessary to predict the next step for 300 h obtained from the weather station.

Figure 10.5 presents the results for the root mean square error (RMSE) of the topologies studied for next-step prediction of solar radiation. We observe that the MLP network presented the smallest error in estimating this climatic variable.

In Fig. 10.6, we observe the results for averages of learning times (ETL) for the six topologies. Again, the MLP network presented the best result.

Figure 10.7 shows the results for the mean processing time (ETP). In this case, the complete network (CP) was the one that processed the data faster during next-step prediction of the solar radiation.

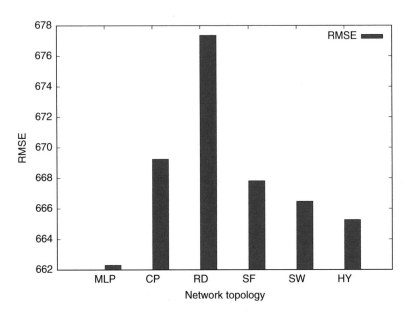

Fig. 10.5 The Root Mean Square Errors (RMSE) of solar radiation for networks with MLP, Complete (CP), Random (RD), Scale-free (SF), Small World (SW), and Hybrid (HY) topologies, after 100 simulations

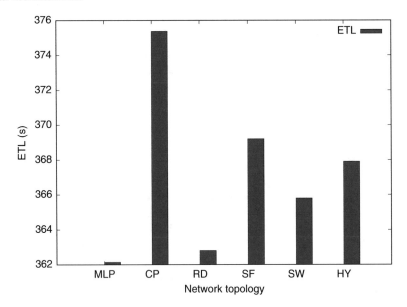

Fig. 10.6 Average of time to learn (ETL) of solar radiation for networks with MLP, Complete (CP), Random (RD), Scale-free (SF), Small World (SW), and Hybrid (HY) topologies, after 100 simulations

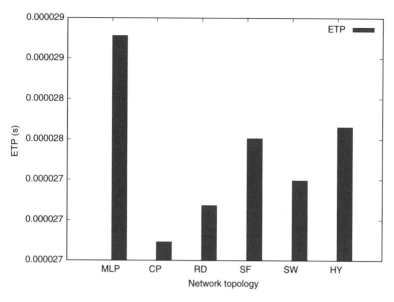

Fig. 10.7 Average next-step prediction times (ETP) of solar radiation for networks with MLP, Complete (CP), Random (RD), Scale-free (SF), Small World (SW), and Hybrid (HY) topologies, after 100 simulations

These results indicate that if we want a higher accuracy in the result, we must use an MLP network for training, while if we need speed to predict solar radiation values, we must use a complete network for this purpose.

In Fig. 10.8, we can see a comparison between the values of solar radiation obtained from the meteorological station and the average of those produced by the neural network when we use the MLP topology.

A procedure like that used in the next-step prediction of the solar radiation was performed to predict the temperature.

Figure 10.9 presents the results for the root mean square error (RMSE) of the topologies studied to predict the next temperature step. We observed that the network with small world topology (SW) was the one that presented the smallest error in estimating this climatic variable.

In Fig. 10.10, we present the results for averages of learning times (ETL) for the six topologies. Again, the MLP network presented the best result.

Figure 10.11 shows the results for the mean processing time (ETP). Here, the random network (RD) was the one that processed the data faster during next-step prediction of temperature.

These results indicate that if we want a better accuracy in the result, we must use a small world network (SW) for the training, while if we need speed to predict the temperature values, we must use a random (RD) network for this purpose.

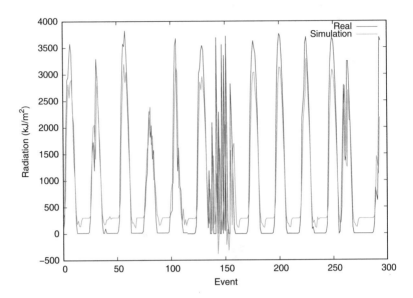

Fig. 10.8 Comparison of real signals of solar radiation and simulated with the MLP neural network, using 300 h of data

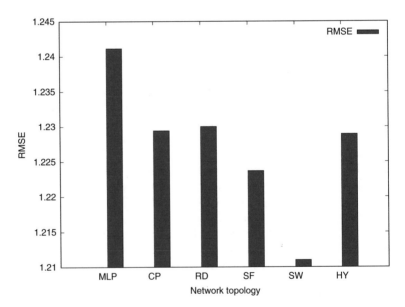

Fig. 10.9 The Root Mean Square Errors (RMSE) of temperature for networks with MLP, Complete (CP), Random (RD), Scale-free (SF), Small World (SW), and Hybrid (HY) topologies, after 100 simulations

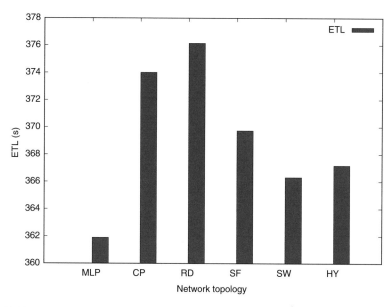

Fig. 10.10 Average of time to learn (ETL) of temperature for networks with MLP, Complete (CP), Random (RD), Scale-free (SF), Small World (SW), and Hybrid (HY) topologies, after 100 simulations

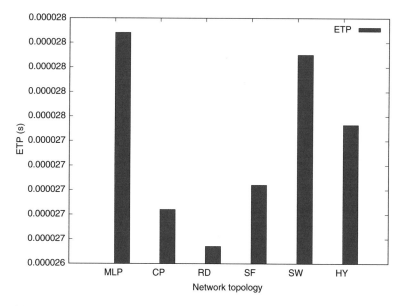

Fig. 10.11 Average next-step prediction times (ETP) of temperature for networks with MLP, Complete (CP), Random (RD), Scale-free (SF), Small World (SW), and Hybrid (HY) topologies, after 100 simulations

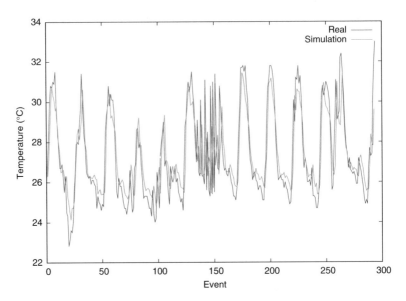

Fig. 10.12 Comparison of real signals of temperature and simulated with the MLP neural network, using 300 h of data

In Fig. 10.12, we see a comparison between the temperature values obtained from the meteorological station and the average of those produced by the neural network when we use the small world topology (SW).

10.5   Concluding Remarks

In this study, we performed next-step prediction simulations with neural networks of six different topologies—multilayer perceptron (MLP), complete network (CP), random network (RD), scale-free network (SF), small world network (SW), and hybrid network (HY)—to predict solar radiation and temperature. For each network, we repeated the experiment 100 times, and we calculated the means of the root mean square error (RMSE), the means of the learning times (ETL), and the average processing times (ETP). We observed that for each climatic variable studied, one network topology performed better than the others.

We also observed that different topology networks presented different performances for the same climatic variable, when considering learning time and processing time. This makes the optimization of these indicators mutually exclusive.

Based on the results obtained, we conclude that to predict solar radiation, we must use an MLP topology network, while to predict the temperature, we must use a small world topology (SW) network if we want the smallest possible error in the estimates. On the other hand, based on the graphs of the processing times, that is, of the times to predict the next step, we realize that if we want the highest possible processing speed, we must use a complete topology network to predict solar radiation and a random network to predict the temperature.

Acknowledgements The authors acknowledge SENAI CIMATEC University Center and CNPq—Fundação Carlos Chagas Filho de Amparo à Pesquisa do Estado do Rio de Janeiro for their financial support.

References

1. Abhishek, K., Singh, M., Ghosh, S., Anand, A.: Weather forecasting model using artificial neural network. Procedia Technol. **4**, 311–318 (2012). http://dx.doi.org/10.1016/j.protcy.2012.05.047
2. Barabási, A.L., Albert, R.: Emergence of scaling in random networks. Science **286**(5439), 509–512 (1999)
3. Bohland, J., Minai, A.: Efficient associative memory using small-world architecture. Neurocomputing **38**, 38–40 (2001)
4. Coutinho, E.R., Silva, R.M., Delgado, A.R.S.: Use of computational intelligence techniques in meteorological data prediction (In Portuguese). Revista Brasileira de Meteorologia **31**(1), 24–36 (2016). http://dx.doi.org/10.1590/0102-778620140115
5. Do Hoai, N., Udo, K., Mano, A.: Downscaling global weather forecast outputs Using ANN for flood prediction. J. Appl. Math. **2011**, 1–14 (2011). http://dx.doi.org/10.1155/2011/246286
6. Emmert-Streib, F.: Influence of the neural network topology on the learning dynamics. Neurocomputing **69**(10–12), 1179–1182 (2006). http://dx.doi.org/10.1016/j.neucom.2005.12.070
7. Erdös, P., Rényi, A.: On the evolution of random graphs. Publ. Mat. Inst. Hung. Acad. Sci. **5**, 17–61 (1960)
8. Fadigas, I.S., Pereira, H.B.B.: A network approach based on cliques. Physica A **392**(10), 2576–2587 (2013)
9. Hayati, M., Mohebi, Z.: Application of artificial neural networks for temperature forecasting. World Acad. Sci. Eng. Technol. **28**(2), 275–279 (2007)
10. Hayati, M., Mohebi, Z.: Temperature forecasting based on neural network approach. World Appl. Sci. J. **2**(6), 613–620 (2007)
11. Maqsood, I., Khan, R., Abraham, A.: An ensemble of neural networks for weather forecasting. Neural Comput. Appl. **13**(2), 112–122 (2004). http://dx.doi.org/10.1007/s00521-004-0413-4
12. Monteiro, R.L.S.: An evolutionary model for the simulation of affinity networks (in Portuguese). Ph.D. thesis, Federal University of Bahia, Salvador (2012)
13. Monteiro, R.L.S., Carneiro, T.K.G., Fontoura, J.R.A., da Silva, V.L., Moret, M.A., Pereira, H.B.B.: A model for improving the learning curves of artificial neural networks. PLOS One **11**(2), 1–11 (2016). http://dx.doi.org/10.1371/journal.pone.0149874
14. Pereira, H.B.B., Fadigas, I.S., Senna, V., Moret, M.A.: Semantic networks based on titles of scientific papers. Physica A **390**(6), 1192–1197 (2011)
15. Rosenblatt, F.: The Perceptron: a probabilistic model for information storage and organization in the brain. Psychol. Rev. **65**(6), 386–408 (1958)
16. Rumelhart, D.E., Hinton, G.E., Williams, R.J.: Learning representations by back-propagating errors. Nature **323**(6088), 533–536 (1986)
17. Shrivastava, G., Karmakar, S., Kowar, M.K., Guhathakurta, P.: Application of artificial neural networks in weather forecasting: a comprehensive literature review. Int. J. Comput. Appl. **51**(18), 17–29 (2012)

18. Voyant, C., Muselli, M., Paoli, C., Nivet, M.L.: Numerical weather prediction (NWP) and hybrid ARMA/ANN model to predict global radiation. Energy **39**(1), 341–355 (2012). http:// dx.doi.org/10.1016/j.energy.2012.01.006
19. Watts, D.J., Strogatz, S.H.: Collective dynamics of 'small-world' networks. Nature **393**, 440–442 (1998)
20. Werbos, P.J.: Backpropagation through time: what it does and how to do it. Proc. IEEE **78**(10), 1550–1560 (1990). http://dx.doi.org/10.1109/5.58337

Chapter 11
Constructal Design Associated with Genetic Algorithm to Maximize the Performance of H-Shaped Isothermal Cavities

Emanuel da Silva Dias Estrada, Elizaldo Domingues dos Santos, Liércio André Isoldi, and Luiz Alberto Oliveira Rocha

11.1 Introduction

The heat transfer realm has been investigated exhaustively to enhance the performance of the thermal systems. Extended surfaces also known as fins or assembly of fins have played an important role in this search [17, 34]. Special interest deserves recent works dedicated to the study of the geometry of fins. These articles used constructal design method [3, 6, 12, 13, 18–20, 23, 30] to understand how the configuration, architecture, shape, or structure affects the performance of the system.

Constructal design method is the method used to apply the Constructal law: "For a finite size flow system to persist in time (survive) its currents should evolve in time to make easy the access of the currents that flow through it." [1, 2, 4, 5]. Using Constructal design one should first identify what is flowing (heat, fluid, mass, electricity, goods, etc.). As the system is finite, it must obey some constraints, for example, the volume that it occupies. Now, to be alive (to flow, to move, to survive) the system must be free to move.

Sometimes, to improve the performance of engineering systems, the designer associates constructal design with an optimization method [11]. This association allows the designer to elect an objective function to perform the optimization.

E. d. S. D. Estrada
Centre for Computational Sciences, Federal University of Rio Grande, Rio Grande, Brazil
e-mail: emaenuelestrada@furg.com

E. D. dos Santos · L. A. Isoldi
School of Engineering, Federal University of Rio Grande, Rio Grande, Brazil
e-mail: elizaldosantos@furg.br; liercioisoldi@furg.br

L. A. O. Rocha (✉)
UNISINOS, São Leopoldo, Brazil

© Springer Nature Switzerland AG 2019
G. Mendes Platt et al. (eds.), *Computational Intelligence, Optimization and Inverse Problems with Applications in Engineering*,
https://doi.org/10.1007/978-3-319-96433-1_11

Now, the search for best shapes can take place based on the following: objective function, constraints (volumes, areas), and the degrees of freedom, normally aspect ratios between the lengths that are important in order to design the domain.

Genetic Algorithm [10, 14–16, 33] is an important optimization method. Recently, it has been commonly used associated to constructal design method [29, 31]. It allows the designer to save time of simulation, when this time is compared to the time used by the exhaustive search optimization method. The accuracy of genetic algorithm method has been tested and the results are validated frequently [29].

Sometimes it is not possible to use extended surfaces to improve the heat transfer rate of the thermal systems. Walls, or other obstacles, do not allow the use of fins. Therefore, the design adopts an intelligent solution: the use of cavities. Cavities can be understood as negative or inverted fins [7, 13, 24]. If the heat generation in the body is uniform, isothermal cavities are important to cool the body. To improve its performance, for a fixed area of the cavity and total area of the body, several geometries of cavities have been studied. Elemental cavity (C-shaped Cavity) [7], T- (first construct) [25, 35], X- [27, 28], Y- [22, 31], T-Y- [21, 26], and H-shaped cavity (second construct) [8] are examples of studied cavities. These cavities were designed with the help of constructal design associated with the exhaustive search method.

The H-shaped cavity [8], because of its large number of degrees of freedom, was not optimized completely. The work presented in this chapter goes further: optimizes its six degrees of freedom. This was possible because constructal design was associated with the genetic algorithm. This is an important contribution once these results can be used in several industrial applications.

11.2   H-Shaped Construct: Constructal Design and Numerical Formulation

Consider the two-dimensional H-shaped conducting body shown in Fig. 11.1. The external dimensions (H, L) vary, where L is the length and H the height of the body. The figure is symmetric with respect to the abscissa axis. The length of the superior and the inferior branches of the H is $2L0$ meaning the whole figure is symmetric. The third dimension, W, is perpendicular to the plane of the figure. The total volume occupied by this body is fixed,

$$V = HLW \tag{11.1}$$

Once the heat transfer occurs in a two-dimensional domain it is considered here that thermal gradients in the normal direction is negligible. Then, the volume

Fig. 11.1 Isothermal
H-shaped intrusion into a
two-dimensional conducting
body with uniform heat
generation

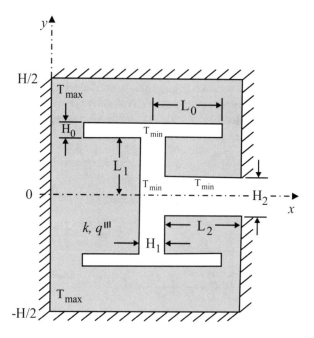

constraint can be expressed as an area constraint $(A = HL)$. The variable
dimensions of the solid and the cavity are given by: H, L, H_0, L_0, H_1, L_1,
H_2, L_2.

The cavity constraint can be written by

$$V = (4H_0L_0 + 2H_1L_1 + H_2L_2)W \qquad (11.2)$$

This second constraint may be replaced by the statement that the volume fraction
occupied by the cavity is fixed,

$$\phi = \frac{V_0}{V} = \frac{(4H_0L_0 + 2H_1L_1 + H_2L_2)}{HL} \qquad (11.3)$$

The solid is isotropic with the constant thermal conductivity k. It generates
heat uniformly at the volumetric rate q''' [W/m^3]. The outer surfaces of the heat
generating body are perfectly insulated. The generated heat current $(q'''A)$ is
removed by cooling the wall of the cavity. The cavity wall is maintained at the
minimal domain temperature (T_{min}), that is, temperatures in the solid are higher
than T_{min}. The hot spot of temperature T_{max} occurs at one or more points inside the
solid.

An important thermal design constraint is the requirement that temperatures
must not exceed a certain level. This makes the hot spot temperature T_{max} a
constraint. However, the location of T_{max} is not a constraint. In the present problem

statement, this design objective is represented by the minimization of maximum excess temperature.

The numerical optimization of the geometry consisted of simulating the temperature field in a large number of configurations, calculating the maximum excess temperature for each configuration, and selecting the configuration with the smallest maximum excess temperature. Symmetry allowed us to perform calculations in only half of the domain, $y \geq 0$. The conduction equation for the solid region is

$$\frac{\partial^2 \tilde{T}}{\partial x^2} + \frac{\partial^2 \tilde{T}}{\partial y^2} + 1 = 0 \qquad (11.4)$$

where the dimensionless variables are

$$\tilde{T} = \frac{T - T_{min}}{q''' A / k} \qquad (11.5)$$

$$(\tilde{x}, \tilde{y}, \tilde{H}, \tilde{L}, \tilde{H}_0, \tilde{L}_0, \tilde{H}_1, \tilde{L}_1, \tilde{H}_2, \tilde{L}_2) = \frac{(x, y, H, L, H_0, L_0, H_1, L_1, H_2, L_2)}{A^{1/2}} \qquad (11.6)$$

The boundary conditions are indicated in Fig. 11.1. The maximum dimensionless excess temperature, T_{max}, is also the dimensionless global thermal resistance of the construct,

$$\tilde{T}_{max} = \frac{T_{max} - T_{min}}{q''' A / k} \qquad (11.7)$$

Note that the corresponding dimensionless representations for Eqs. (11.1) and (11.3) are given by

$$\tilde{H} \tilde{L} = 1 \qquad (11.8)$$

and

$$\phi = 4 \tilde{H}_0 \tilde{L}_0 + 2 \tilde{H}_1 \tilde{L}_1 + \tilde{H}_2 \tilde{L}_2 \qquad (11.9)$$

Equation (11.4) was solved with a finite element code based on triangular elements, developed in MATLAB environment and using the PDE (partial-differential-equations) toolbox [32]. The domain is symmetric at the x-axis. Therefore, for the sake of simplicity, only half of the domain was used to perform the simulations. The grid was non-uniform in both x and y directions, and varied for different geometries. The appropriate mesh size was determined by successive refinements, increasing the number of elements four times from one mesh size to the next, until the criterion $|(\tilde{T}_{max}^j - \tilde{T}_{max}^{j+1})/\tilde{T}_{max}^j| < 5 \times 10^{-3}$ is satisfied. Here $\tilde{T}_{max}^j$ represents the maximum temperature calculated using the current mesh size, and $\tilde{T}_{max}^{j+1}$ corresponds to the maximum dimensionless excess temperature using the next mesh size, where the

Table 11.1 Numerical tests showing the achievement of grid independence ($H/L = 1$, $\phi = 0.1$, $H_2/L_2 = 0.15$, $L_1/L_2 = 0.6$, $L_0/L_2 = 0.6$, $H_1/H_2 = 0.75$, $H_0/H_2 = 1.0$)

Iteration	# of elements	$\tilde{T}_{max}$	$\lvert(\tilde{T}_{max}^j - \tilde{T}_{max}^{j+1})/\tilde{T}_{max}^j\rvert$
1	174	0.089752	2.8×10^{-2}
2	696	0.092265	1.11×10^{-2}
3	2784	0.093291	4.4×10^{-3}
4	11,136	0.093699	–

Table 11.2 Comparison between the results obtained for an isothermal C-cavity [7], and the present numerical work using Matlab and Comsol ($H/L = 1$, $\phi = 0.3$)

H_0/L_0	Present work (Matlab)	Present work (Comsol)	Ref. [7]
1.875	0.1873	0.187368	0.1873
1.3	0.1436	0.151486	0.1436
0.83334	0.10865	0.108689	0.10865
0.4686	0.06574	0.065747	0.0657

Table 11.3 Comparison between exhaustive search and GA, when $H/L = 1$ and $\phi = 0.1$, considering five degrees of freedom

Study	H_2/L_2	L_1/L_2	L_0/L_2	H_1/H_2	H_0/H_2	$\tilde{T}_{max}$
GA	0.02	0.446	0.925	0.1	4.728	0.02465
Exhaustive search	0.01	0.5	0.9	0.1	5	0.02447

number of elements was increased by four times. Table 11.1 shows an example on how grid independence was achieved.

The accuracy of the numerical method was also tested by reproducing with very good agreement the results for the same cavity developed with Comsol Multiphysics [9]. Table 11.2 shows some examples of this comparison.

Once the accuracy and grid independence of the numerical model is assured, it is time to search for the optimal configuration. The computational domain has six degrees of freedom. Therefore, the optimization procedure would take a huge number of simulations if the exhaustive search was used [7]. Therefore, we rely on genetic algorithm (GA) to investigate the optimal shapes in the search space. The parameters and configurations used by GA were defined based on the literature, and adjusted by comparing the thermal performance of the problem with the results obtained by exhaustive search [7], for five degrees of freedom. Table 11.3 shows this comparison. In this sense, the GA adopted the configuration below for all performed simulations:

- Population type: bitstream;
- Population size: 200 individuals;
- Selection function: uniform stochastic sampling;
- Elitism: 5% of population size;
- Crossover fraction: 0.8;
- Mutation fraction: 0.05;
- Stopping criteria : 50 generations without $\tilde{T}_{max}$ change.

11.3 Optimal H-Shaped Cavities

The numerical work consisted of determining the temperature field in a large number of configurations of the type shown in Fig. 11.1. Figure 11.2 shows an example of the calculated temperature distribution, and where the maximum temperature is, i.e. the hot spots are located in the rim of Fig. 11.2.

This same process of determining the temperature field was repeated in the present study for several values of the degrees of freedom, H_0/H_2, H_1/H_2, L_0/L_2, L_1/L_2, H_2/L_2, and H/L. Also, the influence of the parameter ϕ, as defined by Eq. (11.9), is evaluated. The genetic algorithm handles the order that each degree of freedom is varied and its corresponding range of search: $0.1 \leq H/L \leq 10$, $0.005 \leq H_0/L_2 \leq 20$, $0.01 \leq H_1/H_2 \leq 20$, $0.1 \leq L_0/L_2 \leq 3$, $0.01 \leq L_1/L_2 \leq 15$, $0.001 \leq H_2/L_2 \leq 2$. Initially, the degree $H/L = 1$ and the parameter $\phi = 0.1$ are fixed, and it is allowed that all the other degrees of freedom vary freely. The temperature field is calculated for each set of the degrees of freedom and genetic algorithm returns the maximum excess temperature. Later, the procedure is repeated for each new value of H/L. A curve corresponding to the value of $\phi = 0.1$ is then built, and the results are shown in Fig. 11.3. This figure shows that the minimum value of the maximum excess temperature in the studied range was found for $H/L = 0.1$. The maximum excess temperature has also a local minimum in the other extreme of the studied range. The worst configuration is the one corresponding to the degree of freedom $H/L = 2$, where it is found the maximum value of the maximum excess temperature. These results indicate that the configurations that better distribute the hot spots (imperfections) are the ones as slender as possible in both directions, horizontal and vertical. The worst configurations are the ones close to the square shape. Figure 11.3 shows also the curves corresponding to $\phi = 0.2$ and

Fig. 11.2 Dimensionless
temperature distribution of a
typical configuration of
H-shaped cavity ($\phi = 0.1$,
$H/L = 10$, $H_2/L_2 = 0.0517$,
$L_1/L_2 = 10.294$,
$L_0/L_2 = 0.7973$,
$H_1/H_2 = 4.563$,
$H_0/H_2 = 0.4338$)

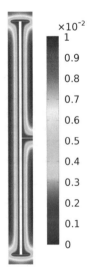

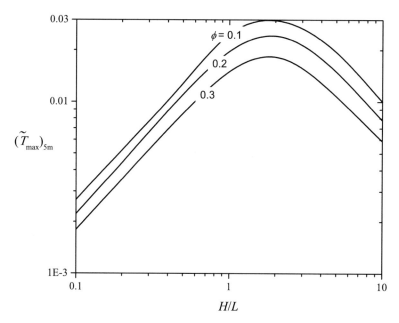

Fig. 11.3 Maximum dimensionless temperature as function of ϕ and H/L

0.3. Both curves have the same behavior shown by the curve presented for $\phi = 0.1$, indicating that minimum and maximum values of the maximum excess temperature correspond to the same values for the degree of freedom H/L shown by the curve $\phi = 0.1$.

The optimal values corresponding to the curves presented in Fig. 11.3 are shown in Tables 11.4, 11.5, and 11.6. Note that the subscripts associated with each degree of freedom indicate the order of the simulation used by the genetic algorithm. For example, the degree of freedom $(H_0/H_2)_{5o}$ was the first optimized and, in the end of the simulation, it was optimized five times, receiving the subscript $5o$. According to the order of the simulation, the other degrees of freedom received its corresponding subscript. The higher the subscript, the sooner the degree of freedom was used in the optimization procedure.

The results presented in Fig. 11.3 are further detailed in Fig. 11.4. Such figure emphasizes that the dimensionless maximum excess temperature decreases as the volume fraction increases and the ratio $H/L = 0.1$ is the best for each value of the volume fraction studied. Note that the configuration $H/L = 2$ has the worst performance. This happens because this configuration does not allow the body enough freedom to reach slender geometries. In addition, one can observe that as ϕ increases there is a decrease in maximum temperature. These results show how important is the investigation of the effect of the geometry in the performance of the thermal systems.

Table 11.4 The best performances for $\phi = 0.1$ and several values of H/L

H/L	$(H_0/H_2)_{5o}$	$(H_1/H_2)_{4o}$	$(L_0/L_2)_{3o}$	$(L_1/L_2)_{2o}$	$(H_2/L_2)_{1o}$	$(\tilde{T}_{max})_{5m}$
0.1	1.1751	0.9878	0.9957	0.0461	0.0069	0.0027
0.2	2.5216	0.2521	0.9953	0.0912	0.0072	0.0052
0.5	7.0945	0.0138	0.9675	0.2233	0.0069	0.01275
1	6.9683	0.041	0.9444	0.4576	0.0147	0.0246
2	3.7071	6.576	0.7444	1.2245	0.0236	0.0297
5	0.01	16.634	1.0667	5.0246	0.0147	0.0194
10	0.4338	4.563	0.7937	10.294	0.0517	0.0100

Table 11.5 The best performances for $\phi = 0.2$ and several values of H/L

H/L	$(H_0/H_2)_{5o}$	$(H_1/H_2)_{4o}$	$(L_0/L_2)_{3o}$	$(L_1/L_2)_{2o}$	$(H_2/L_2)_{1o}$	$(\tilde{T}_{max})_{5m}$
0.1	1.3131	0.9259	1.0004	0.0417	0.0128	0.0022
0.2	1.1746	0.1264	1.0016	0.0841	0.0283	0.0044
0.5	1.2176	0.5025	1.0016	0.2124	0.0674	0.0110
1	11.5079	0.0586	0.9667	0.4035	0.0176	0.0196
2	0.7180	3.6407	0.09286	1.4987	0.1290	0.0241
5	0.005	5.0458	1.1732	5.6514	0.1125	0.0149
10	0.2053	13.944	0.9528	10.902	0.0385	0.0078

Table 11.6 The best performances for $\phi = 0.3$ and several values of H/L

H/L	$(H_0/H_2)_{5o}$	$(H_1/H_2)_{4o}$	$(L_0/L_2)_{3o}$	$(L_1/L_2)_{2o}$	$(H_2/L_2)_{1o}$	$(\tilde{T}_{max})_{5m}$
0.1	1.2133	0.1964	1.0016	0.0379	0.0205	0.0018
0.2	1.4197	0.011	0.9835	0.0742	0.0361	0.0035
0.5	1.7731	0.2243	0.9796	0.1845	0.0752	0.0085
1	14.204	0.2870	0.8851	0.3380	0.0225	0.0149
2	0.0887	14.364	1.4349	2.2431	0.0845	0.0182
5	0.0246	13.1070	1.4244	6.5916	0.0793	0.0107
10	0.005	13.9448	1.0425	13.853	0.0666	0.0059

    The best configurations obtained, being fixed the values of H/L and ϕ, in Figs. 11.3 and 11.4 are shown in scale in Fig. 11.5. Note that in these best shapes ($H/L = 0.1$) the temperature is approximately uniform, i.e. the gradients are very small. The gradients increase, but the temperature is still almost uniform for the configuration $H/L = 10$. This temperature distribution becomes more uniform as the volume fraction increases. The larger gradients in Fig. 11.5 are shown when the ratio $H/L = 2$, the worst configuration. However, this configuration is important because it shows how the hot spots (imperfections) are optimally distributed within the body. The more the hot spots are well distributed in the body, the smaller is the maximum dimensionless excess temperature. Considering these optimal configurations for $H/L = 2$, the maximum dimensionless excess of temperature decreases as the volume fraction increases.

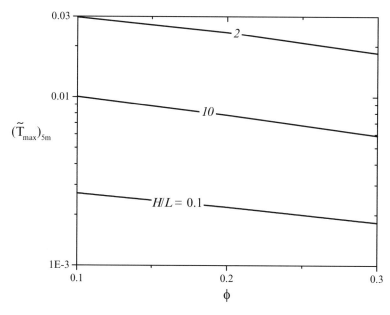

Fig. 11.4 Dimensionless maximum excess temperature as function of ϕ when $H/L = 0.1, 2$, and 10

It is important to notice that the maximum dimensionless excess temperature calculated for the best H-shaped cavity with ratio $H/L = 0.1$ is approximately only 30% of the maximum dimensionless excess temperature calculated for the elemental C-shaped cavity studied by Biserni et al. [7].

11.4 Conclusions

This chapter studied how to discover the geometry that mostly improves the performance of a H-shaped cavity intruded into a solid body. We rely on constructal design method associated with the genetic algorithm to perform the optimization procedure. The dimensionless maximum excess temperature was minimized with respect to six degrees of freedom, while the cavity volume fraction was evaluated for few values. The results showed that there is no universal geometry that always is the best configuration. The best shapes depend on the degrees of freedom and the cavity volume fraction. In general, as expected, the higher is the cavity volume fraction, the smaller is the maximum dimensionless excess temperature. It was also shown that the dimensionless maximum excess temperature of the H-shaped cavity (for the same thermal conditions) is only approximately 30% of the maximum dimensionless excess temperature of the elemental C-shaped cavity for the studied cases in this work.

Fig. 11.5 Dimensionless
temperature distribution for
the best shapes observed in
Figs. 11.3 and 11.4: (**a**)
$\phi = 0.1$, $H/L = 0.1$; (**b**)
$\phi = 0.2$, $H/L = 0.1$; (**c**)
$\phi = 0.3$, $H/L = 0.1$; (**d**)
$\phi = 0.1$, $H/L = 2$; (**e**)
$\phi = 0.2$, $H/L = 2$; (**f**)
$\phi = 0.3$, $H/L = 2$; (**g**)
$\phi = 0.1$, $H/L = 10$; (**h**)
$\phi = 0.2$, $H/L = 10$; (**i**)
$\phi = 0.3$, $H/L = 10$

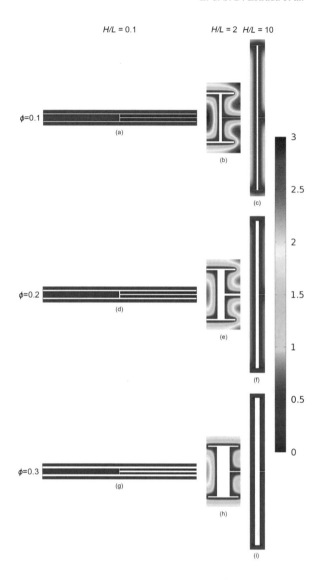

Acknowledgements Professors Elizaldo D. dos Santos, Liércio A. Isoldi and Luiz A.O. Rocha acknowledge the sponsorship from CNPq—Conselho Nacional de Desenvolvimento Científico e Tecnológico. Professor Emanuel S. D. Estrada acknowledges the financial support from CAPES—Coordenação de Aperfeiçoamento de Pessoal de Nível Superior.

References

1. Bejan, A.: Shape and Structure, from Engineering to Nature. Cambridge University Press, New York (2000)
2. Bejan, A.: The Physics of Life: The Evolution of Everything. St. Martins Press, New York City (2016)
3. Bejan, A., Almogbel, M.: Constructal T-shaped fins. Int. J. Heat Mass Transf. **43**(12), 2101–2115 (2000). http://dx.doi.org/10.1016/S0017-9310(99)00283-5
4. Bejan, A., Lorente, S.: Design with Constructal Theory. Wiley, Hoboken (2008)
5. Bejan, A., Zane, J.P.: Design in Nature: How the Constructal Law Governs Evolution in Biology, Physics, Technology, and Social Organization, 1st edn. Doubleday, New York (2012)
6. Bello-Ochende, T., Meyer, J., Bejan, A.: Constructal multi-scale pin–fins. Int. J. Heat Mass Transf. **53**(13–14), 2773–2779 (2010). http://dx.doi.org/10.1016/j.ijheatmasstransfer.2010.02.021
7. Biserni, C., Rocha, L.A.O., Bejan, A.: Inverted fins: geometric optimization of the intrusion into a conducting wall. Int. J. Heat Mass Transf. **47**(12–13), 2577–2586 (2004). http://dx.doi.org/10.1016/j.ijheatmasstransfer.2003.12.018
8. Biserni, C., Rocha, L.A.O., Stanescu, G., Lorenzini, E.: Constructal H-shaped cavities according to Bejan's theory. Int. J. Heat Mass Transf. **50**(11–12), 2132–2138 (2007). http://dx.doi.org/10.1016/j.ijheatmasstransfer.2006.11.006
9. COMSOL multiphysics: COMSOL multiphysics reference manual (2014). www.comsol.com
10. Goldberg, D.E.: Genetic Algorithms in Search, Optimization and Machine Learning, 1st edn. Addison-Wesley Longman Publishing Co., Inc., Boston (1989)
11. Gonzales, G., Estrada, E.S.D., Emmendorfer, L., Isoldi, L., Xie, G., Rocha, L., Santos, E. D.: A comparison of simulated annealing schedules for constructal design of complex cavities intruded into conductive walls with internal heat generation. Energy **93**, 372–382 (2015). http://dx.doi.org/10.1016/j.energy.2015.09.058
12. Hajmohammadi, M.R., Poozesh, S., Nourazar, S.S.: Constructal design of multiple heat sources in a square-shaped fin. Proc. Inst. Mech. Eng. E: J. Process Mech. Eng. **226**, 324–336 (2012). http://dx.doi.org/10.1177/0954408912447720
13. Hajmohammadi, M.R., Poozesh, S., Campo, A., Nourazar, S.S.: Valuable reconsideration in the constructal design of cavities. Energy Convers. Manag. **66**, 33–40 (2013). http://dx.doi.org/10.1016/j.enconman.2012.09.031
14. Haupt, R.L.: Practical Genetic Algorithms, 2nd edn. Wiley, Hoboken (2004)
15. Holland, J.H.: Adaptation in Natural and Artificial Systems: An Introductory Analysis with Applications to Biology, Control, and Artificial Intelligence. University of Michigan Press, Ann Arbor (1975)
16. Jong, K.A.D., Spears, W.M.: An analysis of the interacting roles of population size and crossover in genetic algorithms. In: Schwefel, H.P., Männer, R. (eds.) Parallel Problem Solving from Nature, pp. 38–47. Springer, Berlin (1991)
17. Kraus, A.D.: Developments in the analysis of finned arrays. Int. J. Transp. Phenom. **1**, 141–164 (1999)
18. Kundu, B., Bhanja, D.: Performance and optimization analysis of a constructal T-shaped fin subject to variable thermal conductivity and convective heat transfer coefficient. Int. J. Heat Mass Transf. **53**(1–3), 254–267 (2010). http://dx.doi.org/10.1016/j.ijheatmasstransfer.2009.09.034
19. Lorenzini, G., Rocha, L.A.O.: Constructal design of Y-shaped assembly of fins. Int. J. Heat Mass Transf. **49**(23–24), 4552–4557 (2006). http://dx.doi.org/10.1016/j.ijheatmasstransfer.2006.05.019
20. Lorenzini, G., Rocha, L.A.O.: Constructal design of T-Y assembly of fins for an optimized heat removal. Int. J. Heat Mass Transf. **52**(5–6), 1458–1463 (2009). http://dx.doi.org/10.1016/j.ijheatmasstransfer.2008.09.007

21. Lorenzini, G., Rocha, L.A.O.: Geometric optimization of T-Y-shaped cavity according to constructal design. Int. J. Heat Mass Transf. **52**(21–22), 4683–4688 (2009). http://dx.doi.org10. 1016/j.ijheatmasstransfer.2009.06.020

22. Lorenzini, G., Biserni, C., Isoldi, L.A., Santos, E.D., Rocha, L.A.O.: Constructal design applied to the geometric optimization of Y-shaped cavities embedded in a conducting medium. J. Electron. Packag. **133**(4), 41008–41015 (2011). http://dx.doi.org/10.1115/1.4005296

23. Lorenzini, G., Corrêa, R.L., Santos, E.D., Rocha, L.A.O.: Constructal design of complex assembly of fins. J. Heat Transf. **133**(8), 81902–81908 (2011). http://dx.doi.org/10.1115/1. 4003710

24. Lorenzini, G., Rocha, L.A.O., Biserni, C., Santos, E.D., Isoldi, L.: Constructal design of cavities inserted into a cylindrical solid body. J. Heat Transf. **134**(7), 71301–71306 (2012). http://dx.doi.org/10.1115/1.4006103

25. Lorenzini, G., Biserni, C., Garcia, F., Rocha, L.: Geometric optimization of a convective t-shaped cavity on the basis of constructal theory. Int. J. Heat Mass Transf. **55**(23–24), 6951–6958 (2012). http://dx.doi.org/10.1016/j.ijheatmasstransfer.2012.07.009

26. Lorenzini, G., Garcia, F.L., Santos, E.D., Biserni, C., Rocha, L.A.O.: Constructal design applied to the optimization of complex geometries: T-Y-shaped cavities with two additional lateral intrusions cooled by convection. Int. J. Heat Mass Transf. **55**(5–6), 1505–1512 (2012). http://dx.doi.org/10.1016/j.ijheatmasstransfer.2011.10.057

27. Lorenzini, G., Biserni, C., Link, F., Santos, D., Isoldi, L., Rocha, L.A.O.: Constructal design of isothermal X-shaped cavities. Therm. Sci. **18**(2), 349–356 (2014). http://dx.doi.org/10.2298/ TSCI120804005L

28. Lorenzini, G., Biserni, C., Rocha, L.: Geometric optimization of X-shaped cavities and pathways according to Bejan's theory: comparative analysis. Int. J. Heat Mass Transf. **73**, 1–8 (2014). http://dx.doi.org/10.1016/j.ijheatmasstransfer.2014.01.055

29. Lorenzini, G., Biserni, C., Estrada, E.S.D., Santos, E.D., Isoldi, L.A., Rocha, L.A.O.: Genetic algorithm applied to geometric optimization of isothermal Y-shaped cavities. J. Electron. Packag. **136**(3), 31011–31019 (2014). http://dx.doi.org/10.1115/1.4027421

30. Lorenzini, G., Biserni, C., Correa, R.L., Santos, E.D., Isoldi, L.A., Rocha, L.A.O.: Constructal design of T-shaped assemblies of fins cooling a cylindrical solid body. Int. J. Therm. Sci. **83**, 96–103 (2014). http://dx.doi.org/10.1016/j.ijthermalsci.2014.04.011

31. Lorenzini, G., Biserni, C., Estrada, E.D., Isoldi, L.A., Santos, E. D., Rocha, L.A.O.: Constructal design of convective Y-shaped cavities by means of genetic algorithm. J. Heat Transf. **136**(7), 71702–71702 (2014). http://dx.doi.org/10.1115/1.4027195

32. MATLAB: version 7.10.0 (R2010a). The MathWorks Inc., Natick (2010)

33. Renner, G., Ekárt, A.: Genetic algorithms in computer aided design. Comput. Aided Des. **35**(8), 709–726 (2003). http://dx.doi.org/10.1016/S0010-4485(03)00003-4

34. Snider, A.D., Kraus, A.D.: The quest for the optimum longitudinal fin profile. Heat Transfer Eng. **8**(2), 19–25 (1987). http://dx.doi.org/10.1080/01457638708962790

35. Xie, Z., Chen, L., Sun, F.: Geometry optimization of T-shaped cavities according to constructal theory. Math. Comput. Model. **52**(9–10), 1538–1546 (2010). http://dx.doi.org/10.1016/j.mcm. 2010.06.017

Chapter 12
Co-design System for Tracking Targets Using Template Matching

Yuri Marchetti Tavares, Nadia Nedjah, and Luiza de Macedo Mourelle

12.1 Introduction

With the development and enhancement of sensors and the advent of intelligent equipment capable of capturing, storing, editing and transmitting images, the acquisition of information, which is extracted from images and videos, became possible and led to a blossoming of an important research area. In defense and security fields of expertise, this kind of research is very relevant as it allows target recognition and tracking in image sequences. It can provide solutions for the development of surveillance and monitoring systems [13], firing control [2], guidance [5], navigation [8], remote biometrics authentication [4], control of guided weapons [16], among many other applications.

In general, a pattern is an arrangement, or a collection of objects that are similar, and thus it is identified by its arrangement of disposition. One of the most used techniques for finding and tracking patterns in images is generally identified as template matching [1, 12]. It consists basically in finding a small image, considered as a template, inside a larger image. Among the methods used to evaluate the

Y. M. Tavares
Department of Weapons, Navy Weapons Systems Directorate, Brazilian Navy, Rio de Janeiro, Brazil

N. Nedjah (✉)
Department of Electronic Engineering and Telecommunications, Rio de Janeiro State University, Rio de Janeiro, Brazil
e-mail: nadia@eng.uerj.br

L. d. M. Mourelle
Department of Systems Engineering and Computing, Rio de Janeiro State University, Rio de Janeiro, Brazil
e-mail: ldmm@eng.uerj.br

© Springer Nature Switzerland AG 2019
G. Mendes Platt et al. (eds.), *Computational Intelligence, Optimization and Inverse Problems with Applications in Engineering*,
https://doi.org/10.1007/978-3-319-96433-1_12

227

matching process, the normalized cross correlation is very known and widely used. The underlying task is computationally very expensive, especially when using large templates with an extensive image set [18].

Aiming at improving the performance, we propose an implementation of the template matching as a software/hardware co-design system, using global best PSO, implemented in software, while the required computation of PCC, implemented in hardware via a dedicated co-processor. In order to evaluate the proposed design, the processing time as required by the software-only and the co-design systems are evaluated and compared.

The rest of this paper is organized in six sections. First, in Sect. 12.2, we define the template matching and correlation concepts as they are used in this work. After that, in Sect. 12.3, we briefly present the methods and evaluate the performance of GAs, global best PSO and local best PSO as computational intelligence techniques to be used as software sub-system. Subsequently, in Sect. 12.4, we describe in detail the proposed hardware used as co-processor. Then, in Sect. 12.5, we present and analyze the obtained performance results. Finally, in Sect. 12.6, we draw some useful conclusions and point out one direction for future works.

12.2   Template Matching

Template matching is used in image processing, basically, to find an object given via its image inside a bigger image. The object to be recognized is compared to a predefined template. Figure 12.1 illustrates the dynamics of the used process for a gray scale image, wherein the pixels are represented by bytes of value between 0 and 255 (higher values are represented by squares with color closer to white). The template, identified by a red square in the figure, slides over all the pixels. In each position, the images are compared using a similarity measure. Once the similarity evaluation is completed, regarding all pixels, the pixel that provides the highest correlation degree is identified as the location of the template within the image [15].

Among existing similarity measures for template matching, the normalized cross correlation is well known and often used. The Pearson's Correlation Coefficient (PCC) is used as a measure of similarity between two variables and can be interpreted as a dimensionless measure. When applied to images, the PCC can be computed as defined in Eq. (12.1):

$$PCC(A, P) = \frac{\sum_{i=1}^{N}(p_i - \overline{p}) \times (a_i - \overline{a})}{\sqrt{\sum_{i=1}^{N}(p_i - \overline{p})^2 \times \sum_{i=1}^{N}(a_i - \overline{a})^2}} \qquad (12.1)$$

wherein p_i is the intensity of pixel i in template P; $\overline{p}$ is the average intensity of all pixels of the template; a_i is the intensity of pixel i in patch A from the analyzed image; $\overline{a}$ is the average intensity of all pixels in the patch of image A. Note that template P and patch A of the image must have the same dimensions. The

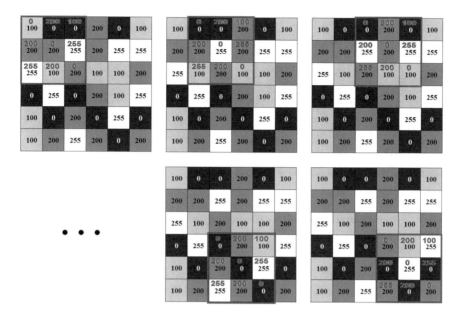

Fig. 12.1 Template matching process

PCC always assumes real values within $[-1, +1]$. Coefficient $+1$ implies a perfect positive correlation of the compared variables while coefficient -1 represents a perfect negative correlation of the compared variables.

The advantage of template matching is that the template stores particular characteristics of the object (color, texture, shape, edges, centroid), differentiating it from the others, allowing more accuracy. Besides, the object detection does not depend on how the objects are classified or represented. The disadvantage is the underlying high computational cost necessary to perform the computation regarding all possible positions.

12.3   Software Development

Figure 12.2b shows an example where the PCC is calculated for all pixels of the Fig. 12.2a. The template is highlighted by a green square. We can note that the correlation graphic has a visible maximum peak, which corresponds to the center of the object. So, the tracking problem is related to an optimization problem, where the global maximum must be discovered and the objective function is PCC.

Optimization algorithms are search methods in which the objective is to find out an optimal solution, by associating values to variables, for a given optimization problem. For this, the algorithm examines the space of solutions to discover possible solutions that meet the problem constraints [7]. Many computational intelligence

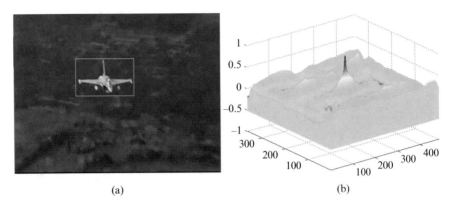

(a)   (b)

Fig. 12.2 Aircraft image—361 × 481 pixels. (**a**) Template highlighted—137 × 85. (**b**) Correlation aspect

techniques comply with this purpose. In the work presented in this chapter, Genetic Algorithms and two topologies of Particle Swarm Optimization were evaluated, as described in the next sections.

12.3.1 Genetic Algorithms

Genetic Algorithms is one of the most established methods of computational intelligence [9]. It is inspired by the evolution theory, which states that the best solution to a problem is found from the combination of possible solutions that are improved iteratively.

At each iteration (generation), a new population of candidate solutions (individuals) is created from the genetic information of the best individuals of the previous generation. The algorithm represents a possible solution using a simple structure, called chromosome. Genetic inspired operators are applied (selection, crossover, and mutation) to chromosomes in order to simulate the process of solution evolution. Each chromosome is comprised of genes that are usually represented by bits, depending on codification. Figure 12.3 shows the flowchart of GA execution.

In the basic genetic algorithms, an initial population of possible solutions is generated randomly. Each individual is composed by a chromosome, which is represented internally by a binary array. The population is evaluated by a fitness function, which assigns a value proportional to the quality of the solution to the problem.

In order to create the population of the next generation, a selection process is used. During this process, individuals are selected to generate the offspring based on their fitness. An individual has a chance to reproduce and thus pass its genetic information to the individuals of the next generation proportional to its fitness.

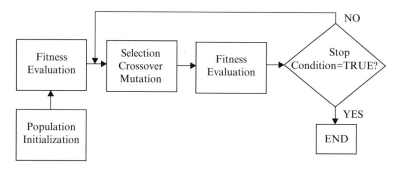

Fig. 12.3 Typical steps of GA

In GA, the selection process is based on an addicted roulette, where each individual is represented by a space proportional to its fitness [9]. The selection process is repeated, with replacement, until it reaches the desired number of individuals required to form the new population. Some genetic algorithms implementations use the principle of elitism to prevent losing the best solution. It consists in an unconditional repetition of the best-fitted individual in the next generation.

After the selection process, the crossover is carried out. The parents are separated, two by two, to participate on the crossover. If the crossover does not occur, the two descendants are exact copies of their parents. The canonical genetic algorithms generally use one-point crossover that consists in swapping the chromosomes of each parent, at a random point.

After some generations, the population will have similar individuals, located in the same region of the search space, which may not be the region where the global optimum is located, indicating a premature convergence. In order to overcome this problem, the mutation operator is used. It provides an exploratory behavior, inducing the algorithm to sample new points of the search space. In the proposed operator in Ref. [9], the mutation is characterized by flipping, with a certain probability, some bits of the chromosome.

All the process is repeated, with the creation and evaluation of new populations, until a stopping criterion is achieved. This criterion can be a target fitness value, considered acceptable, or a maximum number of generations.

12.3.2 Particle Swarm Optimization

The PSO algorithm resulted from the observation of social behavior of flock birds and fish schools [10]. The particles, considered as possible solutions, behave like birds looking for food using their own learning (cognitive component) and the flock learning (social component). The problem is expressed via an objective function. The quality of the solution represented by a particle is the value of the objective function in the position of that particle. The term particle is used, in analogy to the physical concept, because it has a well-defined position and velocity, but it has no mass or volume. The term swarm represents a set of possible solutions. At each iterative cycle, the position of each particle of the swarm is updated according to Eq. (12.2):

$$x_i^{(t+1)} = x_i^{(t)} + v_i^{(t+1)} \tag{12.2}$$

where $x_i^{(t)}$ is the position of the particle i at time t, and $x_i^{(t+1)}$ and $v_i^{(t+1)}$ are the position and the velocity of the particle i at time $t + 1$, respectively.

The velocity is the sum of three components: inertia, memory, and cooperation. Inertia keeps the particle in the same direction. The memory conducts the particle towards the best position discovered by itself during the process so far. The cooperation conducts the particle towards the best position discovered by the swarm or neighborhood, depending on the topology used. The neighborhood of a particle is directly linked to the topology considered.

The social structure of the PSO is determined by the way particles exchange information and exert influence on each other [7]. In the star topology, the neighborhood of the particle consists of all other particles of the swarm. In this case, the velocity vector is updated according to Eq. (12.3):

$$v_i^{(t+1)} = wv_i^{(t)} + c_1r_1(pbest_i - x_i^{(t)}) + c_2r_2(gbest - x_i^{(t)}) \tag{12.3}$$

where w is a constant that represents the inertia of the particle; c_1 and c_2 are constants that give weights to the components cognitive and social, respectively; r_1 and r_2 are random numbers between 0 and 1; $pbest_i$ is the best position found by the particle i and $gbest$ is the best position found by the swarm. The positions of the particles are confined into the search space, and the maximum speed can be set for each dimension of the search space.

In the ring topology, the neighborhood of each particle consists of a subset of two neighboring particles. In this case, the velocity vector is updated according to Eq. (12.4):

$$v_i^{(t+1)} = wv_i^{(t)} + c_1r_1(pbest_i - x_i^{(t)}) + c_2r_2(lbest - x_i^{(t)}) \tag{12.4}$$

where $lbest$ is the best position found by the neighborhood of the particle. So, the social component of the velocity is influenced only by a limited neighborhood, and the information flows more slowly through the swarm.

The aforementioned steps are repeated until the stopping criterion is reached. This criterion can be a target fitness value, considered acceptable, or a maximum number of iterations. The steps used in PSO are summarized in Fig. 12.4.

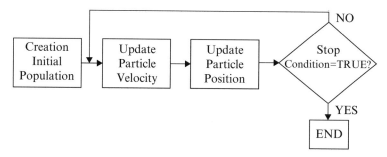

Fig. 12.4 Typical steps of PSO

12.3.3   Comparison Between GA and PSO

In order to perform the comparisons between the computational intelligence algorithms, we used, initially, MATLAB software 7.10.0 (R2010a) installed on a computer ASUS, Intel Core i7-2630 QM 2 GHz, 6 GB memory and Windows 7 Home Premium 64-bit. For the implementation of the genetic algorithms, an external toolbox for MATLAB was used.

For the Genetic Algorithm, a script was written using the classical algorithm, as described in the previous section. PCC is considered as the fitness function and chromosomes correspond to random positions in the main image. The following steps were taken into account:

1. First, transform the main image and the template into gray scale.
2. Then, generate the initial population, with chromosomes corresponding to random positions in the main image;
3. For each individual, extract a patch of the same size as the template, and compute the corresponding PCC at its center that corresponds to its fitness. It is noteworthy to point out that the limits of the main image are completed with zeros.
4. After that, generate the new population, using selection by roulette and elitism, one-point crossover and mutation. Considering that correlation has negative values, the fitness has been normalized before applying the roulette selection.
5. Repeat steps 3 and 4 until an acceptable PCC value is reached or a maximum limit for the generation number is exceeded.

For Particle Swarm Optimization, another script was written. We evaluated two topologies, based on previous section: star and ring. For the star topology, hereinafter called global best PSO (PSO-G), the neighborhood of the particles are all particles of the swarm. For the ring topology, hereinafter called local best PSO (PSO-L), the neighborhood of the particle are only two neighboring particles, considering the index of the particles. The PCC was used as fitness function, and the particles corresponded to random positions in the main image. So, the following steps were taken into account:

1. First, transform the main image and the template into gray scale.
2. Then, generate the initial particle swarm, with random positions and velocities.
3. For each particle, extract a patch of the same size as the template, and compute the corresponding PCC at its center. It is noteworthy to point out that the limits of the main image are completed with zeros.
4. After that, store the best value of PCC found for each particle and also that related to the whole swarm of particles in the case of PSO-G or in its neighborhood in the case of PSO-L.
5. Update the particle positions with Eq. (12.2), and the particle velocities with Eqs. (12.3) and (12.4) for the PSO-G and PSO-L, respectively.
6. Repeat steps 3, 4, and 5 until an acceptable PCC value is reached or a maximum limit for the iteration number is exceeded.

An aircraft video was downloaded from YouTube [22], and a frame, identified hereinafter as aircraft was extracted. Figure 12.2 shows the aircraft frame and its corresponding correlation behavior. Also, the frame number 715 of the benchmark video EgTest05 downloaded from a website [6] was used. Hereinafter, this frame is identified as pickup715, and is shown in Fig. 12.5, together with its corresponding PCC behavior. Furthermore, the first frame of the benchmark video EgTest02, downloaded from the same website [6], was also used for performance evaluation. The resolution of this frame was reduced to 320 × 240 pixels and, hereinafter, is identified as cars320. It is shown in Fig. 12.6, together with the corresponding PCC behavior.

For comparison purposes, the brute force search considering all pixels of the image was also performed, hereinafter called Exhaustive Search (ES). All algorithms are used to find the pixel localization of the template center within the main image. In Table 12.1 are shown the center position of the template, the PCC and computational time for the different images and optimization methods. In this

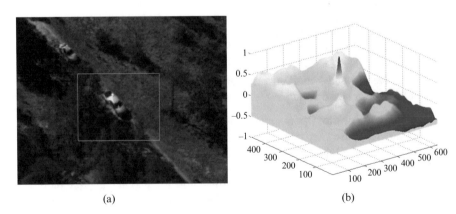

(a)   (b)

Fig. 12.5 Pickup715 image—640 × 480 pixels. (**a**) Template highlighted—249 × 193. (**b**) Correlation aspect

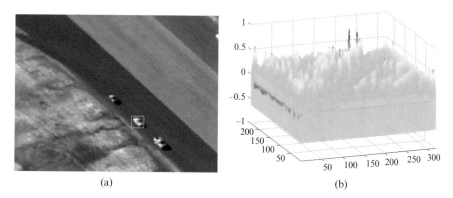

Fig. 12.6 Cars320 image—320 × 240 pixels. (**a**) Template highlighted—21 × 19. (**b**) Correlation aspect

Table 12.1 Some results for ES, GA, and PSO

		aircraft config$_1$	pickup715 config$_1$	cars320 config$_1$	cars320 config$_2$
Position (line,column)	ES	(145,214)	(259,310)	(164,194)	(164,194)
	GA	(144,214)	(260,310)	(164,194)	(164,194)
	PSO-G	(144,215)	(260,312)	(164,194)	(164,194)
	PSO-L	(146,214)	(258,310)	(164,194)	(164,194)
PCC	ES	0.9997	0.9994	0.9987	0.9987
	GA	0.9596	0.9850	0.9987	0.9987
	PSO-G	0.9699	0.9556	0.9987	0.9987
	PSO-L	0.9597	0.9849	0.9987	0.9987
Time (s)	BE	34.5	188.53	4.67	4.67
	GA	0.7318	0.6814	0.6882	2.1382
	PSO-G	0.3075	0.5209	0.2472	0.4705
	PSO-L	0.6683	1.1117	0.2852	0.9458

case, the parameters for the GA, PSO-G, and PSO-L were set empirically as follows ($config_1$):

- $GA_{config1}$: population of 30 individuals; selection by roulette and elitism; crossover rate 80%; mutation rate 12%.
- $PSO_{config1}$: swarm of 100 particles; maximum velocity $V_{max} = 10$, $w = 1$, $c_1 = 1.5$, and $c_2 = 2$.

The stopping criteria were also set empirically. We considered a threshold of 0.95 for the PCC or a maximum number of iterations. For the genetic algorithms, this number was set to 300 generations while for PSO-G and PSO-L as 50 iterations.

It is possible to observe in the correlation graphic of the image cars320 (Fig. 12.6) that it includes many local maxima and thus it entails a complex optimization

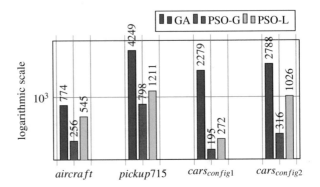

Fig. 12.7 Average processing time for the images over 100 searches (ms)

process in order to identify the target. During this case study, the algorithms were reconfigured to avoid local maxima. The results of this configuration, called $config_2$, are shown in the last column of Table 12.1. This configuration is set as follows:

- $GA_{config2}$: population of 35 individuals; selection by roulette and elitism; crossover rate 80%; mutation rate 15%; maximum 400 generations.
- $PSO_{config2}$: swarm of 200 particles; $w = 1.4$; $c_1 = c_2 = 1.5$; maximum velocity $V_{max} = 10$; the maximum number of iteration for PSO-G was set to 50, while for PSO-L, this number was set to 150 iterations.

Because of the stochastic characteristic of GA and PSO, it is important to evaluate convergence. To do this, Monte Carlo simulations were used, and the process was repeated 100 times (we performed 100 tracking simulations). The results are shown and compared in the bar diagrams of the Figs. 12.7, 12.8, and 12.9. Note that the PSO-G is the fastest, and is more accurate than both PSO-L and GA. PSO-L is faster than GA. For the image cars320, using $config_2$, the speedup obtained with GA is only 2, with 71% success rate.

We can conclude that the global best PSO is the most appropriate tool to perform the optimization. Thus, this technique was implemented as software of the proposed co-design system in order to track targets using template matching and correlation. More detailed results can be verified in Ref. [20], where the authors also performed a comparison between GA and PSO.

12.4 Hardware Architecture

The most expensive part of template matching is the PCC related computations, as verified in Ref. [19]. In order to improve the processing time, and thus allow real-time execution, this computation is implemented via a dedicated co-processor,

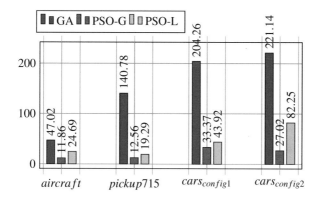

Fig. 12.8 Average iterations/generations number for the images over 100 searches

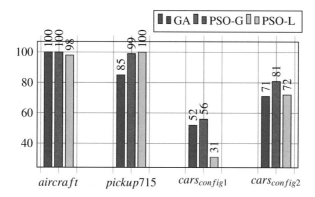

Fig. 12.9 Success rate for the images over 100 searches (%)

taking advantage of the inherent parallelism of the hardware. Furthermore, the search for the location with the maximum correlation is assisted by global best PSO, executed as a software, by a general purpose processor. Unlike brute force methods, this intelligent technique optimizes the search process.

This hardware/software approach, usually called co-design, is a methodology to develop an integrated system using hardware and software components, to satisfy performance requirements and cost constraints [14]. The final target architecture usually has software components executed by a soft processor that is aided by some dedicated hardware components developed especially for the application.

The macro-architecture of the proposed integrated system is presented in Fig. 12.10. The system includes: a processor (PS), in order to execute the PSO iteration; a co-processor to compute the required PCC; two dedicated memory blocks (BRAM IMG and BRAM TMP) that store the original image and the template, respectively; and the drivers to access these memories (GET IMG and GET TMP).

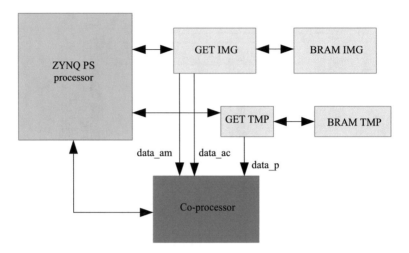

Fig. 12.10 Macro-architecture of the proposed system

12.4.1 Dedicated Co-processor

Figure 12.11 presents the proposed architecture for the co-processor that is responsible for performing the correlation computation for two given images, as defined in Eq. (12.1). The architecture is designed so as it can operate in a pipeline way. Each of the three blocks corresponds to one of the three pipeline stages. In each rising edge of the clock signal, the co-processor requires three data:

- `data_p`: one pixel from the template, consisting of 8 bits;
- `data_ac`: one pixel from the image to be compared, consisting of 8 bits;
- `data_am`: one pixel from the next image, consisting of 8 bits.

All the images considered during the comparison are composed by 64×64 pixels, consisting of a total of 4096 pixels that are represented by 4 KBytes. The computations are performed on a block, and the obtained results are transferred to the next one, at every synchronism pulse. This pulse is generated by component `sincro` at every 4103 clock cycles. As output, the co-processor provides the value of the PCC (`result`), as 32 bits two's complement together with a flag, indicating the operation end (`flag_end`) as well as an error signal (`error`), indicating that the result is not valid, which usually occurs when there is a division by zero.

12.4.1.1 First Stage

Figure 12.12 shows the architecture of block 1, which represents the first stage of the pipeline and is responsible for the computation of the average of the pixels of the images to be compared. It has output registers that are loaded only when the

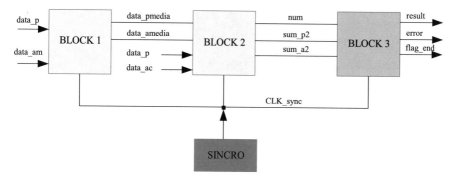

Fig. 12.11 Macro-architecture of the co-processor

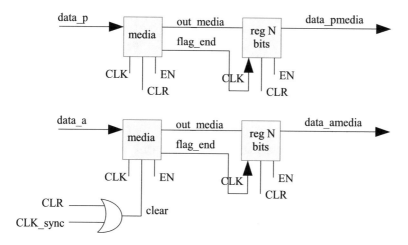

Fig. 12.12 Micro-architecture of block 1

stage task is completed. At a syncronism pulse, only component `media`, which is responsible for computing the main image average, is restarted.

12.4.1.2   Second Stage

Figure 12.13 shows the architecture of the block 2, which represents the pipeline second stage, and is responsible for computing the 3 sums of Eq. (12.1). It consists of two components `subt_A2`, three components `mult_CLK` and three components `sum_A2`. Component `subt_A2` performs, in two's complement, the subtraction of image pixels of the averages obtained by the block 1. Component `mult_CLK` performs, in one clock pulse, the multiplication of the results provided by components `subt_A2`. Component `sum_A2` performs the sum, in two's complement, of the multiplications provided by components `mult_CLK`. Like block 1, block 2

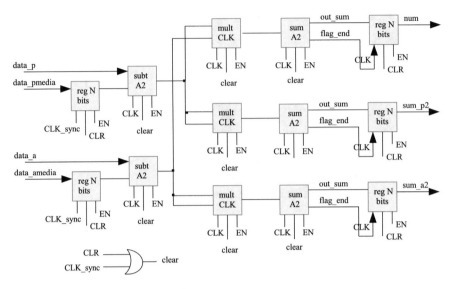

Fig. 12.13 Micro-architecture of block 2

has output registers that are loaded only when the stage task is completed. At the syncronism pulse, components `subt_A2`, `mult_CLK`, and `subt_A2` are reset.

12.4.1.3 Third Stage

Figure 12.14 shows the architecture of the block 3, which represents the pipeline third stage, and is responsible for computing the main multiplication, the square-root and the division of Eq. (12.1). It consists of component `mult_CLK` that performs, in one clock pulse, the multiplication of the denominator sum of Eq. (12.1), component `SQRT` that calculates the square-root and component `div_frac_A2` that performs the division, providing a result with 2^{-24} precision. This last component provides the output signals of the co-processor. The operation of this block is controlled by a state machine (`FSM`), which is responsible for coordinating the block's components. At a syncronism pulse, the FSM returns to its initial state. Similarly to block 1 and block 2, block 3 has output registers that are loaded only when the stage task is completed.

12.4.1.4 Square-Root Computation

Historically, one of the first methods to calculate the square root was developed by ancient Babylonians [11]. Based on this numerical method, so-called *Babylonian*, the component `SQRT` in block 3 has been implemented, in hardware, using an iterative process of N steps, based on Eq. (12.5):

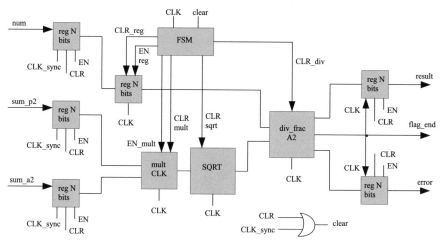

Fig. 12.14 Micro-architecture of block 3

$$x_{i+1} = 0,5 \times \left(x_i + \frac{S}{x_i}\right) \tag{12.5}$$

wherein x is the square root, S is the radicand ($x = \sqrt{S}$), x_i is the value of x in the current iteration, and x_{i+1} is the value of the next iteration. The number of iterations N is set according to the required result precision.

12.4.1.5   Division Computation

Algorithm 12.1 is used to compute the required division result. This algorithm is implemented, in hardware, as component `div_frac_A2` in block 3.

12.4.2   Memory Controllers

The dedicated memory blocks `BRAM_TMP` and `BRAM_IMG` store the template and the main image, respectively. They are implemented in the programmable logic of the Zynq chip (PL). Component `BRAM_TMP` can store up to 4096 8-bit pixels, totalizing 4KBytes corresponding to the template size. On the other hand, component `BRAM_IMG` can store up to 573×463 8-bit pixels, totalizing 260 KBytes. The maximum size of main image is 510×400 pixels. Note that the image is padded with zeros.

Drivers `GET_TMP` and `GET_IMG` are responsible for providing access to dedicated memories `BRAM_TMP` and `BRAM_IMG`, respectively. They deliver the data to

---

Algorithm 12.1 Division $Q = A/B$

$N_B := 0$

$RA_{\langle N-1 \ldots \frac{N}{2} \rangle} := A$

$RA_{\langle \frac{N}{2}-1 \ldots 0 \rangle} := 0$

while $RA_{\langle N-1 \ldots \frac{N}{2} \rangle} > B$ **do**

   $RA := $ shift right RA

   $N_B := N_B + 1$

end while

$Q := 0$

for $j = 1, N_B$ **do**

   $RA := $ shift left RA

   $Q := $ shift left Q

   if $RA_{\langle N-1 \ldots \frac{N}{2} \rangle} > B$ **then**

      $RA_{\langle N-1 \ldots \frac{N}{2} \rangle} := RA_{\langle N-1 \ldots \frac{N}{2} \rangle} - B$

      $Q_0 := 1$

   else

      $Q_0 := 0$

   end if

end for

for $j = 1, 24$ **do**

   $RA := $ shift left RA

   $Q := $ shift left Q

   if $RA_{\langle N-1 \ldots \frac{N}{2} \rangle} > B$ **then**

      $RA_{\langle N-1 \ldots \frac{N}{2} \rangle} := RA_{\langle N-1 \ldots \frac{N}{2} \rangle} - B$

      $Q_0 := 1$

   else

      $Q_0 := 0$

   end if

end for

---

the co-processor, on the right time. The read and write cycles are synchronized by the clock signal (CLK) and the syncronism signal (CLK_sync). Moreover, these memory controllers also allow the processor to access the dedicated memory.

12.5 Performance and Results

The proposed design is implemented using the *Smart Vision Development Kit* rev 1.2 (SVDK) of *Sensor to Image* [17]. This board has a Xilinx module called *PicoZed 7Z015 System-On Module* Rev.C that provides a hardware environment for developing and evaluating designs targeting machine vision applications. Among the various features, common to many embedded processing systems, this board has a UART interface, a USB3 interface, 1GB DDR3 memory ($\times 32$), 33.333 MHz oscillator, a HDMI video encoder, and a chip XC7Z015 Xilinx [3, 17]. This chip is part of Zynq-7000 family, with architecture *All Programmable System on Chip*. It integrates a dual-core ARM Cortex-A9 based processing system (PS) and a Xilinx

Fig. 12.15 Reference images used in the performance tests. (**a**) Pickup—I_1. (**b**) Truck—I_2. (**c**) Rcar—I_3. (**d**) Cars—I_4. (**e**) Sedan—I_5. (**f**) IR1—I_6. (**g**) IR2—I_7. (**h**) IR3—I_8

programmable logic (PL), in a single device [21]. This component provides the flexibility and scalability of an FPGA, while providing a performance, power, and ease of use typically of an ASIC. So, it is perfect for co-design implementations.

The implementation in SVDK board used 11% of flip-flops, 39% of LUTs, 25% of buffers, and 69% of BRAMs (Block RAMs). Because of synthesis time constraints associated with the project, the co-processor runs at 25 MHz.

Figure 12.15 shows the images used in this section, and the corresponding templates, highlighted by the inner squares. These images are considered benchmarks for computational vision and are available in [6].

The performance of the proposed system was evaluated using three scenarios:

1. ES: which is the exhaustive search of all main image pixels, executed by the main processor;
2. PSO_{SO}: which is the intelligent search using global best PSO implemented only in software and executed by the main processor; and
3. PSO_{HP}: which is the intelligent search using global best PSO implemented in software whereby PCCs are calculated by the co-processor, working in pipeline mode.

The correlation for each pixel is calculated extracting a patch, of the same size as the template, and computing the corresponding PCC at its center. It is noteworthy to point out that the limits of the main image are completed with zeros.

The canonical PSO algorithm is used, and the PCC is the cost function. The parameters were set, empirically, as follows: 18 particles, inertial coefficient $w = 0.6$, cognitive coefficient $c_1 = 0.6$, social coefficient $c_2 = 2$ with a maximum velocity of 10. As stopping criteria, we combined either an acceptable PCC threshold of 0.95 or maximum of 10 iterations. In terms of search space, we limited the search in a window of 101×101 pixels. Note that in real situations, the object

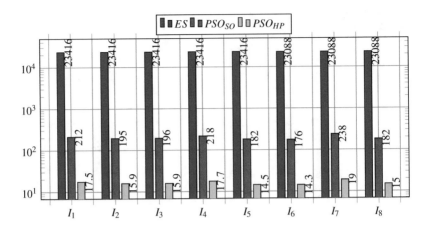

Fig. 12.16 Average processing time for the images over 1000 searches (ms)

does not change position abruptly, and it remains nearby its position in the previous frame. All templates considered have 64×64 pixels.

The results, in terms of average processing time, as obtained for the compared scenarios are given in Fig. 12.16. These average results are for 1000 template searches performed in each image. We can observe for the image IR1, with the best result, the time processing has been improved $131\times$, comparing the PSO_{SO} with the ES, and has been improved $1614\times$, comparing the PSO_{HP} with the ES. In order to track an object in a usual video of 30 frames/s, it is necessary to process each frame in a time of at most 33 ms. All results for PSO_{HP} are lower than 19 ms, confirming that the proposed design is a viable way to achieve real-time execution.

Because of the stochastic behavior of the PSO, the success rates on finding the object for the PSO scenarios are also evaluated. The results are depicted in Fig. 12.17 for 1000 template searches in each image. It is possible to observe that the success rate is above 89.4% for all the images, which is acceptable in real-world object tracking.

12.6 Conclusions

The proposed co-design system, wherein the PCC computation is implemented in hardware keeping the global best PSO iterative process in software, showed to be a viable way to achieve real-time execution in template matching. The average processing time for the images was improved up to $1614\times$, comparing with the exhaustive search, and is lower than 19 ms. As a future work, and in order to further improve the performance of the process, the co-processor design could be examined as to increase the operation frequency.

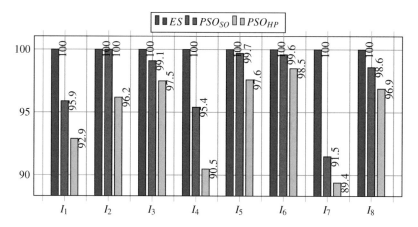

Fig. 12.17 Average success rate for the benchmark images over 1000 searches

Acknowledgements Y. M. Tavares acknowledges the Brazilian Navy for the support given during the development of his research work. We are also grateful to FAPERJ (Fundação Carlos Chagas Filho de Amparo à Pesquisa do Estado do Rio de Janeiro, http://www.faperj.br) and CNPq (Conselho Nacional de Desenvolvimento Científico e Tecnológico, http://www.cnpq.br) for their continuous financial support.

References

1. Ahuja, K., Tuli, P.: Object recognition by template matching using correlations and phase angle method. Int. J. Adv. Res. Comput. Commun. Eng. **2**(3), 1368–1373 (2013)
2. Ali, A., Kausar, H., Khan, M.I.: Automatic visual tracking and firing system for anti aircraft machine gun. In: 6th International Bhurban Conference on Applied Sciences and Technology (IBCAST), pp. 253–257. IEEE, Piscataway (2009)
3. Avnet: PicoZed 7Z015 / 7Z030 System-On Module Hardware User Guide, version 1.3 (2015)
4. Benfold, B., Reid, I.: Stable multi-target tracking in real-time surveillance video. In: IEEE Conference on Computer Vision and Pattern Recognition (CVPR), pp. 3457–3464. IEEE, Piscataway (2011)
5. Choi, H., Kim, Y.: UAV guidance using a monocular-vision sensor for aerial target tracking. Control Eng. Pract. **22**, 10–19 (2014)
6. Collins, R., Zhou, X., Teh, S.K.: An open source tracking testbed and evaluation web site. In: IEEE International Workshop on Performance Evaluation of Tracking and Surveillance, vol. 2, p. 35 (2005)
7. Engelbrecht, A.P.: Fundamentals of Computational Swarm Intelligence. Wiley, West Sussex (2006)
8. Forlenza, L., Fasano, G., Accardo, D., Moccia, A.: Flight performance analysis of an image processing algorithm for integrated sense-and-avoid systems. Int. J. Aerosp. Eng. **2012**, 1–8 (2012)
9. Holland, J.H.: Adaptation in Natural and Artificial Systems: An Introductory Analysis with Applications to Biology, Control and Artificial Intelligence. MIT Press, Cambridge (1992)
10. Kennedy, J., Eberhart, R.: Particle swarm optimization. In: IEEE International Conference on Neural Network, vol. 4, pp. 1942–1948. EUA (1995)

11. Kosheleva, O.: Babylonian method of computing the square root: justifications based on fuzzy techniques and on computational complexity. In: Fuzzy Information Processing Society, NAFIPS, pp. 1–6. IEEE, Piscataway (2009)
12. Mahalakshmi, T., Muthaiah, R., Swaminathan, P., Nadu, T.: Review article: an overview of template matching technique in image processing. Res. J. Appl. Sci. Eng. Technol. **4**(24), 5469–5473 (2012)
13. Narayana, M.: Automatic tracking of moving objects in video for surveillance applications. Ph.D. Thesis, University of Kansas (2007)
14. Nedjah, N., Mourelle, L.M.: Co-design for System Acceleration: A Quantitative Approach. Springer Science & Business Media, Berlin (2007)
15. Nixon, M.S., Aguado, A.S.: Feature Extraction and Image Processing, 1st edn. Academic, Great Britain (2002)
16. Olson, T.L., Sanford, C.W.: Real-time multistage IR image-based tracker. In: AeroSense'99, pp. 226–233. International Society for Optics and Photonics, Bellingham (1999)
17. SensorToImage: SVDK Hardware User Guide, revision 1.1 (2015)
18. Sharma, P., Kaur, M.: Classification in pattern recognition: a review. Int. J. Adv. Res. Comput. Sci. Softw. Eng. **3**(4), 1–3 (2013)
19. Tavares, Y.M., Nedjah, N., Mourelle, L.M.: Embedded Implementation of template matching using correlation and particle swarm optimization. In: International Conference on Computational Science and Its Applications, pp. 530–539. Springer, Beijing (2016)
20. Tavares, Y.M., Nedjah, N., Mourelle, L.M.: Tracking patterns with particle swarm optimization and genetic algorithms. Int. J. Swarm Intell. Res. **8**(2), 34–49 (2017)
21. Xilinx: UG585 Zynq-7000 AP SoC Technical Reference Manual, version 1.10 (2015)
22. YouTube: Rafale - High Technology Hunting Plane (Brazil) (in Portuguese) (2015)

Chapter 13
A Hybrid Estimation Scheme Based on the Sequential Importance Resampling Particle Filter and the Particle Swarm Optimization (PSO-SIR)

Wellington Betencurte da Silva, Julio Cesar Sampaio Dutra,
José Mir Justino da Costa, Luiz Alberto da Silva Abreu,
Diego Campos Knupp, and Antônio José Silva Neto

13.1 Introduction

The increasing availability of modern high performance computers has made possible the solution of problems of state estimation in several knowledge fields, such as propagation of fire, combustion, tumor growth, chemical reactors, and flow in pipe-in-pipe system, among others [17]. Such problems are generally treated within state space representation [17], which comprises an evolution model and a measurement model for the system under observation. One of the most attractive features of this approach is that estimation and prediction can be applied sequentially when new information (measurement) is obtained [20]. In the Bayesian framework, one of the main interests is the estimation of the a posteriori probability of the conditional distribution of the states, given the measurements from the system. However, depending on the problem, the solution requires dealing with intractable integrals so that the use of numerical methods is required [14].

Sequential Monte Carlo methods are widely used techniques for this purpose, and, in particular, particle filters are very important. For well-behaved problems (that is, with smooth changes), particle filters have been extensively used showing good

W. B. da Silva · J. C. S. Dutra (✉)
Chemical Engineering Program, Center of Agrarian Sciences and Engineering, Federal University of Espírito Santo, Alegre, Brazil

J. M. J. da Costa
Statistics Department, Federal University of Amazonas, Manaus, Brazil
e-mail: zemir@ufam.edu.br

L. A. d. S. Abreu · D. C. Knupp · A. J. Silva Neto
Department of Mechanical Engineering and Energy, Polytechnic Institute, IPRJ-UERJ, Nova Friburgo, Brazil
e-mail: diegoknupp@iprj.uerj.br; ajsneto@iprj.uerj.br

© Springer Nature Switzerland AG 2019
G. Mendes Platt et al. (eds.), *Computational Intelligence, Optimization and Inverse Problems with Applications in Engineering*,
https://doi.org/10.1007/978-3-319-96433-1_13

results [4, 6–10, 15–19, 21–26]. However, when the dynamics show abrupt changes (that is, large discontinuities), the usually applied filters (Sequential Importance Sampling-SIS, Sampling Importance Resampling-SIR and Auxiliary Sampling Importance Resampling—ASIR) may face difficulty in directing the particles (also called samples) to high probability regions [13]. As the a priori distribution at a given time is related to a posterior distribution of the previous time, it is advisable the use of an a priori informative distribution if the previous state is well known. In this case, when abrupt changes occur, and the uncertainties associated with the evolution model are small, the new particles generated to evolve a step forward are not able to be distributed in the more credible regions [13].

A typical solution normally taken to avoid this situation is to increase both the number of particles and the level of uncertainty associated with the model evolution (that is, assigning a high standard deviation to the evolution model) [16, 18, 22]. Proposals embedding some optimization methods into the usual particle filters have been tested with good results, such as PSO (Particle Swarm Optimization) [2, 27]. However, the optimization step is performed at each time instant, even if the problem dynamics has not undergone significant changes. In this situation, the application of the optimization method is unnecessary, since it increases considerably the computational cost.

In this chapter, we propose a new approach for state and parameter estimation using a particle filter algorithm, in which an evolutionary optimization method is embedded to direct particles to high probability regions. Unlike previous works in the literature, the optimization step is performed only when a warning signal is triggered. In this proposal, the warning signal is based on the effective sample size $\hat{N}_{eff}$ [9, 21]. When $\hat{N}_{eff}$ is below a threshold proportional to the specified sample size, it is assumed that the system has undergone a behavior change. If so, the estimation scheme considers switching on the optimization step to direct the particles to the new search region with the highest probability, where new samples are drawn.

This new estimation scheme was applied to a heat transfer problem in a thin plate to estimate sequentially the unknown boundary heat flux, as well as for filtering temperature measurements and estimating the convective heat transfer coefficient. To do so, non-intrusive temperature measurements were taken at the exposed surface. The proposal allowed obtaining very accurate results. Among the main remarks, it is worth to highlight that the computational time and the amount of particles were much lower when compared to the standard SIR approach for the estimation problem.

13.2 Physical Problem and Mathematical Formulation

Consider a thin plate, with thickness L_{X_1}, to which a spatially uniform, but time varying, heat flux, $q(t)$, is applied at the surface $X_1 = 0$. The opposite surface,

Fig. 13.1 Schematic representation of the physical problem

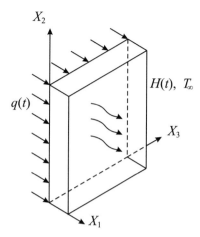

$X_1 = L_{X_1}$, exchanges heat with the environment at temperature T_∞, with a heat transfer coefficient $H(t)$, as schematically presented in Fig. 13.1.

As the applied heat flux is spatially uniform in the problem here described, if one further considers that the heat losses through the lateral borders are negligible, no significant temperature variations are expected in the directions X_2 and X_3. Thus, performing a lumped analysis across the plate thickness, X_1, the general three-dimensional heat conduction equation can be simplified to [5]:

$$C_p \frac{dT(t)}{dt} = \frac{q(t)}{L_{X_1}} - \frac{H(t)}{L_{X_1}}(T(t) - T_\infty), \quad t > 0 \tag{13.1a}$$

$$T(0) = T_\infty \tag{13.1b}$$

where T is the plate spatial average temperature, and C_p the volumetric heat capacity.

In this chapter, the parameters C_p and L_{X_1} appearing in the model are considered known, and the goal is to estimate the applied heat flux $q(t)$ and the heat transfer coefficient $H(t)$, employing non-intrusive temperature measurements at the exposed surface, $X_1 = L_{X_1}$, as obtained from an infrared thermography system [12], for instance. The inverse problem formulation and solution are presented in detail in the next section.

13.3   Inverse Problem Formulation and Solution

The solution of the inverse problem within the Bayesian framework is recast in the form of statistical inference from the *posterior probability density*, which is the statistical model for the conditional probability distribution of the unknown parameters given the measurements [10, 15, 18, 26]. In the present chapter, two different approaches were addressed: the Bayesian filter known as SIR (Sampling Importance Resampling), and the proposed hybrid estimation scheme based on SIR and PSO (Particle Swarm Optimization).

The measurement model incorporating the related uncertainties is called the *likelihood function*, i.e., the conditional probability of the measurements $\mathbf{z}_k$ given the unknown parameters $\mathbf{x}_k$, noted as $\pi(\mathbf{z}_k|\mathbf{x}_k)$. By assuming that the measurement errors are Gaussian random variables, with zero mean and known covariance matrix $\mathbf{W}$, and that the measurement errors are additive and independent of the parameters $\mathbf{x}_k$, the likelihood function can be expressed as given by Eq. (13.2) [10, 15, 18, 26]. In this representation, $\mathbf{h}_k(\mathbf{x}_k)$ is the solution of the direct problem.

$$\pi(\mathbf{z}_k|\mathbf{x}_k) = (2\pi)^{-1/2}|\mathbf{W}|^{-1/2}\exp\left\{-\frac{1}{2}[\mathbf{z}_k - \mathbf{h}_k(\mathbf{x}_k)]^T\mathbf{W}^{-1}[\mathbf{z}_k - \mathbf{h}_k(\mathbf{x}_k)]\right\}$$
(13.2)

The model for the unknowns that reflects the uncertainty of the parameters, without the information conveyed by the measurements, is called the prior model [10, 15, 18, 26]. It is noted $\pi(\mathbf{x}_k)$, where the subscript k is the discrete time instants.

In this context, Bayes' theorem is stated as:

$$\pi_{posterior}(\mathbf{x}_k) = \pi(\mathbf{x}_k|\mathbf{z}_k) = \frac{\pi(\mathbf{x}_k)\pi(\mathbf{z}_k|\mathbf{x}_k)}{\pi(\mathbf{z}_k)}$$
(13.3)

In this equation, $\pi_{posterior}(\mathbf{x}_k)$ represents the posterior probability density, $\pi(\mathbf{x}_k)$ the prior density, $\pi(\mathbf{z}_k|\mathbf{x}_k)$ the likelihood function, and $\pi(\mathbf{z}_k)$ the marginal probability density of the measurements, which plays the role of a normalizing constant.

13.3.1 Principle of the Sequential Monte Carlo Based Estimation

In order to estimate the unknown boundary heat flux sequentially, progressing iteratively from $k = 1$ up to $k = NP$, a non-stationary inverse problem must be solved [10, 15, 18, 26]. This is of great interest in innumerable practical applications, especially those regarding online monitoring, virtual inference, identification, model updating, and control [4, 7, 10, 15–19, 22–24, 26]. In such type of problems, the available measured data is used along with prior knowledge about the physical phenomena involved and the measuring devices, in order to produce estimates of the desired dynamic variables.

The sequential Monte Carlo problem considers a set of equations for [10, 15]:

(a) *state space model*—Eq. (13.4), which represents the dynamic evolution of the state vector $\mathbf{x}_k \in \mathbf{R}^n$ considering the input vector $\mathbf{u}_k \in \mathbf{R}^m$ and the state uncertainty vector $\mathbf{v}_k \in \mathbf{R}^n$:

$$\mathbf{x}_k = \mathbf{f}_k(\mathbf{x}_{k-1}, \mathbf{u}_{k-1}, \mathbf{v}_{k-1}) \qquad (13.4)$$

(b) *observation model*—Eq. (13.5), which provides the solution of the direct problem accounting for the state vector and the measurement uncertainty $\mathbf{n}_k \in \mathbf{R}^{nz}$:

$$\hat{\mathbf{z}}_k = \mathbf{h}_k(\mathbf{x}_k, \mathbf{n}_k) \qquad (13.5)$$

Both functions $\mathbf{f}$ and $\mathbf{h}$ are generally nonlinear, and the subscript $k = 1, 2, \ldots, NP$ denotes a time instant for a dynamic problem. The state vector and the input vector, which consider parameters and external signals, contain the variables to be dynamically reconstructed. It can be assumed, without loss of generality, that the state vector is the actual state vector augmented with the input vector, in such a way that $\mathbf{x} = [\mathbf{x}_{actual}, \mathbf{u}]$. In order to perform the sequential Monte Carlo procedure, the estimation error must be minimized statistically [10, 15, 18, 26].

In this regard, the evolution and observation models are based on the assumptions that the sequence $\mathbf{x}_k$ depends on the past observations only through its own history, $\pi(\mathbf{x}_k|\mathbf{x}_{k-1})$, and the sequence $\hat{\mathbf{z}}_k$ is a Markovian process with respect to the history of $\mathbf{x}_k$, $\pi(\hat{\mathbf{z}}_k|\mathbf{x}_k)$. For the state and observation noises, it is assumed that the noise vectors $\mathbf{v}_i$ and $\mathbf{v}_j$, as well as $\mathbf{n}_i$ and $\mathbf{n}_j$, are mutually independent, for $i \neq j$, and also mutually independent of the initial state $\mathbf{x}_0$, and the noise vectors $\mathbf{v}_i$ and $\mathbf{n}_j$ are mutually independent for all i and j. Under this framework, different applications can be considered as prediction; filtering, fixed-lag smoothing, and whole-domain smoothing problems [10, 15, 18, 26].

The objective here is the accurate estimation of the temperature field, in addition to the heat flux input and the convective heat transfer coefficient. Thus, the augmented state vector is given by $\mathbf{x}(t) = [\mathbf{T}(t), \mathbf{q}(t), \mathbf{H}(t)]^T$, where the vectors $\mathbf{T}$, $\mathbf{q}$ and $\mathbf{H}$ represent the timewise distribution of the temperature, the heat flux, and the convective heat transfer coefficient, respectively. For linear and Gaussian systems, the application of Bayes' theorem and the state space framework can form a recursive solution for estimating the posterior distribution probability. However, it is computationally intractable for nonlinear and non-Gaussian systems. For this reason, Particle Filter Methods were proposed.

13.3.2 The Particle Filter Method

The *Particle Filter Method* [8, 9, 21] is a Sequential Monte Carlo technique for the solution of the state estimation problem. The key idea is to represent the required posterior density function by a set of N random samples called particles $(\mathbf{x}_i^k)$ with associated weights ($\mathbf{w}_k^i$), given by the set $\{\mathbf{x}_k^i, \mathbf{w}_k^i\}$ with $i = 1, \ldots, N$ and $k = 1, \ldots, NP$. Afterwards, one can compute the estimates based on these samples and weights. As the number of samples becomes very large, this Monte Carlo characterization becomes an equivalent representation of the posterior probability

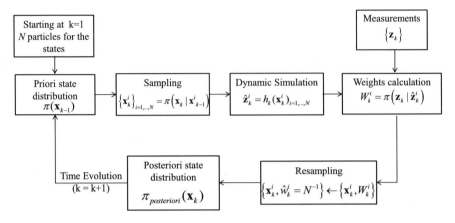

Fig. 13.2 The SIR algorithm [16]

function, and the solution approaches the optimal Bayesian estimate. The particle filter algorithms generally make use of an *importance density*, which is a density proposed to represent the sought posterior density. In this regard, particles are drawn from the importance density instead of the actual density.

The sequential application of the particle filter might result in the *degeneracy phenomenon*, where after a few time iterations all, but very few particles, have negligible weight [8, 9, 21]. The degeneracy implies that a large computational effort is devoted to updating particles whose contribution to the approximation of the posterior density function is almost zero. This problem can be overcome with a resampling step in the application of the particle filter. Resampling involves a mapping of the random measure $\{\mathbf{x}_k^i, \mathbf{w}_k^i\}$ into $\{\mathbf{x}_k^{i*}, N^{-1}\}$ with uniform weights, leading, therefore, to the elimination of particles originally with low weights, and the replication of particles with high weights. Resampling can be performed if the number of effective particles (particles with large weights) falls below a certain threshold number [8, 9, 21]. Alternatively, resampling can also be applied indistinctively at every instant t_k, as in the *Sampling Importance Resampling* (SIR) algorithm described in [4, 6–9, 16, 17, 21–25].

Such algorithm can be summarized in the steps presented in Fig. 13.2, as applied to the system evolution iteratively from k to $k + 1$. In the first step of the SIR algorithm, it should be noticed that the weights are given directly by the likelihood function $\pi(\mathbf{z}_k|\hat{\mathbf{z}}_k^i)$. Subsequently, in this algorithm, the resampling step is applied at each time instant, and then the weights $w_k^i = N^{-1}$ are uniform.

The greatest advantage of Particle Filter Methods is the easy computational implementation. However, usual formulation of these methods may not provide good results in problems with large discontinuities (or irregular/abrupt changes), because the available information may not be enough to draw suitable samples for the estimation. In this regard, it is proposed here to move the a priori information to a new search region by means of an evolutionary optimization method, namely Particle Swarm Optimization (PSO), as explained in the following section.

13.4 The Scheme Based on PSO and SIR Particle Filter

13.4.1 The Particle Swarm Optimization Method (PSO)

In 1995, James Kennedy and Russell Eberhart [11] developed the Particle Swarm Optimization method (PSO) as an alternative to the Genetic Algorithm methods. PSO approaches the social behavior of various species and tries to equilibrate the individuality and sociability of the individuals in order to locate the optimum of interest. The original idea of Kennedy and Eberhart came from the observation of birds looking for a nesting place. The development of the algorithm is based on the theory of swarms, that consists in the fact that birds or particles make use of past experiences, either personal or from the group, to help them find the nest or food (or optimal point). PSO searches a space by adjusting the trajectories of individual vectors, also called particles, which are conceptualized as moving points in a multidimensional problem space with two associated vectors, position vector ($\mathbf{P}_j^{ik}$) and velocity vector ($\mathbf{V}_j^{ik}$), $j = 1, \ldots, M$ for the current evolutionary iteration ik. The velocity of the individual particle in PSO is dynamically adjusted according to its own flying experience and its companions' flying experience, in the moves around the search space. The former was termed cognition-only model, and the latter was termed social-only model [3]. Therefore, one gets:

$$\mathbf{v}_j^{ik+1} = \alpha \mathbf{v}_j^{ik} + \beta \mathbf{r}_{1j}(\mathbf{B}_j - \mathbf{P}_j^{ik}) + \beta \mathbf{r}_{2j}(\mathbf{B}_g - \mathbf{P}_j^{ik}) \tag{13.6}$$

$$\mathbf{P}_j^{ik+1} = \mathbf{P}_j^{ik} + \mathbf{v}_j^{ik+1} \tag{13.7}$$

In this formulation, $\mathbf{r}_{1j}$ and $\mathbf{r}_{2j}$ are random numbers between 0 and 1, and two positive real numbers $0 < \alpha < 1$ and $1 < \beta < 2$, denoted learning parameters, must be chosen. Besides that, $\mathbf{P}_j$ is the j-th individual of the vector of parameters; $\mathbf{v}_j^{ik} = 0$ for $ik = 0$; $\mathbf{B}_j$ is the best value found by the j-th individual, $\mathbf{P}_j$; and $\mathbf{B}_g$ is the best value found by the entire population.

A general description of the PSO algorithm follows.

Step 1: Setting of initial conditions for the swarm; i.e. for each particle, the position ($\mathbf{P}_j^0$) and velocity ($\mathbf{V}_j^0$) are randomly generated, given suitable ranges;
Step 2: Evaluation of the objective function $F(\mathbf{P})$ for each particle of the swarm; and the positions $\mathbf{B}_j$ and $\mathbf{B}_g$ are eventually updated; where

$$F(\mathbf{x}_{k-1}) = \left(\frac{\mathbf{z}_k - \hat{\mathbf{z}}_k}{\sigma}\right)^2 \tag{13.8}$$

and σ represents the standard deviation of the measurements.

Step 3: Update of the velocity of each particle of the swarm using Eq. (13.6);

Step 4: Update of the position of each particle of the swarm using Eq. (13.7), in order to obtain the new position $\mathbf{P}_j^{ik+1}$;

Step 5: Check of the stopping criteria; if it is not verified, return to Step 2 for the next iteration.

13.4.2 The PSO-SIR Filter Algorithm

It is well known from the literature that sequential importance sampling provides Monte Carlo estimates whose computational cost increases exponentially with the sample size. In practice, while the algorithm iterates, very few particles may dominate the entire sample, resulting in a very small number of particles with nonzero weights to approximate the filtering distribution [8, 9, 21]. Therefore, adding resampling steps [1, 8] has been used in literature as an efficient mechanism to minimize the weight degeneracy.

Due to the simplicity of the implementation, resampling steps have been used by many researchers and practitioners. The resampling algorithms produce multiple copies of particles with high weights, and then eliminate those with small weights. However, the excessive use of resampling implies the sample impoverishment [1, 8]. Therefore, resampling should be performed only when needed, typically when we do not obtain a low variance of weights. A typical statistical metric to evaluate if resampling should be used is given by the effective sample size [1]:

$$\widehat{N}_{eff} = \frac{1}{\sum_{i=1}^{N}(W_k^i)^2} \tag{13.9}$$

The maximum value for the effective sample size is N, which is user specified. Resampling can be applied whenever $\widehat{N}_{eff}$ is below a threshold (N_T), e.g. 50% or 40% of the simulated sample. In this chapter, it is assumed the threshold $N_T = 0.4N$. The scheme based on PSO and SIR Particle Filter (PSO-SIR) is similar to the standard SIR filter method; however, our approach uses the effective sample size to evaluate if the generated sample represents the posterior probability density. In other words, whenever $\widehat{N}_{eff} < N_T$, it is evident that a priori information used in the generation of particles was not suitable in that instant of time. If so, there is a need to move the prior information to a higher probability region. Therefore, the idea of this new method is to use the PSO to find this better region, and improve prior information to draw a set of particles with acceptable effective size. Such algorithm can be summarized in the steps presented in Fig. 13.3.

The objective function $F(\mathbf{x}_{k-1})$ used in this chapter is given by Eq. (13.8). This metric represents the least square weighted by measurement error covariance.

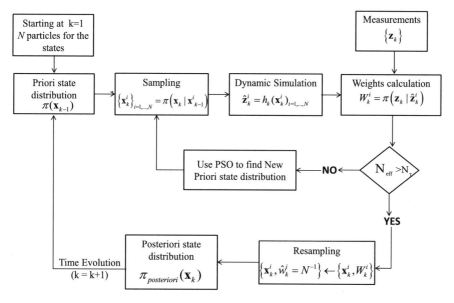

Fig. 13.3 The PSO-SIR algorithm

13.5 Results and Discussions

In order to validate the technique discussed and presented in this chapter, it was considered an actual experimental apparatus, similar to the one presented in Ref. [12], which corresponds to a set up with a 40×40 mm plate, with 1.5 mm thickness, made of aluminum with the following properties: thermal conductivity $k = 237$ W/mK, specific mass $\rho = 2702$ kg/m^3, and thermal capacity $c_p = 903$ J/kg K. An electric potential difference of 9 V was applied to the heater (39.8 Ω) during an interval of 240 s beginning at two distinct time instants $t_1 = 240$ s and $t_2 = 720$ s, resulting in a power dissipation of 1.75 W in each interval, and a heat flux of 549 W/m^2. The initial measurements, taken until 240 s, were used to calculate the standard deviation of the measurements $\sigma = 0.16\,^\circ$C. The temperature measurements at the plate surface were then taken at a frequency of 1 measurement per second, during 1200 s, and the experiment was carried out with room temperature $T_\infty = 22.7\,^\circ$C.

In the framework of the particle filters, two auxiliary models are needed: (1) an evolution model, given by Eqs. (13.1), and (2) an observation model, given by Eq. (13.5). The evolution model for the spatially averaged temperature was integrated by means of the Runge-Kutta method, and added a Gaussian uncertainty with standard deviation of 0.05 $^\circ$C. For the heat flux, it was used a random walk model according to:

$$q(t) = q(t - 1)\sigma_q \epsilon_q \qquad (13.10)$$

where σ_q is the standard deviation related to the model evolution for the heat flux, taken as 10 and 300 W/m^2 in all cases here presented, and ϵ_q are random numbers with uniform distribution between $[-1, 1]$. The first value for ϵ_q means that the a priori information is good, and the second standard deviation expresses that there is not much trust on the value of the parameters.

For the construction of the credible interval, it was considered as the lower and upper limits of the interval, the quantis 0.005 and 0.995, respectively, of the posterior distribution at each time.

For the convective heat transfer coefficient estimation, it was used a uniform distribution between 15 and 22 W/m^2 K, typical values for natural convection.

In order to allow for a close comparison regarding the estimates obtained with the different methodologies, it is also calculated the Mean Absolute Percentage Error (MAPE) of the estimated temperatures with respect to the experimental data, defined as [20]:

$$\text{MAPE} = \frac{100}{NP} \sqrt{\sum_{i=1}^{NP} \left| \frac{T_{\text{exp},i} - T_{est,i}}{T_{\text{exp},i}} \right|} \tag{13.11}$$

where $T_{\text{exp},i}$ are the experimental data employed, and $T_{est,i}$ are the estimated temperatures. NP is the number of sampling time instants.

In order to assess the proposed approach, it was also used a usual SIR filter with different number of particles (50, 100, and 500), and standard deviation for heat flux (10 and 300 W/m^2) for the samples generation. In addition to MAPE, it was also computed the CPU time by the filters. Table 13.1 presents a comparison between the two methodologies employed (standard SIR and PSO-SIR).

It can be observed that the best result for the SIR filter was found when applied the highest value of particles number ($NP = 500$), and heat flux standard deviation ($q = 300$ W/m^2). With this filter setting, the initial information about the system parameters is supposed not to be well known, what allows improving the filter exploitation ability. In other words, even if the researchers doing the estimation are aware of the problem, there is not much trust on the a priori information, in such a way that the initial guess is estimated at each time. On the other hand, using $N = 50$ and $q = 100$ W/m^2 means, respectively, that the search field for the

Table 13.1 MAPE errors and computational times for SIR and PSO-SIR filters

Particle filter	Number of particle, N	Heat flux standard deviation, σ_q (W/m^2)	MAPE (%)	Elapsed time (s)
SIR	50	10	1.2001	227.51
	100		1.1436	449.04
	500		1.0761	2226.20
	50	300	0.5208	254.04
	100		0.5008	449.04
	500		0.4860	2379.52
PSO-SIR	**50**	**10**	**0.3362**	**536.85**

particle is limited, and the knowledge degree about the parameters is less uncertain (in such a way that there would be no need for exploring much beyond the a priori information). For this reason, the worst values of MAPE are observed for the SIR filter in this condition, as the ability to follow the actual behavior of the heat transfer is diminished. Since this setting is stringent, the PSO-SIR was only addressed in this condition to highlight the redirection feature of the particles.

Despite the strict setting, PSO-SIR showed the best performance (lines in bold in Table 13.1), since the estimates obtained led to lower values of MAPE, with a low computational burden. In relation to the SIR filter with the same number of particles and heat flux standard deviation, PSO-SIR took more time due to the activation of the optimization step. As the index MAPE is related to prediction accuracy, the filtering of the plate temperature is very satisfactory with PSO-SIR. Specifically, our approach, when compared to SIR, presented performance results for MAPE 1.5 times lower, and for elapsed time 4.4 times shorter. This shows that the PSO-SIR filter is much more efficient and faster than the traditional SIR particle filter, since it provided better estimation accuracy and lower computational cost, even with a strict setting.

This reveals that the PSO-SIR feature to redirect particle sample allows the estimation procedure to capture external changes in the heat flux. Even in face of a strict search field, the step of particle swarm optimization in the present approach embedded an adaptive characteristic to the standard SIR filter. According to the effective sample size, the proposed PSO-SIR algorithm considers to switch on the optimization step in the search for new a priori information, to draw particle samples, taking into account the measurement covariance.

In this regard, Fig. 13.4 shows the effective sample size during the estimation procedure with PSO-SIR for the test case, considering 50 particles and the threshold with 20 particles. The black stars in the graph indicate the monitoring of the effective sample size along the time. The status on and off for the optimization is also plotted, in such a way that the optimization was switched on when the red asterisk is nonzero and switched off elsewhere.

It can be seen that the optimization step was switched on at just a few discrete time instants, which are related to changes in the heat flux. As the case test considered a double pulse, the optimization step was switched on four times. Whenever the effective sample size (black star) is below the threshold value (NT = 20), the a priori information of the system is moved around the search space in order to account for external changes. The effect is that the estimation procedure is able to find the actual system input and parameters. In addition, since the optimization step is performed a few times, this did not increase significantly the required computational time as shown in Table 13.1.

The graphical results for the estimation of boundary heat flux, plate temperature, and heat transfer coefficient are presented in Figs. 13.5 and 13.6. These results correspond to a double pulse test, using the SIR filter with $N = 500$ and σ_q of 10 and 300 W/m², as well as the PSO-SIR with $N = 50$ and $\sigma_q = 10$ W/m². These filters led to estimates with MAPE, respectively, equal to 1.0761%, 0.4860%, and 0.3362%.

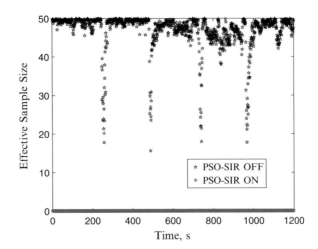

Fig. 13.4 Effective sample size during the estimation procedure for the analyzed test cases, considering 50 particles in the SIR filter and threshold equal to 20 particles to switch on PSO

In Fig. 13.5a, we can observe that the SIR filter with $\sigma_q = 10\,\text{W/m}^2$, despite the high number of particles, was not able to estimate the heat flux input because the a priori information was taken as certain. Even with the credible interval being narrow, the estimated profile does not represent the experimental test double pulse heat flux. For the SIR filter with $\sigma_q = 300\,\text{W/m}^2$, see Fig. 13.5b, the heat flux could be estimated for the sake of the broad search field. However, the estimates present a very wide credible interval, with a positive bias value when the heat flux should be zero (since the heat source was turned off), and an oscillatory behavior when the heat flux was set at the maximum value. Possibly, if σ_q was decreased, and the particle number increased, one could improve the result for this filter.

The heat flux estimation with PSO-SIR can be seen in Fig. 13.5c, d the comparison with the previous filters. Our approach provided very accurate estimates with narrow credible interval. The only problem was observed when the external heat source is turned off after the first pulse. The algorithm should estimate a null value for the heat flux, but there was a systematic deviation as it approached the input of the last (second) pulse. This may be related to the noises present in the measurements and to the fact that the change in the heat source is implemented analogically.

Finally, Fig. 13.6 presents the estimation results of: (a) temperature; and (b) heat transfer coefficient. The SIR filter with $\sigma_q = 10\,\text{W/m}^2$ showed a major temperature deviation in relation to the other tested filters, since its MAPE value is the highest. Unexpectedly, the result for heat transfer coefficient with PSO-SIR is diffuse, probably due to the uniform distribution; even though, it is located in the middle of the interval considered.

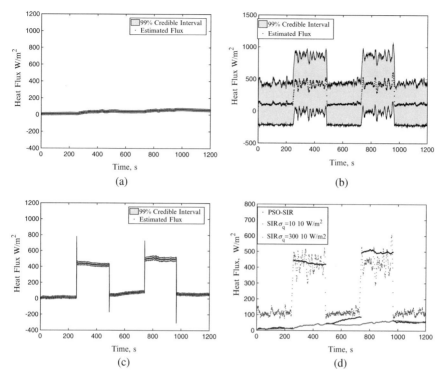

Fig. 13.5 Estimation of boundary heat flux for the double pulse test. (**a**) SIR filter ($\sigma_q = 10$ W/m^2 and $N = 500$). (**b**) SIR filter ($\sigma_q = 300$ W/m^2 and $N = 500$). (**c**) PSO-SIR ($\sigma_q = 10$ W/m^2 and $N = 50$). (**d**) Comparison of the heat flux estimates

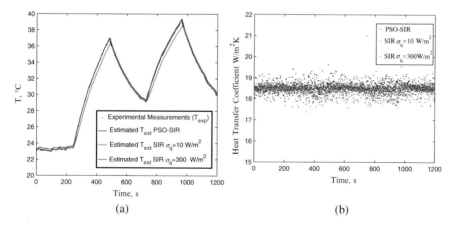

Fig. 13.6 Estimation of (**a**) temperature, and (**b**) heat transfer coefficient, for the double pulse test case, considering $N = 500$ particles

13.6 Conclusion

In this chapter, a new particle filter was proposed to deal with parameter and state estimation problems involving large discontinuities. To improve the estimates in these cases, the proposed approach combines classic filters with stochastic optimization methods.

In the methodology presented in this chapter, a combination of the SIR filter and the Particle Swarm Optimization (PSO) method was used.

The results obtained with the new approach were compared with the solution obtained with a classic SIR filter to solve an inverse problem of heat transfer estimation of contour heat flux in a plate.

The results, using real temperature measurements obtained from a previous work by the authors, showed that the new method is more effective and accurate than the classical methodologies, thus showing that the method is very promising.

Acknowledgements The authors acknowledge the financial support provided by FAPERJ–Fundação Carlos Chagas Filho de Amparo à Pesquissa do Estado do Rio de Janeiro, CNPq–Conselho Nacional de Desenvolvimento Científico e Tecnológico, and CAPES–Coordenação de Aperfeiçoamento de Pessoal de Nível Superior, research supporting agencies from Brazil.

References

1. Cappé, O., Moulines, E., Rydén, T.: Inference in Hidden Markov Models. Springer, New York (2005)
2. Cheng, W.C.: PSO algorithm particle filters for improving the performance of lane detection and tracking systems in difficult roads. Sensors **12**, 17168–17185 (2012)
3. Colaço, M.J., Dulikravich, G.S.: A survey of basic deterministic, heuristic, and hybrid methods for single-objective optimization and response surface generation. In: Orlande, H.R.B., Fudym, O., Maillet, D., Cotta, R.M. (eds.) Thermal Measurements and Inverse Techniques, vol. 1, pp. 355–405. CRC Press, Boca Raton (2011)
4. Costa, J.M.J., Orlande, H.R.B., Campos Velho, H.F., Pinho, S.T.R., Dulikravich, G.S., Cotta, R.M., Cunha Neto, S.H.: Estimation of tumor size evolution using particle filters. J. Comput. Biol. **22**(7), 1–17 (2015)
5. Cotta, R.M., Milkhailov, M.D.: Heat Conduction: Lumped Analysis, Integral Transforms, Symbolic Computation. Wiley, Chichester (1997)
6. Demuner, L.R., Rangel, F.D., Dutra, J.C.S., Silva, W.B.: Particle filter application for estimation of valve behavior in fault condition. J. Chem. Eng. Chem. **1**, 73–87 (2015) (in Portuguese)
7. Dias, C.S.R., Demuner, L.R., Rangel, F.D., Dutra, J.C.S., Silva, W.B.: Online state estimation through particle filter for feedback temperature control. In: XXI Brazilian Congress of Chemical Engineering, Fortaleza (2016)
8. Doucet, A., Freitas, N., Gordon, N.: Sequential Monte Carlo Methods in Practice. Springer, New York (2001)
9. Gordon, N., Salmond, D., Smith, A.F.M.: Novel approach to nonlinear and non-Gaussian Bayesian state estimation. Proc. Inst. Elect. Eng. **140**, 107–113 (1993)
10. Kaipio, J., Somersalo, E.: Statistical and Computational Inverse Problems, Applied Mathematical Sciences. Springer, New York (2004)

11. Kennedy, J., Eberhart, R.C.: Particle swarm optimization. In: Proceedings of the 1995 IEEE International Conference on Neural Networks vol. 4, pp. 1942–1948 (1995)
12. Knupp, D.C., Naveira-Cotta, C.P., Ayres, J.V.C., Orlande, H.R.B., Cotta, R.M.: Space-variable thermophysical properties identification in nanocomposites via integral transforms, Bayesian inference and infrared thermography. Inverse Probl. Sci. Eng. **20**(5), 609–637 (2012)
13. Li, T., Sun, S., Sattar, T.P., Corchado, J.M.: Fight sample degeneracy and impoverishment in particle filters: a review of intelligent approaches. Expert Syst. Appl. **41**, 3944–3954 (2014)
14. MacKay, D.J.C.: A practical Bayesian framework for back-propagation networks. Neural Comput. **4**, 448–472 (1992)
15. Maybeck, P.: Stochastic Models, Estimation and Control. Academic, New York (1979)
16. Orlande, H.R.B., Colaço, M.J., Dulikravich, G.S., Vianna, F.L.V., Silva, W.B., Fonseca, H.M., Fudym, O.: Kalman and particle filters. In: METTI5 Advanced Spring School: Thermal Measurements & Inverse Techniques (2011)
17. Orlande, H.R.B., Colaço, M.J., Dulikravich, G.S., Vianna, F., Silva, W.B., Fonseca, H., Fudym, O.: State estimation problems in heat transfer. Int. J. Uncertain. Quantif. **2**(3), 239–258 (2012)
18. Orlande, H.R.B., Dulikravich, G.S., Neumayer, M., Watzenig, D., Colaço, M.J.: Accelerated Bayesian inference for the estimation of spatially varying heat flux in a heat conduction problem. Numer. Heat Transf. A Appl. **65**(1), 1–25 (2014)
19. Pacheco, C.C., Orlande, H.R.B., Colaço, M.J., Dulikravich, G.S.: Estimation of a location-and time-dependent high-magnitude heat flux in a heat conduction problem using the Kalman filter and the approximation error model. Numer. Heat Transf. A Appl. **68**(11), 1198–1219 (2015)
20. Petris, G., Petrone, S., Campagnoli, P.: Dynamic Linear Models with R. Springer, New York (2009)
21. Ristic, B., Arulampalam, S., Gordon, N.: Beyond the Kalman Filter. Artech House, Boston (2004)
22. Silva, W.B., Orlande, H.R.B., Colaço, M.J.: Evaluation of Bayesian filters applied to heat conduction problems. In: 2nd International Conference on Engineering Optimization, Lisboa (2010)
23. Silva, W.B., Orlande, H.R.B., Colaço, M.J., Fudym, O.: Application of Bayesian filters to a one-dimensional solidification problem. In: 21st Brazilian Congress of Mechanical Engineering, Natal (2011)
24. Silva, W.B., Rochoux, M., Orlande, H.R.B., Colaço, M.J., Fudym, O., El Hafi, M., Cuenot B., Ricci, S.: Application of particle filters to regional-scale wildfire spread. High Temp. High Pressures **43**, 415–440 (2014)
25. Silva, W.B., Dutra, J.C.S., Abreu, L.A.S., Knupp, D.C., Silva Neto, A.J.: Estimation of timewise varying boundary heat flux via Bayesian filters and Markov Chain Monte Carlo method. In: II Simposio de Modelación Matemática Aplicada a la Ingeniería, Havana (2016)
26. Wang, J., Zabaras, N.: A Bayesian inference approach to the inverse heat conduction problem. Int. J. Heat Mass Transf. **47**(17), 3927–3941 (2004)
27. Zhao, J., Li, Z.: Particle filter based on particle swarm optimization resampling for vision tracking. Expert Syst. Appl. **37**, 8910–8914 (2010)

Chapter 14
Fault Detection Using Kernel Computational Intelligence Algorithms

Adrián Rodríguez-Ramos, José Manuel Bernal-de-Lázaro, Antônio José Silva Neto, and Orestes Llanes-Santiago

14.1 Introduction

During the last two decades, the kernel methods have been established as a valuable alternative tool for numerous areas of research [28]. In fact, they have played a significant role in reducing dimensionality, removing noise, and extracting features from the huge databases, including the historical data sets obtained from the complex industrial processes [2, 16, 37]. Many current publications incorporate kernel methods in the fault diagnosis tasks because they allow the mapping of input data into a feature space where it is possible the use of linear algorithms, by avoiding the nonlinearity in the original data. However, the aforementioned operation, and the structure underlying the data are totally determined by the kernel function selected. This means that the inappropriate parameter settings of these kernel methods may result in non-satisfactory diagnosis results. Both, the choosing of an appropriate kernel and the proper setting of its parameters, are open problems in the current fault diagnosis applications.

The present chapter addresses the use of kernel methods for the fault detection in complex industrial systems with nonlinear relationships between variables, and with non-Gaussian characteristics, such as those involving chemical and biological

A. Rodríguez-Ramos · J. M. Bernal-de-Lázaro (✉) · O. Llanes-Santiago
Department of Automation and Computing, Universidad Tecnológica de La Habana José Antonio Echeverría, Cujae, Habana, Cuba
e-mail: adrian.rr@automatica.cujae.edu.cu; jbernal@automatica.cujae.edu.cu; orestes@tesla.cujae.edu.cu

A. J. Silva Neto
Department of Mechanical Engineering and Energy, Polytechnic Institute, IPRJ-UERJ, Nova Friburgo, Brazil
e-mail: ajsneto@iprj.uerj.br

© Springer Nature Switzerland AG 2019
G. Mendes Platt et al. (eds.), *Computational Intelligence, Optimization and Inverse Problems with Applications in Engineering*,
https://doi.org/10.1007/978-3-319-96433-1_14

processes. The proposed procedures are focused on the detection of incipient faults with small effects on the monitored system, which can be hidden by the disturbances in the systems and abrupt faults. Besides, it is investigated how the indirect kernel optimization criteria can improve the performance of the kernel classification algorithms used in this stage.

The procedures evaluated in this study involve the data preprocessing with the Kernel Independent Component Analysis (KICA) algorithm, and the fault detection by using a classifier based on the Kernel Fuzzy C-means (KFCM) algorithm. The KICA algorithm allows to eliminate information that can overload the system as redundant variables, noises, and outliers. Moreover, it is very useful in nonlinear processes, with complex interactions between its variables and with a non-Gaussian distribution [2, 16, 37]. On the other hand, the KFCM algorithm is used to obtain a better separability among classes, and, therefore, the classification results are improved. In addition, optimization algorithms are used to adjust the kernel parameters in the KFCM algorithm. In this chapter an evolutionary algorithm (Differential Evolution) and an algorithm based on group intelligence (Particle Swarm Optimization) are used. The study is evaluated using the Tennessee Eastman (TE) process benchmark.

The remaining of the chapter is organized as follows. In Sect. 14.2 the techniques used in the fault detection tasks and the optimization algorithms employed for the adjustment of the KFCM classifier are described. In Sect. 14.3, the study case is presented, and the performance of the proposed fault detection scheme is discussed. Finally, based on the analysis of the results, some conclusions and future potential research lines are presented.

14.2 Preprocessing and Classification Tasks for Fault Diagnosis

When fault diagnosis is performed based on historical data, it can be seen as a pattern recognition procedure that includes the preprocessing of the data, extraction of the characteristic patterns and their classification [34].

14.2.1 Preprocessing by Using Kernel ICA

Several papers, as for example Refs. [1, 2], have shown that when a fault diagnosis system incorporates a stage of data preprocessing, the results in the classification process are improved.

Kernel ICA (KICA) is an advanced version of the Independent Component Analysis (ICA) algorithm. The aforementioned technique is mainly used for non-Gaussian processes in order to transform multivariate data into statistically independent components [21]. The basic idea of kernel ICA is to perform a nonlinear mapping of the data into a hyper-dimensional feature space $\mathcal{H}$, and then to extract the useful information by using the ICA algorithm [18]. With this purpose, the KICA procedure is divided into two steps. Firstly, a whitened Kernel Principal Components Analysis (KPCA) is carried out, and secondly the linear ICA is applied in the kernel transformed space [2]. In this sense, in order to obtain the whitened KPCA score vector, can be considered the initial step of the kernel ICA.

It is worth noting that here the goal of the kernel PCA, as a first step, is to eliminate the correlation between the variables, not to make them independent. Unlike PCA, ICA does not have orthogonality constraints, and implies the use of higher order statistics [19]. This means that it eliminates the correlation between the data, and the dependence between the variables is also reduced by considering the higher order statistics.

14.2.2   Kernel Fuzzy C-Means (KFCM)

The kernel clustering methods have been widely used in several fields, including genetic classification, handwritten digits recognition, as well as image processing [6, 23, 30, 32]. Nevertheless, their applications are still an innovative topic in the field of fault diagnosis [4, 17]. Specifically, KFCM is a kernelized version of the Fuzzy C-means algorithm with a high potential for fault detection tasks [39]. The algorithm starts employing a kernel function K to map the input data $\mathbf{x}$ into a hyper-dimensional feature space $\mathcal{H}$. Afterwards, it performs the conventional FCM in the transformed space. The KFCM algorithm can be formalized as follows, by first defining

$$J_{\mathrm{KFCM}} = \sum_{i=1}^{c} \sum_{k=1}^{N} (\mu_{ik})^m \, \|\Phi(\mathbf{x_k}) - \Phi(\mathbf{v_i})\|^2 \tag{14.1}$$

where $\|\Phi(\mathbf{x_k}) - \Phi(\mathbf{v_i})\|^2$ is the square of the distance between the mapping data $\Phi(\mathbf{x_k})$ and $\Phi(\mathbf{v_i})$, $\mathbf{v_i}$ is the center of each cluster, c is the total number of clusters, μ indicates the pertinence of each data point to each cluster, and m is a control parameter, that may be adjusted. The distance in the feature space is calculated through the kernel in the input space as follows:

$$\|\Phi(\mathbf{x_k}) - \Phi(\mathbf{v_i})\|^2 = \mathbf{K}(\mathbf{x_k}, \mathbf{x_k}) - 2\mathbf{K}(\mathbf{x_k}, \mathbf{v_i}) + \mathbf{K}(\mathbf{v_i}, \mathbf{v_i}) \tag{14.2}$$

Using a Radial Basis Function (RBF) as the kernel function results

$$\|\Phi(\mathbf{x_k}) - \Phi(\mathbf{v_i})\|^2 = \mathbf{2}\,(\mathbf{1} - \mathbf{K}(\mathbf{x_k}, \mathbf{v_i})) \tag{14.3}$$

where

$$\mathbf{K}(\mathbf{x_k}, \mathbf{v_i}) = e^{-\|\mathbf{x_k}-\mathbf{v_i}\|^2/\sigma^2} \tag{14.4}$$

As a result, Eq. (14.1) may be rewritten as:

$$J_{\mathrm{KFCM}} = 2 \sum_{i=1}^{c} \sum_{k=1}^{N} (\mu_{ik})^m \|1 - \mathbf{K}(\mathbf{x_k}, \mathbf{v_i})\|^2 \tag{14.5}$$

Then, minimizing the above expression under the conditions for local extreme allows to find the center of each cluster and the pertinence of each data point to each cluster as follows:

$$\mathbf{v}_i = \frac{\sum_{k=1}^{N} \left(\mu_{ik}^m \mathbf{K}(\mathbf{x_k}, \mathbf{v_i})\mathbf{x_k}\right)}{\sum_{k=1}^{N} \mu_{ik}^m \mathbf{K}(\mathbf{x_k}, \mathbf{v_i})} \tag{14.6}$$

$$\mu_{ik} = \frac{1}{\sum_{j=1}^{c} \left(\frac{1 - \mathbf{K}(\mathbf{x_k}, \mathbf{v_i})}{1 - \mathbf{K}(\mathbf{x_k}, \mathbf{v_j})}\right)^{1/(m-1)}} \tag{14.7}$$

14.2.3 Optimization Algorithms and Kernel Function

The metaheuristics are stochastic global search methods. They are able to efficiently locate the vicinity of the global optimum even against tough conditions such as multimodality, time-variance, and noise [3]. This permits a remarkable level of flexibility with respect to other optimization techniques, hence their popularity [29].

In many scientific areas, and in particular in the fault diagnosis field, metaheuristic algorithms have been widely used, with excellent results in the solution of optimization problems [9, 25, 26]. They can locate efficiently the neighborhood of the global optimum in most of the occasions, with an acceptable computational time.

In this chapter, the Differential Evolution (DE) and the Particle Swarm Optimization (PSO) algorithms are employed to adjust the parameters for the KFCM-based classifier, with the goal to obtain the best results in the classification task.

14.2.3.1 Optimization with Differential Evolution

Differential Evolution (DE) is one of the most popular optimization algorithms due to its good convergence and easy implementation [11, 22, 33]. This algorithm is based on three operators: Mutation, Crossover, and Selection, for which must be defined the population size NP, the number of parameters to be optimized, and the scale factor F. The crucial idea behind DE is the combination of these operators at each j-th iteration using vector operations to obtain a new solution candidate. The configuration of DE can be summarized using the following notation:

$$DE/\mathbb{X}^j/\gamma/\lambda^*$$

where $\mathbb{X}^j$ denotes the solution to be perturbed in the j-th iteration; γ is the number of pair of vectors used for disturbing $\mathbb{X}^j$, and λ^* indicates the distribution function that will be used in the crossover. In this chapter it has been considered the configuration $DE/\mathbb{X}^{j\,(best)}/1/Bin$, where $\mathbb{X}^{j\,(best)}$ indicates the best individual of the population, and Bin the Binomial Distribution function. The mutation operator is expressed in the following way:

$$\mathbb{X}^{j+1} = \mathbb{X}^{j\,(best)} + F_S(\mathbb{X}^{j\,(\alpha)} - \mathbb{X}^{j\,(\beta)}) \tag{14.8}$$

where $\mathbb{X}^{j+1}$, $\mathbb{X}^{j\,(best)}$, $\mathbb{X}^{j\,(\alpha)}$, $\mathbb{X}^{j\,(\beta)} \in \mathbb{R}^n$, $\mathbb{X}^{j\,(\alpha)}$ and $\mathbb{X}^{j\,(\beta)}$ are elements of the population Z, i.e. one pair of vectors, and F_S is a scaling factor. For complementing the mutation operator, the crossover operator is defined for each component $\mathbb{X}_n$ of the solution vector:

$$\mathbb{X}_n^{j+1} = \begin{cases} \mathbb{X}_n^{j+1}, & \text{if } R < C_R \\ \mathbb{X}_n^{j\,(best)}, & \text{otherwise} \end{cases} \tag{14.9}$$

where $0 \le C_R \le 1$ is the crossover constant that is another control parameter in DE, and R is a random number generated by the distribution λ^*, which in this case is the binomial distribution.

Finally, the selection operator results as follows:

$$\mathbb{X}^{j+1} = \begin{cases} \mathbb{X}^{j+1}, & \text{if } F\left(\mathbb{X}^{j+1}\right) \le F\left(\mathbb{X}^{j\,(best)}\right) \\ \mathbb{X}^{j\,(best)}, & \text{otherwise} \end{cases} \tag{14.10}$$

where F is the objective function.

14.2.3.2 Selecting Parameters with Particle Swarm Optimization

Particle Swarm Optimization (PSO) is an algorithm inspired by the social behavior of different species [8]. The underlying idea of this algorithm is based on the collaborative work of individual organisms to reduce energy at the time of migration or to find food in nature. In general, the application of PSO requires a balance between its capacity (Individual search capacity) and its exploitation capacity (ability to learn from the neighbors). If there is little exploration, the algorithm will converge to the first optimum found, possibly a local minimum. On the other hand, if there is little exploitation, it will never converge. In the methodology presented in this chapter, the estimation of the kernel parameters is performed in the off-line stage. Therefore, it is not of interest to analyze the computational cost. It should also be emphasized that there are many variants of this algorithm. In the present chapter, the conventional PSO version developed in Ref. [8] is used, given its simplicity and easy implementation for parameter estimation problems, with kernel methods.

PSO works with a group or population (swarm) of $\mathbf{Z}$ agents (particles), which are interested in finding a good approximation to the global minimum or maximum x_0 of the objective function $f : \mathbb{D} \subset \mathbb{R}^n \rightarrow \mathbb{R}$. Each agent moves throughout the search space $\mathbb{D}$. The position of the z-th particle is identified with a solution for the optimization problem. On each l-th iteration, its value is updated and it is represented by a vector $\mathbf{X}_z^l \in \mathbb{R}^n$.

Each particle accumulates its historical best position $\mathbf{X}_z^{pbest}$, which represents the best achieved individual experience. The best position that was achieved along the iterative procedure, a among all the agents in the population, i.e. $\mathbf{X}^{gbest}$ represents the collective experience.

The generation of the new position needs the current velocity of the particle $\mathbf{V}_z^l \in \mathbb{R}^n$ and the previous position $\mathbf{X}_z^{l-1}$

$$X_z^l = X_z^{l-1} + V_z^l \tag{14.11}$$

The vector $\mathbf{V}_z^l$ is updated according to the following expression:

$$V_z^l = V_z^{l-1} + c_1 R(X_z^{pbest} - X_z^{l-1}) + c_2 R(X^{gbest} - X_z^{l-1}) \tag{14.12}$$

where V_z^{l-1} is the previous velocity of the z-th particle; R denotes a diagonal matrix with random numbers in the interval $[0,1]$; and c_1, c_2 are the parameters that characterize the trend during the velocity updating [20], balancing the individual and group experiences. They are called cognitive and social parameters, respectively. They represent how the individual and social experiences influence the next agent decision. Some studies have been made in order to determine the best values for c_1 and c_2. The values $c_1 = c_2 = 2$, $c_1 = c_2 = 2.05$ or $c_1 > c_2$ with $c_1 + c_2 \leq 4.10$ are recommended [10].

Some variants of the algorithm have been developed with the objective of improving some characteristics of PSO, e.g. velocity, stability, and convergence.

Equations (14.11) and (14.12) represent the canonical implementation of PSO. Another well-known variant is the one with inertial weight, which considers either constant inertial weight or inertial weight with reduction. The idea behind this variant is to add an inertial factor ω for balancing the importance of the local and global search [20]. This parameter ω affects the updating of each particle velocity by the expression

$$V_z^l = \omega V_z^{l-1} + c_1 R(X_z^{pbest} - X_z^{l-1}) + c_2 R(X^{gbest} - X_z^{l-1}) \tag{14.13}$$

Nowadays, the most accepted strategy for ω is to establish $\omega \in [\omega_{min}; \omega_{max}]$, and reduce its value according to the number of the current iteration l by means of

$$\omega = \omega_{max} - \frac{\omega_{max} - \omega_{min}}{Itr_{max}} l \qquad (14.14)$$

where Itr_{max} is the maximum number of iterations to be reached. The basic PSO is recognized as a particular case for the alternative that considers the inertial weight $\omega = 1$ along all the execution of the algorithm [20].

14.2.3.3   Choosing a Kernel

In general, the selection of a kernel depends on the application. However, the Gaussian, or RBF, kernel is one of the most popular [2, 28, 38]. This function is a homogeneous kernel, which maps the input space to a higher dimension space. The RBF kernel is mathematically defined as:

$$K(\mathbf{x}_i, \mathbf{x}_j) = \exp\left(-\frac{\|\mathbf{x}_i - \mathbf{x}_j\|^2}{2\sigma^2}\right) \qquad (14.15)$$

where σ is called bandwidth, and indicates the degree of smoothness of the function. If σ is overestimated, the exponential tends to show a linear behavior, and its projection in a higher dimensional space loses its ability to separate nonlinear data. Meanwhile, if σ is underestimated, the result will be highly sensitive to noise in the training step of the algorithm.

14.2.3.4   Fitness Function

The fitness function used in this chapter is the partition coefficient (PC) [24, 31, 35] which measures the fuzziness degree of the partition U. This expression, shown in Eq. (14.16), is a validity index to evaluate quantitatively the result of a clustering method, and comparing its behavior when its parameters vary.

$$PC = \frac{1}{N} \sum_{i=1}^{c} \sum_{k=1}^{N} (\mu_{ik})^2 \qquad (14.16)$$

The parameters N, c, and μ_{ik} are defined in Sect. 14.2.2.

If the partition U is less fuzzy, the clustering process is better. Being analyzed in a different way, it allows to measure the degree of overlapping among the classes. In this case, the optimum comes up when PC is maximized, i.e., when each pattern belongs to only one class. Likewise, the minimum comes up when each pattern belongs to all classes.

Therefore, the optimization problem is defined as:

$$max\{PC\} = max\left\{\frac{1}{N}\sum_{i=1}^{c}\sum_{k=1}^{N}(\mu_{ik})^m\right\} \qquad (14.17a)$$

subject to:

$$m_{min} < m \leq m_{max} \qquad (14.17b)$$

$$\sigma_{min} \leq \sigma \leq \sigma_{max} \qquad (14.17c)$$

where σ is shown in Eq. (14.15). When PC is maximized, a better performance is achieved in the classification process, because confusion is reduced in determining to which class an observation belongs.

14.3 Results and Discussion

In this section, the techniques described previously are applied in the design of a fault detection system for the Tennessee Eastman Process benchmark. The fault detection system designed shows an improvement in the classification process, as a result of the best parameters determination to the KFCM algorithm, using optimization algorithms.

14.3.1 Study Case: Tennessee Eastman Process

The Tennessee Eastman (TE) process is widely used as a chemical plant benchmark to evaluate the performance of new control and monitoring strategies [1, 36]. The process consists of five major units interconnected, as shown in Fig. 14.1.

The TE process contains 21 preprogrammed faults, and one normal operating condition data set. The data sets from the TE are generated during a 48 h operation simulation, with the inclusion of faults after 8 h of simulation. The control objectives and general features of the process simulation are described in Refs. [5, 7]. All data sets used to test the procedure hereby proposed were given in Ref. [7], and it can be downloaded from http://web.mit.edu/braatzgroup/TE_process.zip. Table 14.1 shows the faults considered in this chapter, in order to evaluate the advantages of the fault diagnosis proposal presented in this chapter.

According to the specialized literature, the Faults 3, 5, 9, 10, and 11, as well as Fault 15 have small magnitudes, and therefore their detection is very difficult. Fault 3 is generated from one step in the D feed temperature, but it has a quite close behavior to the normal data in terms of the mean and variance. Beyond that,

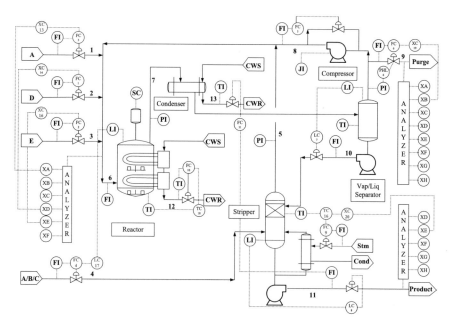

Fig. 14.1 Piping diagram of the Tennessee Eastman process [5]

Table 14.1 Description of the TE process faults

Fault	Process variable	Type	Fault	Process variable	Type
F1	A/C feed ratio, B composition constant	Step	F9	D feed temperature	Random
F2	B composition, A/C ration constant	Step	F10	C feed temperature	Random
F3	D feed temperature	Step	F11	Reactor cooling water inlet temperature	Random
F4	Reactor cooling water inlet temperature	Step	F12	Condenser cooling water inlet temperature	Random
F5	Condenser cooling water inlet	Step	F13	Reaction kinetics	Slow drift
F6	A feed loss	Step	F14	Reactor cooling water valve	Sticking
F7	C header pressure loss-reduced availability	Step	F15	Condenser cooling water valve	Sticking
F8	A, B, and C feed composition	Random			

Fault 5 is due to one step in the condenser cooling water inlet temperature. This variation causes a mean shift on the condenser cooling flow, and a chain reaction in other variables, which produces an out-of-control operation. In this case, the control loops are able to compensate such changes. In consequence of this, the variables return to their set-point, except the condenser cooling water inlet temperature [5]. As a matter of fact the fault does not disappear, it is only hidden.

On the other hand, Fault 9 is a result of one random variation in the feed D temperature. It is hard to detect too. Fault 10 appears when the feed C temperature, of stream 4, is randomly changed. It is interesting to observe that as a result of this fault, the temperature and pressure on the stripper also change. Then, the stripper steam valve is manipulated by the control loops to compensate the changes by means of the stripper steam flow rate, which makes difficult the detection of this fault [15]. Fault 15 is a sticking in the condenser cooling water valve. Similarly to Fault 3, the historical data set of Fault 15 has little difference with respect to the normal data. Therefore, Fault 15 is also hard to detect.

14.3.2 Experimental Results

Figure 14.2 describes the experimental scheme used in this section. Note that to evaluate the proposed diagnostic scheme, two tests were conducted. First, the classifier based on the KFCM algorithm was trained without considering the preprocessing stage, generating the False Alarm Rate (FAR) and the Fault Detection Rate (FDR) indicators. Thereafter, the KICA and the KFCM algorithms were employed together in the fault diagnosis process, generating also the FAR and FDR indicators. In general, a total of 320 observations (samples) for each class (operating

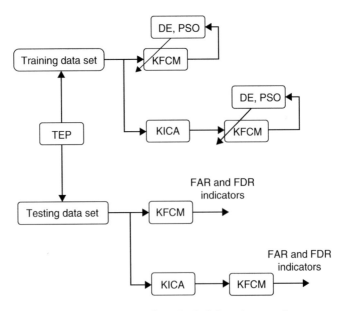

Fig. 14.2 Experimental scheme used to evaluate the fault detection procedure

stages) were used in the training data set, while 800 observations for each class were used in the testing data set. As an outstanding aspect, it should be highlighted that the dimension of feature space was significantly reduced to $\mathbf{R}^{33} \rightarrow \mathbf{R}^{24}$ by the preprocessing with KICA. The number of independent components herein used represents 73% of the information contained in the data set of the TE process.

The DE algorithm implemented for the kernel parameter optimization was executed using the following specifications: population size NP $= 10$, maximum iteration $MaxIter = 100$, difference vector scale factor $F = 0.1$, and crossover criterion $C_R = 0.9$. Moreover, the following search ranges for the parameters to be estimated were considered: $m \in [1, 2]$, and $\sigma \in [1, 150]$. The PSO algorithm was also configured with such ranges. However, for the PSO algorithm the estimated parameters were searched by using the following specifications: population $size = 20$, $w_{mix} = 0.9$, $w_{min} = 0.4$, $c_1 = 2$, $c_2 = 2$ and $Itr_{max} = 100$. In this context, were obtained the results for experiment 1 (without data preprocessing) and 2 (with data preprocessing using KICA algorithm).

For the implementation of the DE and PSO algorithms the following stopping criteria were considered:

- Criterion 1: Maximum number of iterations (100).
- Criterion 2: Value of the objective function (0.9999). See Eq. (14.16).

The value of the σ parameter for the KICA algorithm used in experiment 2 was 492.53, and it was taken from Ref. [2].

The behavior of the objective function (PC) for the experiments 1 and 2 is shown in Fig. 14.3a, b, respectively. It can be observed that when the DE algorithm is applied, the value of the objective function converges to one faster than when the algorithm PSO is used.

In Tables 14.2 and 14.3 are shown the values of the parameters m and σ estimated for each experiment. In Figs. 14.4 and 14.5 is shown the evolution of the best estimated values for such parameters at each iteration for the experiments 1 and 2. In both experiments it can be seen a greater exploitation capacity of the algorithm DE, since the estimated values for parameters converge faster.

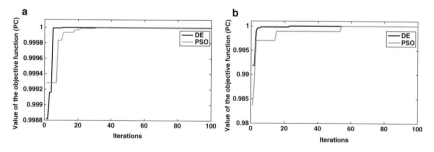

Fig. 14.3 Value of the objective function (PC), Eq. (14.16). (**a**) Experiment 1 (without data preprocessing). (**b**) Experiment 2 (with data preprocessing using KICA algorithm)

Table 14.2 Values of the parameters estimated for experiment 1 (without data preprocessing)

Parameters	DE	PSO
m	1.7148	1.7150
σ (KFCM)	94.9676	85.9322

Table 14.3 Values of the parameters estimated for experiment 2 (with data preprocessing using KICA algorithm)

Parameters	DE	PSO
m	1.3832	1.4284
σ (KFCM)	55.5045	37.7942

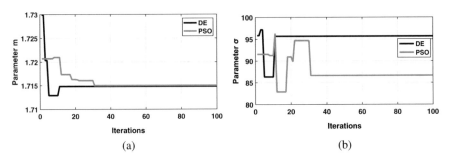

Fig. 14.4 Evolution of the parameters (experiment 1). (**a**) Parameter m. (**b**) Parameter σ

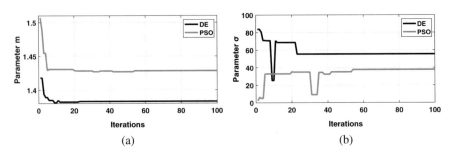

Fig. 14.5 Evolution of the parameters (experiment 2). (**a**) Parameter m. (**b**) Parameter σ

14.3.2.1 Results of the Classification

On the other hand, the information provided by the confusion matrix $\mathbf{C}$ associated with the classification process was used to evaluate the performance of the fault diagnosis procedures [12]. In the confusion matrix, the main diagonal represents the number of observations successfully classified. In the first row, outside the main diagonal represents the false alarms are represented (i.e., $i = 1, j = 2, 3, \ldots, k$). The number of missing alarms are shown at the first column (i.e., $j = 1, i = 2, 3, \ldots, k$). Then, all general information about the fault diagnosis stage is available in the confusion matrix. For example, the detectability of the fault detection schemes can be studied in detail through the False Alarm Rate (FAR) and the Fault Detection Rate (FDR), given by

$$\text{FAR} = \frac{\text{No. of samples } (J > J_{lim} | f = 0)}{\text{total samples } (f = 0)} \times 100\% \qquad (14.18)$$

$$\text{FDR} = \frac{\text{No. of samples } (J > J_{lim} | f \neq 0)}{\text{total samples } (f \neq 0)} \times 100\% \qquad (14.19)$$

where J is the output for the used discriminative algorithms by considering the fault detection stage as a binary classification process, and J_{lim} is the threshold that determines whether one sample is classified as a fault or normal operation.

The results shown in Tables 14.4 and 14.5 were obtained using a cross validation process. The cross validation involves partitioning a sample of data into complementary subsets (d), by performing the analysis on $d - 1$ subsets (called the training set), and validating the analysis on the other subset (called the validation set or testing set). To reduce variability, multiple rounds of cross-validation are performed using different partitions, and the validation results are averaged. In Fig. 14.6 it is shown the cross-validation process for four partitions of the data set. In the experiments implemented for the TE process, the cross-validation was performed with ten partitions of the data set.

Tables 14.4 and 14.5 show the performance of the evaluated procedure in terms of false alarms and missing faults detected. The results summarized in Table 14.4 were obtained without using the data preprocessing with the KICA algorithm. Note that in this case some faults are easily detected (e.g., Faults 1, 2, 4, 6, 7, 8, 12, 13, and 14), with higher values for the FDR measure. Nonetheless, as expected, some faults are difficult to detect (e.g., Faults 3, 5, 9, 10, 11, and 15) due to the fact that they are hidden by the influence of other variables of the process. In general, the performance for these faults is characterized by a high FAR value or a small FDR value. That means a low probability of distinguishing correctly between the normal operating condition (NOC) and the abnormal situations.

Fig. 14.6 Cross-validation process

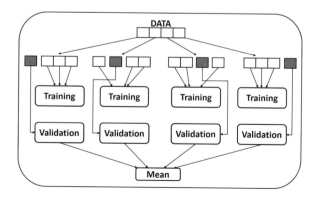

Table 14.4 Fault detection performance without data preprocessing

Faults	KFCM-DE		KFCM-PSO	
	FAR(%)	FDR(%)	FAR(%)	FDR(%)
F1	9.25	89.13	0.00	95.25
F2	6.88	96.13	0.00	98.5
F3	8.25	77.63	6.25	64.88
F4	6.13	92.75	7.25	89.75
F5	8.88	76.25	8.13	77.88
F6	0.00	97.88	6.13	97.25
F7	9.88	93.13	5.25	93.25
F8	7.88	91.38	8.75	97.88
F9	5.25	76.25	0.00	83.13
F10	0.00	75.38	7.25	76.13
F11	4.75	82.25	8.88	78.25
F12	8.5	91.25	8.38	96.75
F13	8.88	95.13	8.38	96.75
F14	10.75	93.13	6.88	94.13
F15	0.00	72.25	10.25	58.88

Table 14.5 Fault detection performance with data preprocessing through KICA

Faults	KFCM-DE		KFCM-PSO	
	FAR(%)	FDR(%)	FAR(%)	FDR(%)
F1	0.00	98.75	0.00	99.13
F2	0.00	99.25	0.00	92.63
F3	4.88	88.38	18.75	61.13
F4	0.00	96.25	0.00	95.38
F5	8.75	81.75	17.25	64.88
F6	0.00	95.13	0.00	99.75
F7	0.00	96.25	0.00	68.38
F8	0.00	93.25	0.00	66.75
F9	5.38	92.63	21.88	74.13
F10	6.25	79.25	9.25	87.88
F11	7.38	80.75	2.75	81.25
F12	0.00	92.13	0.00	94.88
F13	0.00	91.75	0.00	95.63
F14	0.00	90.13	0.00	99.38
F15	6.38	63.25	3.25	57.13

In general, the experiments performed have shown that PSO gives the best results for the Faults 1, 2, 5, 7, 8 , 9, 10, 12, 13 and 15. Meanwhile, the parameters obtained with the DE algorithm allowed to achieve a best performance for the Faults 3, 4, 6, 11, as well as 15.

Table 14.5 summarizes the fault detection performances for the combined work between the KICA and KFCM algorithms. It is interesting to observe that due to preprocessing stage with KICA there is an important reduction in the false alarms with respect to the previous results. Regarding in the small magnitude faults, the higher results for FDR are obtained with the parameters adjusting by using the DE algorithm. Except for Faults 10 and 11, the fault detection scheme presented a worse performance with respect to the small magnitude when the PSO algorithm was used. In particular, using KICA and PSO a worse performance for Faults 5 and 9 is obtained.

In general terms, the performance for Faults 2, 3, 4, 5, 7, 8, 9, and 15 is greater when the classifier uses the DE algorithm to estimate its parameters. Through the results shown in the tables, it is demonstrated that the overall performance of the detection system is better with the data preprocessing based on KICA. However, it should not be forgotten the important role that the configuration of the parameters has for the application of the kernel methods in the fault detection tasks. In fact, it is necessary to emphasize that the detection levels herein achieved, for the small-magnitude faults in this process, are still insufficient for the current industrial standards.

As several algorithms are presented (KFCM-DE, KFCM-PSO, KICA-KFCM-DE and KICA-KFCM-PSO), we must analyze if there are significant differences between them. In order to achieve this purpose, it is necessary to apply statistical tests [13, 14, 27].

14.3.3 Statistical Tests

Table 14.6 shows the global classification (%) obtained for the ten partitions of the data set used in the cross-validation. First, the non-parametric Wilcoxon test is applied to each experiment. The two algorithms used in experiment 1 are compared, and the best is determined. Then, the same analysis is performed for experiment 2.

Table 14.6 Global classification (%) obtained with each algorithm

N	Experiment 1		Experiment 2	
	KFCM-DE	KFCM-PSO	KICA-KFCM-DE	KICA-KFCM-PSO
1	85.82	85.17	88.46	80.63
2	85.54	85.08	89.08	81.22
3	86.71	86.22	88.50	82.65
4	86.30	86.55	89.07	83.18
5	88.19	87.34	90.15	83.42
6	83.64	86.26	87.42	80.84
7	84.97	86.41	88.52	84.34
8	89.04	88.09	89.16	83.47
9	90.17	88.87	92.91	84.08
10	86.23 0	85.78	89.33	81.71

Afterwards, the best algorithms for experiments 1 and 2 are selected, and the non-parametric Wilcoxon test is applied again to determine which one is the winning algorithm.

14.3.3.1 Wilcoxon Test

Table 14.7 shows the results of the comparison in pairs of algorithms, for each experiment, using the Wilcoxon test. The first two rows contain the values of the sum of the positive (R^+) and negative (R^-) rank for each comparison established. The next two rows show the statistical values T and the critical value of T for a level of significance $\alpha = 0.05$. The last row indicates which algorithm was the winner in each comparison.

When the data are not preprocessed by KICA (experiment 1), Table 14.7 shows that there are no significant differences in the results when using the optimization algorithms DE (algorithm 1) and PSO (algorithm 2). However, the DE algorithm was selected as the best one, since it allows a faster convergence of the estimated parameters, as shown in Figs. 14.3a and 14.4.

On the other hand, when KICA is applied (experiment 2), the DE algorithm achieves better results than PSO. For this reason, the KFCM-DE and KICA-KFCM-DE algorithms are compared by applying again the non-parametric Wilcoxon test. Table 14.8 shows the results of this comparison, resulting in that the KICA-KFCM-DE algorithm is the winner.

Table 14.7 Results of the Wilcoxon test for experiments 1 and 2

	Experiment 1 (1 vs 2)	Experiment 2 (3 vs 4)
$\sum R^+$	35	55
$\sum R^-$	20	0
T	20	0
$T_{\alpha=0.05}$	8	8
Winner	–	3

Table 14.8 Results of the Wilcoxon test

	KFCM-ED vs KICA-KFCM-ED (1 vs 3)
$\sum R^+$	0
$\sum R^-$	55
T	0
$T_{\alpha=0.05}$	8
Winner	3

14.4   Conclusions and Future Work

In this chapter was presented a comparative study between two metaheuristic optimization algorithms, Differential Evolution (DE) and Particle Swarm Optimization (PSO). These algorithms were used to estimate the parameters of the Kernel Fuzzy C-means (KFCM) classifier. First, a diagnostic classifier without considering a data preprocessing stage was evaluated. Thereafter, the KICA and the KFCM algorithms were jointly employed in the fault detection process. For the comparative evaluation were established as the comparison criteria the false alarm and fault detection rates. The experiments have shown that the overall performance of the detection scheme is better with the data preprocessing, and higher results are obtained adjusting the kernel parameter by using the DE algorithm.

For future researches, it is necessary to analyze the use of KFCM considering the dynamics of the process to improve the detection of incipient and small-magnitude faults. Furthermore, it would be interesting to investigate the use of other optimization techniques for tuning the kernel parameters, including the algorithms applied in the preprocessing and classification stages.

Acknowledgements   The authors acknowledge the financial support provided by FAPERJ—Fundação Carlos Chagas Filho de Amparo à Pesquisa do Estado do Rio de Janeiro, CNPq—Conselho Nacional de Desenvolvimento Científico e Tecnológico and CAPES—Coordenação de Aperfeiçoamento de Pessoal de Nível Superior, research support agencies from Brazil.

References

1. Bernal-de-Lázaro, J.M., Moreno, A.P., Llanes Santiago, O., Silva Neto, A.J.: Optimizing Kernel methods to reduce dimensionality in fault diagnosis of industrial systems. Comput. Ind. Eng. **87**, 140–149 (2015). https://doi.org/10.1016/j.cie.2015.05.012
2. Bernal-de-Lázaro, J.M., Llanes-Santiago, O., Prieto-Moreno, A., Knupp, D.C., Silva Neto, A.J.: Enhanced dynamic approach to improve the detection of small-magnitude faults. Chem. Eng. Sci. **146**, 166–179 (2016). https://doi.org/10.1016/j.ces.2016.02.038
3. Boussaïd, I., Lepagnot, J., Siarry, P.: A survey on optimization metaheuristics. Inf. Sci. **237**, 82–117 (2013). https://doi.org/10.1016/j.ins.2013.02.041
4. Cao, S.Q., Zuo, X.M., Tao, A.X., Wang, J.M., Chen, X.Z.: A bearing intelligent fault diagnosis method based on cluster analysis. In: Mechanical Engineering and Materials, Applied Mechanics and Materials, vol. 152, pp. 1628–1633. Trans Tech Publications, Durnten-Zurich (2012). https://doi.org/10.4028/www.scientific.net/AMM.152-154.1628
5. Chiang, L.H., Russell, E.L., Braatz, R.D.: Fault Detection and Diagnosis in Industrial Systems, 1st edn. Advanced Textbooks in Control and Signal Processing. Springer, London (2001)
6. Ding, Y., Fu, X.: Kernel-based fuzzy c-means clustering algorithm based on genetic algorithm. Neurocomputing **188**, 233–238 (2016). https://doi.org/10.1016/j.neucom.2015.01.106
7. Downs, J.J., Vogel, E.F.: A plant-wide industrial process control problem. Comput. Chem. Eng. **17**(3), 245–255 (1993). https://doi.org/10.1016/0098-1354(93)80018-I
8. Eberhart, R., Kennedy, J.: A new optimizer using particle swarm theory. In: Proceedings of the Sixth International Symposium on Micro Machine and Human Science, pp. 39–43 (1995). https://doi.org/10.1109/MHS.1995.494215

9. Echevarría, L.C., Santiago, O.L., Neto, A.J.S.: An approach for fault diagnosis based on bio-inspired strategies. In: IEEE Congress on Evolutionary Computation, pp. 1–7 (2010). https://doi.org/10.1109/CEC.2010.5586357

10. Echevarría, L.C., Santiago, O.L., Fajardo, J.A.H., Neto, A.J.S., Sánchez, D.J.: A variant of the particle swarm optimization for the improvement of fault diagnosis in industrial systems via faults estimation. Eng. Appl. Artif. Intell. **28**, 36–51 (2014). https://doi.org/10.1016/j.engappai. 2013.11.007

11. Echevarría, L.C., Santiago, O.L., Neto, A.J.S., Velho, H.F.C.: An approach to fault diagnosis using meta-heuristics: a new variant of the differential evolution algorithm. Comp. Sis. **18**(1), 5–17 (2014). https://doi.org/10.13053/CyS-18-1-2014-015

12. Fawcett, T.: An introduction to ROC analysis. Pattern Recogn. Lett. **27**(8), 861–874 (2006). https://doi.org/10.1016/j.patrec.2005.10.010

13. García, S., Herrera, F.: An extension on "Statistical comparisons of classifiers over multiple data sets" for all pairwise comparisons. J. Mach. Learn. Res. **9**, 2677–2694 (2008)

14. García, S., Molina, D., Lozano, M., Herrera, F.: A study on the use of non-parametric tests for analyzing the evolutionary algorithms' behaviour: a case study on the CEC'2005 special session on real parameter optimization. J. Heuristics **15**(6), 617–644 (2008). https://doi.org/10. 1007/s10732-008-9080-4

15. Ge, Z., Yang, C., Song, Z.: Improved kernel PCA-based monitoring approach for nonlinear processes. Chem. Eng. Sci. **64**(9), 2245–2255 (2009). https://doi.org/10.1016/j.ces.2009.01. 050

16. Ge, Z., Song, Z., Gao, F.: Review of recent research on data-based process monitoring. Ind. Eng. Chem. Res. **52**(10), 3543–3562 (2013). https://doi.org/10.1021/ie302069q

17. Hu, D., Sarosh, A., Dong, Y.F.: A novel KFCM based fault diagnosis method for unknown faults in satellite reaction wheels. ISA Trans. **51**(2), 309–316 (2012). https://doi.org/10.1016/j. isatra.2011.10.005

18. Jiang, L., Zeng, B., Jordan, F.R., Chen, A.: Kernel function and parameters optimization in KICA for rolling bearing fault diagnosis. J. Netw. **8**(8), 1913–1919. https://doi.org/10.4304/ jnw.8.8.1913-1919

19. Jolliffe, I.T.: Principal Component Analysis, 2nd edn. Springer Series in Statistics. Springer, New York (2002)

20. Kameyama, K.: Particle swarm optimization – a survey. IEICE Trans. Inf. Syst. **E92.D**(7), 1354–1361 (2009). https://doi.org/10.1587/transinf.E92.D.1354

21. Kano, M., Tanaka, S., Hasebe, S., Hashimoto, I., Ohno, H.: Monitoring independent components for fault detection. AIChE J. **49**(4), 969–976. https://doi.org/10.1002/aic.690490414

22. Knupp, D.C., Sacco, W.F., Silva Neto, A.J.: Direct and inverse analysis of diffusive logistic population evolution with time delay and impulsive culling via integral transforms and hybrid optimization. Appl. Math. Comput. **250**, 105–120 (2015). https://doi.org/10.1016/j.amc.2014. 10.060

23. Lauer, F., Suen, C.Y., Bloch, G.: A trainable feature extractor for handwritten digit recognition. Pattern Recogn. **40**(6), 1816–1824 (2007). https://doi.org/10.1016/j.patcog.2006.10.011

24. Li, C., Zhou, J., Kou, P., Xiao, J.: A novel chaotic particle swarm optimization based fuzzy clustering algorithm. Neurocomputing **83**, 98–109 (2012). https://doi.org/10.1016/j.neucom. 2011.12.009

25. Liu, Q., Lv, W.: The study of fault diagnosis based on particle swarm optimization algorithm. Comput. Inf. Sci. **2**(2), 87–91 (2009). https://doi.org/10.5539/cis.v2n2p87

26. Lobato, F.S., Steffen, V. Jr., Silva Neto, A.J.: Solution of inverse radiative transfer problems in two-layer participating media with differential evolution. Inverse Prob. Sci. Eng. **18**(2), 183–195 (2010). https://doi.org/10.1080/17415970903062054

27. Luengo, J., García, S., Herrera, F.: A study on the use of statistical tests for experimentation with neural networks: analysis of parametric test conditions and non-parametric tests. Expert Syst. Appl. **36**(4), 7798–7808 (2009). https://doi.org/10.1016/j.eswa.2008.11.041

28. Motai, Y.: Kernel association for classification and prediction: a survey. IEEE Trans. Neural Netw. Learn. Syst. **26**(2), 208–223 (2015). https://doi.org/10.1109/TNNLS.2014.2333664

29. Nanda, S.J., Panda, G.: A survey on nature inspired metaheuristic algorithms for partitional clustering. Swarm Evol. Comput. **16**, 1–18 (2014). https://doi.org/10.1016/j.swevo.2013.11.003

30. Niu, X.X., Suen, C.Y.: A novel hybrid CNN–SVM classifier for recognizing handwritten digits. Pattern Recogn. **45**(4), 1318–1325 (2012). https://doi.org/10.1016/j.patcog.2011.09.021

31. Pakhira, M.K., Bandyopadhyay, S., Maulik, U.: Validity index for crisp and fuzzy clusters. Pattern Recogn. **37**(3), 487–501 (2004). https://doi.org/10.1016/j.patcog.2003.06.005

32. Shen, H., Yang, J., Wang, S., Liu, X.: Attribute weighted mercer kernel based fuzzy clustering algorithm for general non-spherical datasets. Soft Comput. **10**(11), 1061–1073 (2006). https://doi.org/10.1007/s00500-005-0043-5

33. Storn, R., Price, K.: Differential evolution – a simple and efficient heuristic for global optimization over continuous spaces. J. Glob. Optim. **11**(4), 341–359 (1997). https://doi.org/10.1023/A:1008202821328

34. Venkatasubramanian, V., Rengaswamy, R., Kavuri, S.N., Yin, K.: A review of process fault detection and diagnosis, part III: process history based methods. Comput. Chem. Eng. **27**(3), 327–346 (2003). https://doi.org/10.1016/S0098-1354(02)00162-X

35. Wu, K.L., Yang, M.S.: A cluster validity index for fuzzy clustering. Pattern Recogn. Lett. **26**(9), 1275–1291 (2005). https://doi.org/10.1016/j.patrec.2004.11.022

36. Yin, S., Ding, S.X., Haghani, A., Hao, H., Zhang, P.: A comparison study of basic data-driven fault diagnosis and process monitoring methods on the benchmark Tennessee Eastman process. J. Process Control **22**(9), 1567–1581 (2012). https://doi.org/10.1016/j.jprocont.2012.06.009

37. Yoo, C.K., Lee, I.B.: Nonlinear multivariate filtering and bioprocess monitoring for supervising nonlinear biological processes. Process Biochem. **41**(8), 1854–1863 (2006). https://doi.org/10.1016/j.procbio.2006.03.038

38. Zhang, Y.: Enhanced statistical analysis of nonlinear processes using KPCA, KICA and SVM. Chem. Eng. Sci. **64**(5), 801–811 (2009). https://doi.org/10.1016/j.ces.2008.10.012

39. Zheng, Z., Jiang, W., Wang, Z., Zhu, Y., Yang, K.: Gear fault diagnosis method based on local mean decomposition and generalized morphological fractal dimensions. Mech. Mach. Theory **91**, 151–167 (2015). https://doi.org/10.1016/j.mechmachtheory.2015.04.009

Index

A

Anomalous diffusion, 188
Ant Colony Optimization, 3, 144, 167, 190
Ant Colony Optimization with dispersion, 140, 144
Artificial Bee Colonies, 167, 168, 176, 190
Artificial neural networks, 201, 202

B

Bat algorithm, 115
Bioreactor benchmark problem, 142
Bridges, 121

C

Cantilever beam problem, 45
Cantilevered beam, 99
Co-design system, 236, 244
Constructal design, 215, 216
Construction management, 122
Covariance Matrix Adaptation Evolution Strategy, 140, 150
Cross-Entropy Algorithm, 167, 168, 172
Cuckoo search, 114, 167

D

Dew point pressures, 2
Differential Evolution, 2, 3, 13, 14, 18, 25, 36, 50, 53, 147, 266
Differential Evolution with Particle Collision, 140, 147
Double retrograde vaporization, 2, 8, 25

F

Fault detection, 263, 270
Fault diagnosis, 139, 141, 144
Firefly algorithm, 2, 114
Flower Pollination Algorithm, 2, 115
Frame structures, 120

G

Genetic Algorithms, 3, 53, 113, 167, 190, 216, 219, 228, 230
Geotechnics, 125

H

Hardware Architecture, 236
Harmony Search, 2, 3, 13, 15, 18, 25, 114, 167
Highly nonlinear limit state, 41
Hooke Jeeves, 90
Hooke–Jeeves pattern search method, 96
Hydraulics and infrastructures, 123

I

In-Core Fuel Management Optimization, 166
Inverse reliability analysis, 35
Isothermal cavities, 216

K

Kernel function, 265, 266
Kernel Fuzzy C-Means, 264, 265
Kernel methods, 263, 268, 277

© Springer Nature Switzerland AG 2019
G. Mendes Platt et al. (eds.), *Computational Intelligence, Optimization and Inverse Problems with Applications in Engineering*,
https://doi.org/10.1007/978-3-319-96433-1

L

Lightning Optimization Algorithm, 189, 190
Lightning Search Algorithm, 190
Low-discrepancy sequence, 55
Luus-Jaakola Algorithm, 2, 3

M

Mean Effective Concept, 30, 50
Mersenne Twister, 54, 55, 57
Minimum Spanning Tree, 68
Modified Particle Swarm Optimization, 70
Multimodal Particle Swarm Optimization, 3

N

Nonlinear limit state, 40
Nuclear Reactor
    Loading Pattern, 166

O

Opposition-Based Learning, 54, 93

P

Particle Collision Algorithm, 147
Particle Filter Method, 251, 252
Particle Swarm Optimization, 3, 53, 64, 65, 67,
    70, 71, 83, 114, 167, 171, 228, 231, 248,
    253, 264, 267
Population-Based Incremental Learning, 167,
    168, 174
Pseudorandom generator, 54
PSO-SIR Filter Algorithm, 254

R

Reactive azeotrope calculation, 3, 25

Reinforced concrete members, 119
Reliability analysis, 30, 33
Reliability-based Design
    First Order Reliability Method, 33
    Second Order Reliability Method, 34
Robust design, 29
Rotation-Based Multi-Particle Collision
    Algorithm, 90
Rotation-Based Sampling, 90, 94

S

Sequential Monte Carlo Based Estimation,
    250
Sequential optimization and reliability
    assessment, 34
Short column design, 43
Simulated Annealing, 3, 113, 167
Smith's algorithm, 65
Sobol sequence, 55, 57
Steiner Tree Problem, 63, 64
Structural Damage Identification, 87

T

Tabu search, 167
Template matching, 227, 228
Tennessee Eastman process, 264, 270
Three bar truss problem, 48
Transportation engineering, 124
Truss structures, 114
Tuned mass damper, 122
Two phase equilibrium, 25

V

Vibration-based damage identification, 98

Printed in the United States
By Bookmasters